Advances in Metal Matrix Composite Coatings and Layers: Microstructure, Physicochemical and Mechanical Properties

Advances in Metal Matrix Composite Coatings and Layers: Microstructure, Physicochemical and Mechanical Properties

Editors

Dariusz Bartkowski
Aneta Bartkowska

Basel • Beijing • Wuhan • Barcelona • Belgrade • Novi Sad • Cluj • Manchester

Editors
Dariusz Bartkowski
Poznan University of
Technology
Poznan
Poland

Aneta Bartkowska
Poznan University of
Technology
Poznan
Poland

Editorial Office
MDPI AG
Grosspeteranlage 5
4052 Basel, Switzerland

This is a reprint of articles from the Special Issue published online in the open access journal *Coatings* (ISSN 2079-6412) (available at: https://www.mdpi.com/journal/coatings/special_issues/Metal_Matrix_Compos_Coat_Layers).

For citation purposes, cite each article independently as indicated on the article page online and as indicated below:

Lastname, A.A.; Lastname, B.B. Article Title. *Journal Name* **Year**, *Volume Number*, Page Range.

ISBN 978-3-7258-2467-0 (Hbk)
ISBN 978-3-7258-2468-7 (PDF)
doi.org/10.3390/books978-3-7258-2468-7

© 2024 by the authors. Articles in this book are Open Access and distributed under the Creative Commons Attribution (CC BY) license. The book as a whole is distributed by MDPI under the terms and conditions of the Creative Commons Attribution-NonCommercial-NoDerivs (CC BY-NC-ND) license.

Contents

Garima Mittal, Nigar Gul Malik, Arunima Bhuvanendran Nair Jayakumari, David Martelo, Namrata Kale and Shiladitya Paul
Spray Parameters and Coating Microstructure Relationship in Suspension Plasma Spray TiO_2 Coatings
Reprinted from: *Coatings* 2023, 13, 1984, doi:10.3390/coatings13121984 1

Dariusz Bartkowski and Aneta Bartkowska
Fe/TaC Coatings Produced on 145Cr6 Steel by Laser Alloying—Manufacturing Parameters and Material Characterization
Reprinted from: *Coatings* 2023, 13, 1432, doi:10.3390/coatings13081432 16

Christoph Mikulla, Lars Steinberg, Philipp Niemeyer, Uwe Schulz and Ravisankar Naraparaju
Microstructure Refinement of EB-PVD Gadolinium Zirconate Thermal Barrier Coatings to Improve Their CMAS Resistance
Reprinted from: *Coatings* 2023, 13, 905, doi:10.3390/coatings13050905 36

Xiao Shan, Tianmeng Huang, Lirong Luo, Jie Lu, Huangyue Cai, Junwei Zhao, et al.
Automatic Recognition of Microstructures of Air-Plasma-Sprayed Thermal Barrier Coatings Using a Deep Convolutional Neural Network
Reprinted from: *Coatings* 2023, 13, 29, doi:10.3390/coatings13010029 59

Anna Og Dikovska, Daniela Karashanova, Genoveva Atanasova, Georgi Avdeev, Petar Atanasov and Nikolay N. Nedyalkov
Fabrication of Nanostructures Consisting of Composite Nanoparticles by Open-Air PLD
Reprinted from: *Coatings* 2024, 14, 527, doi:10.3390/coatings14050527 76

Yingwei Zhang, Xinpeng Zhou, Weihan Shi, Jiarui Chi, Yan Li and Wenfeng Guo
A Visualization Experiment on Icing Characteristics of a Saline Water Droplet on the Surface of an Aluminum Plate
Reprinted from: *Coatings* 2024, 14, 155, doi:10.3390/coatings14020155 91

André V. Fontes, Patrícia Freitas Rodrigues, Daniela Santo and Ana Sofia Ramos
Exploring the Influence of the Deposition Parameters on the Properties of NiTi Shape Memory Alloy Films with High Nickel Content
Reprinted from: *Coatings* 2024, 14, 138, doi:10.3390/coatings14010138 103

Aikaterini Baxevani, Fani Stergioudi and Stefanos Skolianos
The Roughness Effect on the Preparation of Durable Superhydrophobic Silver-Coated Copper Foam for Efficient Oil/Water Separation
Reprinted from: *Coatings* 2023, 13, 1851, doi:10.3390/coatings13111851 116

Pavels Onufrijevs, Liga Grase, Juozas Padgurskas, Mindaugas Rukanskis, Ramona Durena, Dieter Willer, et al.
Anisotropy of the Tribological Performance of Periodically Oxidated Laser-Induced Periodic Surface Structures
Reprinted from: *Coatings* 2023, 13, 1199, doi:10.3390/coatings13071199 131

Yuanzhe Kou, Jianxiao Bian, Xiaonan Pan and Jinchang Guo
Enhancing Photovoltaic Performance and Stability of Perovskite Solar Cells through Single-Source Evaporation and $CsPbBr_3$ Quantum Dots Incorporation
Reprinted from: *Coatings* 2023, 13, 863, doi:10.3390/coatings13050863 145

Amir Ghiasvand, Alireza Fayazi Khanigi, John William Grimaldo Guerrero, Hamed Aghajani Derazkola, Jacek Tomków, Anna Janeczek and Adrian Wolski
Investigating the Effects of Geometrical Parameters of Re-Entrant Cells of Aluminum 7075-T651 Auxetic Structures on Fatigue Life
Reprinted from: *Coatings* **2023**, *13*, 405, doi:10.3390/coatings13020405 . . . . **157**

Yuhang Du, Qinggang Li, Sique Chen, Deli Ma, Baocai Pan, Zhenyu Zhang and Jinkai Li
Effect of Al_2O_3 Content on High-Temperature Oxidation Resistance of Ti_3SiC_2/Al_2O_3
Reprinted from: *Coatings* **2022**, *12*, 1641, doi:10.3390/coatings12111641 . . . . **173**

Article

Spray Parameters and Coating Microstructure Relationship in Suspension Plasma Spray TiO₂ Coatings

Garima Mittal [1], Nigar Gul Malik [1], Arunima Bhuvanendran Nair Jayakumari [1], David Martelo [2], Namrata Kale [2] and Shiladitya Paul [1,2,*]

1. Materials Innovation Centre, School of Engineering, University of Leicester, Leicester LE1 7RH, UK; garima.nano@gmail.com (G.M.); ngm11@leicester.ac.uk (N.G.M.); abnj1@leicester.ac.uk (A.B.N.J.)
2. Materials Performance and Integrity Group, TWI, Cambridge CB21 6AL, UK; david.martelo@twi.co.uk (D.M.); namrata.kale@twi.co.uk (N.K.)
* Correspondence: shiladitya.paul@twi.co.uk

Abstract: In recent years, there has been growing interest in thermal spray techniques using suspension or solution-based coatings. These techniques offer precise control over particle size and microstructure, improving feedstock flowability and allowing for high-quality coating customization. Spray parameters, such as stand-off distance (SOD) and feedstock flow rate, can alter the performance and characteristics of these coatings. Geothermal power plant heat exchangers often face issues like corrosion, scaling, and fouling. The literature suggests that these issues could be mitigated, at least in part, by the use of spray coatings. In this study, TiO₂ coatings were applied on a carbon steel substrate using suspension plasma spray (SPS) to enhance the performance of geothermal heat exchanger materials. The impact of SOD (50, 75, and 100 mm) and feedstock flow rate (10, 20, and 30 mL/min) on these coatings was examined through various techniques, including scanning electron microscope (SEM), profilometry, X-ray diffraction (XRD), and adhesion testing. The results demonstrated that coatings deposited using a 10 mL/min feedstock flow rate were well adhered to the substrate due to the efficient melting of the coating material, but as the SOD and feedstock flow rate increased due to poor thermal and kinetic energy exchange between the torch and feedstock particles, adhesion between the coating and substrate decreased.

Keywords: suspension plasma spray; TiO₂ coatings; coating microstructures; spray parameters

1. Introduction

Geothermal energy is a potential source of sustainable and renewable energy. However, despite having the ability to provide clean and reliable power generation, geothermal energy has yet to reach its full potential. Because the performance and durability of various components of geothermal power plants are under constant threat of corrosion and scaling due to harsh environmental conditions such as high temperature, varied pH, humidity, silica, and acids [1], these issues adversely affect the stability and efficiency of the power plant, leading to increased operation and maintenance (O&M) costs. Stainless steel and titanium are the most commonly used materials in designing geothermal plants' components. Stainless steel provides corrosion resistance, high-temperature sustainability, excellent mechanical strength, and durability. Titanium is also an excellent choice for geothermal heat exchangers due to its lightweight and outstanding corrosion resistance [2]. Material selection depends on the geothermal fluid chemistry influenced by geographical location and heat exchanger configuration. In the long term, replacing components of the heat exchanger is expensive as compared to applying paints and coatings on the geothermal plant's components. Coatings designed for geothermal power plant components, particularly geothermal heat exchangers, could offer a feasible and economical solution, providing protection against extreme thermal gradients, aggressive fluids, and mechanical stress [3,4].

For instance, F. Zhang et al. developed cermet (WC-CoCr, CrC-NiCr) and alloy (Ni self-fluxing, Fe-based amorphous) coatings using liquid feedstock-based high-velocity oxy-fuel (HVOF) for geothermal drilling components, and found negligible damage to coatings in a simulated geothermal erosion–corrosion environment [5]. Furthermore, coatings, especially surfaces with micro and/or nanostructures play a crucial role in improving the overall efficiency and reliability of geothermal energy generation [6]. Recently, F. Fanicchia et al. presented a detailed review summarizing various coatings and paints used in geothermal power plants [7]. Paints, due to their high thickness, limited stand-alone mechanical properties, and low thermal conductivity, have limited applications in geothermal environments. While coatings (or inorganic coatings), due to sustaining higher mechanical, tribological, and thermal stress than paints, are widely used in high-temperature geothermal applications. High entropy alloys (HEAs) are a new class of alloys that show promising performance in high-temperature geothermal environments due to the good corrosion performance of the CoCrFeNiMo alloy [8,9]. The thermo-mechanical and corrosion performance in a geothermal environment could be altered by adding or removing other alloying elements [10]. However, the high cost associated with HEAs restricts their potential use towards cost-efficient geothermal energy. Ceramic oxides could also be suitable candidates for geothermal applications, as they can be deposited via various coating techniques such as electroplating, thermal spray, dip coating, chemical vapor deposition, and physical vapor deposition [11,12].

Compared to other coating methods, thermal spray coatings have gained attention due to depositing high-quality coatings with tailored properties in a time- and cost-efficient manner. Thermal spray gives the freedom to deposit a wide range of coating materials on varied substrates [13,14]. Conventional powder-based thermal spray is widely accepted and employed in various industries but when it comes to depositing very small particles (<10 microns), poor injection due to low inertia of particles restricts the use of gas as a carrier. Therefore, liquid-carrying submicron particles (suspension) or a solution of chemical precursors of the coating material, forming solid particles during flight, are preferred so that coatings with nano- and micro-scale features can be obtained [15]. In addition, using liquid in the form of a suspension or solution or both provides better flowability of the feedstock as well as enhanced control over coating microstructures [16]. There is no specific microstructure that is suitable for each geothermal heat exchanger as the choice of microstructure is influenced by different factors including geothermal plant location (as the geothermal fluid chemistry and constituents vary with the location of the geothermal plant) and types of heat transfer (boilers, condensers, and evaporators). Generally, dense coatings with controlled cracking or columnar structures are preferred in geothermal environments, providing better protection against corrosive agents along with tolerance for thermal cycling. For better control over coating microstructure and properties, it is essential to understand the relationship between coating deposition/spray parameters and coating properties. The performance and characteristics of liquid feedstock-based thermal spray coatings are influenced by numerous key depositing parameters, for instance, feedstock type, feedstock flow rate, particle size, feedstock concentration, solvent type, plasma current and voltage, primary gases' type and flow rate, stand-off distance, and substrate size, shape, and temperature [17].

In this study, suspension plasma spray (SPS) was used to deposit TiO_2 coatings onto a carbon steel substrate. TiO_2 coatings were chosen to protect the geothermal heat exchangers' components against corrosion and erosion as geothermal fluids contain abrasive particles, and to provide high-temperature stability and chemical stability to geothermal heat exchanger components, protecting them against extreme heat conditions (although a corrosion and heat conductivity study is not performed in this manuscript). In addition, there are many reports available mentioning that the use of TiO_2 coatings in heat exchangers improves their thermal conductivity [18–20]. These coatings are formed using an aqueous suspension of TiO_2 nanoparticles as the feedstock material. The aqueous suspension was chosen due to cost-effectiveness, environmental considerations, safety concerns, and avoiding waste disposal issues. The effect of stand-off distance (SOD; the distance between the injector nozzle and the substrate surface) and feedstock flow rate on the coating microstructure was focused, as it is crucial for optimizing the coating process and attaining desirable coating performance [14].

This work focuses on investigating the effect of feedstock flow rate and stand-off distance on coating microstructures in our experimental setup, which was designed to optimize energy consumption, establishing a foundational understanding. Higher power levels might generate excessive heat input, causing thermal stress or distortion in the substrate. Therefore, an SG100 plasma gun with a low power level (~32 kW) was used considering the low thickness of the prototype plate heat exchanger (0.6 mm). Also, for attaining precise and stable flow rates, a syringe pump (Teledyne ISCO® 260D; Lincoln, NE, USA) connected to the external nebulizer (constructed from a modified RS air brush AB931; Northants, UK) was used, which controlled the flow rate through a combination of motor drive mechanisms, microprocessor control, user input, feedback sensor, and advanced flow rate algorithms. Usually, a high solid content of coating material is used in suspensions. Since nanomaterials are expensive and tend to form agglomerates, making it difficult to formulate a stable and uniform suspension at a higher weight percentage of the coating nanomaterial, a low solid content was used in this study. Comprehensive analyses of coatings deposited via systematically varied SOD and the feedstock flow rate were performed using scanning electron microscopy (SEM), roughness measurements, and tape adhesion tests. The findings of this work will help in optimizing the SPS of TiO_2 coatings on geothermal heat exchangers, enhancing their corrosion resistance, scaling resistance, and fouling resistance.

2. Materials and Methods

2.1. Substrate Material

Coatings were developed on $25 \times 25 \times 6$ mm coupons of carbon steel (S275JR, EN 10025-2). Prior to coating deposition, carbon steel substrates were grit blasted with #100 mesh white alumina to improve coating adhesion. The grit blasting parameters were ~551 kPa set pressure (~482 kPa run pressure) at 80 mm SOD. This process was followed by degreasing with acetone to clean the samples just before the coating deposition. The average surface roughness (R_a) of the substrate after grit blasting was 2.32 µm.

2.2. Coating Material

An aqueous suspension of 5 wt.% TiO_2 nanoparticles, commercially obtained from Promethean Particles Ltd. (Nottingham, UK), was used as feedstock material. Since the feedstock was a homogeneous suspension of nanoparticles, no pre-treatment was required. The particle size of TiO_2 in the suspension was 5–10 nm as characterized by the supplier.

2.3. Coating Development

Suspension plasma spraying was performed using a Praxair®SG-100 plasma gun (Praxair S.T. Inc., Indianapolis, IN, USA) attached to an OTC AII-V20 robot (OTC Daihen Inc., Tipp City, OH, USA). The APS plasma console was 3710 (Praxair S.T., Inc., Indianapolis, IN, USA) with an HF 2210 starter kit. The horizontal speed of the robot was 450 mm/s with a 5 mm vertical increase. The suspensions were fed radially into the plasma through a syringe pump (ISCO® 260D) connected to the external nebulizer (constructed from a modified RS air brush AB931). The diameter of the injector was 0.5 mm. Suspension plasma spraying (SPS) was performed at a plasma current of 700 A and voltage of 46 V. A combination of argon (49 L/min) and hydrogen (0.9 L/min) gases was used as a plasma source, and argon (3 L/min) was used as a carrier gas. The number of passes was kept constant, i.e., 50, during plasma trials. The experimental setup for spraying is shown in Figure 1 and the summary of plasma spraying is given in Table 1.

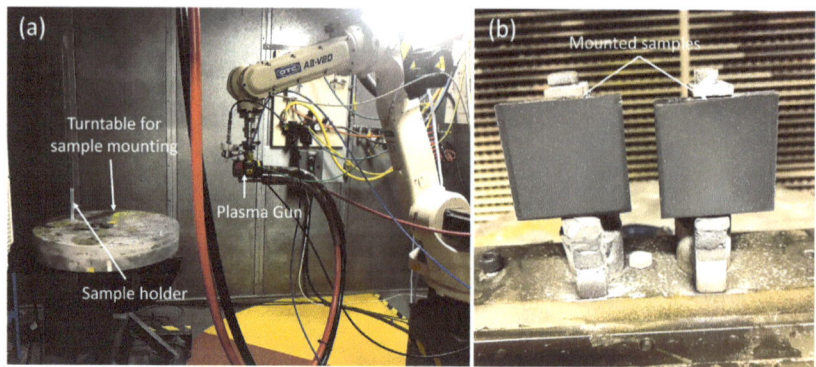

Figure 1. (**a**) The experimental setup for spraying and (**b**) mounted (coated) samples.

Table 1. Summary of suspension plasma spray parameters.

Sample Name	Feedstock Flow Rate (mL/min)	Stand-Off Distance (mm)
Ti-50-10	10	50
Ti-50-20	20	50
Ti-50-30	30	50
Ti-75-10	10	75
Ti-75-20	20	75
Ti-75-30	30	75
Ti-100-10	10	100
Ti-100-20	20	100
Ti-100-30	30	100

2.4. Coating Characterisation

The coating mass was determined by measuring the difference in weight between the sample after and before the coating application. A Sartorius CPA324S four-figure Analytical Balance (Göttingen, Germany) was employed for this weighing process. Deposition efficiency (DE), which indicates the productivity of the process for different spray conditions, was calculated based on ISO 17836:2017 [21].

An EVO LS15 SEM (Zeiss, Jena, Germany) was used to characterise the top-view morphology and the cross-section of the coated samples. An acceleration voltage of 20 kV was used at a working distance of 8.5 mm. EDX analysis of developed coatings was performed using an EDAX spectrometer, and the data were taken at 1K magnification. For cross-sectional SEM imaging, cross-sections were cut using a slow-speed precision saw and mounted in cold EpoFix resin (Buehler, Lake Bluff, IL, USA), followed by grinding (SiC papers P120, P320, P600, P1200, and P2500 from Abrasives, Brighouse, UK) and polishing (with 3 µm and 1 µm clothes). The quantitative evaluation of the porosity of deposited coatings was performed using ImageJ software (version 1.8.0; provided by the National Institutes of Health, Bethesda, MD, USA). For that, 10 images of each coating type along the cross-section were taken randomly at 1000× magnification factor. For image analysis, image filtering was applied to remove noise, followed by image segmentation by thresholding to obtain binary images. The measured pores were categorized based on three area ranges, namely 0.01–1 µm^2, 1–10 µm^2, and >10 µm^2, referring to the fine, medium, and large range pores, respectively. XRD analysis was performed to assess the phase composition of suspension plasma spray TiO$_2$ coatings, using a Bruker AXS D8 diffractometer (Billerica, MA, USA) with a Cu Kα radiation source with λ 0.1542 nm at 40 KV and 100 mA. Scans were taken from 10° to 90° 2θ range with 0.02 steps per second scan rate. The volume percentage of the anatase phase (C_A) in a rutile–anatase mixture can be calculated using the following formula (Equation (1)) [22,23]:

$$C_A = \frac{8I_A}{(8I_A + 13\,I_R)} \qquad (1)$$

where I_A and I_R are X-ray peak intensities of anatase (101) and rutile (110) phases, respectively. Also, anatase and rutile phase crystallite size was calculated from XRD data using the Scherrer equation. A non-contact 3D optical profilometer (Alicona InfinteFocus SL; Alicona Imaging GmbH, Graz, Austria) was used to characterise the surface roughness profile of the deposited specimens using a 5× objective. The analysed area was 25 mm × 25 mm for each coating. A tape adhesion test based on ASTM D3359-17 [24] was performed by cutting a 5 × 5 (at least) grid pattern of 1 × 1 mm squares into the sample surface using an Elcometer 1540 Cross Hatch Cutter (Elcometer Inc., Warren, MI, USA) to ensure that the scraped channels went through the coating and into the steel substrate beneath. Once the pattern was cut, Elcometer 99 Adhesion Test tape (Elcometer Inc.) was carefully adhered to the surface. It was then pulled off at 180° to the sample surface in a single motion, and samples were examined for material loss and delamination.

3. Results and Discussion

The average weight gain of all samples is presented in Figure 2a. It can be seen from the figure that the weight increased along with the increased feedstock flow rate, which can be explained by the fact that when more feedstock material is inserted into the plasma flame, after fragmentation, bigger droplets are formed, which travel in the core region of the plasma and do not deviate from the path [25]. Hence, more material is deposited on the substrate, meaning more weight gain. In the case of SOD, the weight gain fluctuates with the feedstock flow rate. At a low feedstock flow rate (10 mL/min), the weight gain decreases with the increased SOD. This could be because, at a lower feedstock flow rate, smaller droplets with less material are formed, which, after the evaporation of water, give rise to small-size agglomerates. Small droplets/particles tend to follow the gas trajectory and deviate from the central region, leading to less material deposition [25]. Simultaneously, coatings deposited at higher SOD (100 mm) with a higher flow rate (30 mL/min) show more weight gain due to the formation of larger droplets hitting the substrate.

The deposition efficiency (DE) of all samples is shown in Figure 2b, and it was found that all samples show low DE. When spraying suspension feedstock, the solvent evaporates, and only solid content reaches the substrate. It should be noted that DE is the underestimated value due to the loss of feeding material during acceleration and deceleration and overspray of the torch robot [14]. Although a larger substrate can help resolve overspray to some extent (for smaller substrate, due to being carried away with the airflow, sprayed particles tend to overshoot the edges and be lost to the surrounding environment; meanwhile, in the case of the larger substrate, a greatest distance is travelled by sprayed particles before reaching the edges, retaining more material on the substrate), the loss of feedstock material during acceleration and deceleration is unavoidable. The comparison of DEs is still valid as these accuracies are standard for each sample. Low DE is also associated with the radial injection of the feedstock material because, sometimes, the small droplets cannot reach the core region of the plasma plume. Instead, they deflect from their path, leading to insufficient energy exchange, and hence poor deposition occurs. It was observed that at a lower flow rate (10 mL/min), with the increased SOD, DE decreases. This can be elucidated by the fact that when a smaller amount of material is introduced into the plasma after fragmentation, it leads to the formation of smaller droplets. These smaller droplets may not efficiently reach the substrate as they tend to follow the gas trajectory. So, more material is deposited at a smaller SOD (50 mm), where the substrate is comparatively closer to the gun, and the kinetic energy is higher than at longer SODs (100 mm). Meanwhile, DE increases for a higher feedstock flow rate (30 mL/min) at longer SODs because, due to the injection of more feedstock material, larger droplets are formed that have more inertia and higher kinetic energy compared to the smaller droplets. Hence, more material is deposited.

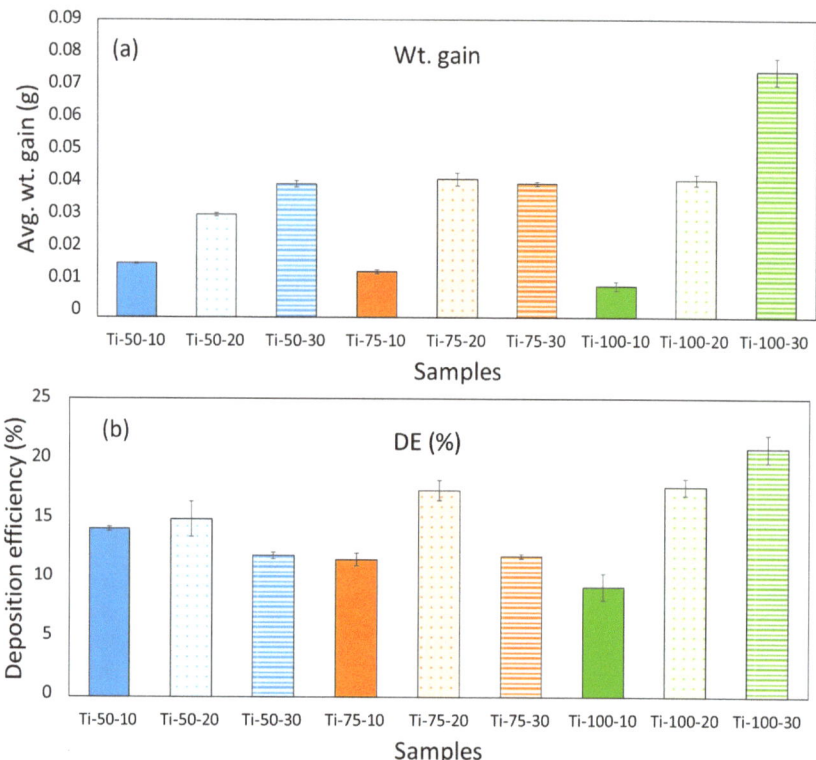

Figure 2. (**a**) Average weight gain and (**b**) deposition efficiency of different TiO$_2$ coatings.

The SEM images of TiO$_2$ coatings at different magnifications are shown in Figures 3 and 4. All coatings show typical cauliflower-like topography due to the deposition of feedstock material in the asperities of the substrate after hitting it at shallow angles. This phenomenon is referred to as the "shadow effect", and it becomes more pronounced when dealing with smaller droplets. These smaller droplets deviate from their intended path as they follow the gas trajectory, often depositing on substrate asperities while moving parallel to the substrate surface [25].

It was found that fully-melted and thinner splats are formed at smaller SODs (Figure 3; circle 1), while thicker splats with partially melted or re-solidified feedstock droplets (Figure 3; circle 7) take place at longer SODs. This happens because, at smaller SODs, the thermal and kinetic energy of formed droplets are higher than the longer SODs. Hence, fully melted droplets hit the substrate with a higher impact, spreading and covering the substrate properly. Meanwhile, at longer SODs, due to the insufficient thermal and kinetic energy transfer, rougher and thicker splats are formed. Also, at longer distances, nanomaterials, because of their high surface area, cool down very quickly and re-solidify even before hitting the substrate or sticking to the surface, giving rise to voids after being trapped between subsequent splats [26].

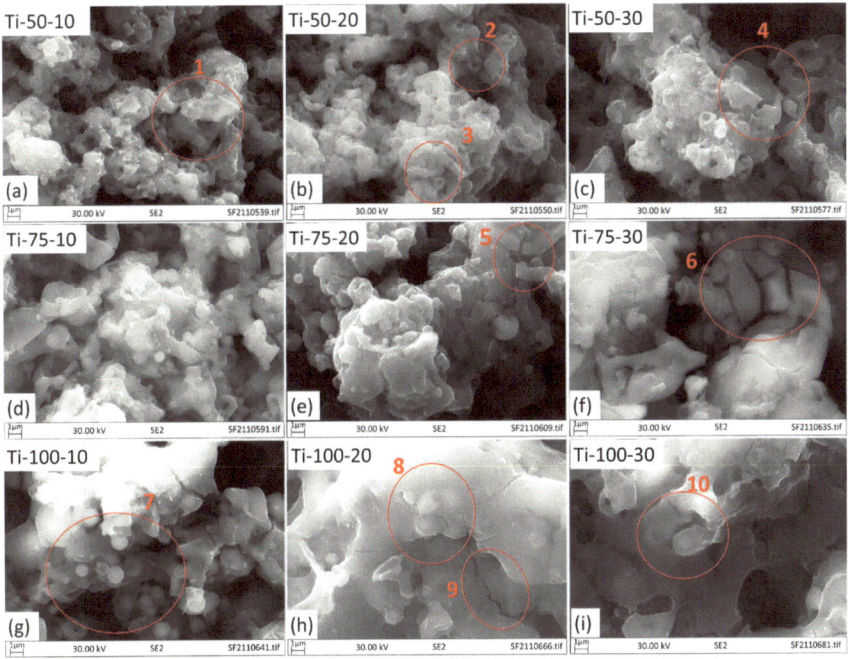

Figure 3. SEM images of different TiO$_2$ coatings at 1000× magnification.

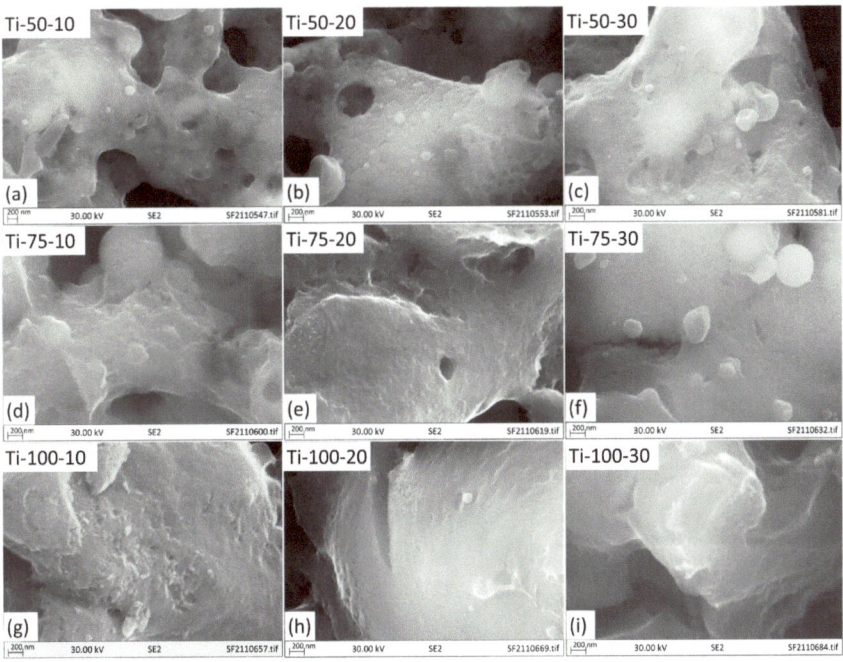

Figure 4. SEM images of different TiO$_2$ coatings at high magnification (10,000×).

With the increased feedstock flow rate, rougher splat formation, partial melting of feedstock droplets (Figure 3; circles 2, 3, 8 & 9), and mud-like crack formation (Figure 3;

circles 4, 5 & 6) also increase due to solvent evaporation. This can be explained by noting that when additional feedstock material is introduced, a greater volume of solvent (water) is also injected into the system. This additional solvent serves to cool down the flame. This leads to a comparatively lower heat transfer from the plasma to feedstock droplets. Hence, rougher and thicker splats and partial or no melting of TiO_2 droplets take place [27]. When these partially molten or unmolten particles deposit on the substrate, there is a difference in their temperature, microstructures, mechanical properties, and bonding characteristics, and the non-uniform distribution of these particles could result in localized stress concentrations, potentially generating mud-crack formation in certain parts of the coating. Figure 4 shows the splat morphology of coatings at higher magnification. As can be seen from the figure, the surface of formed splats is very smooth for coatings deposited at 50 mm SOD with a 10 mL/min feedstock flow rate. As the SOD and flow rate increases, the splat surface becomes comparatively rougher, and the presence of semi- or unmelted particles also increases.

From EDX analysis, it was found that all coatings show peaks related to elements Ti, O, Fe, and C. Also, the Fe peak intensity decreases with the increased flow rate which can be ascribed to the increased thickness of the deposited coating with the increased flow rate. For instance, the EDX pattern of coatings deposited at 75 mm SOD using 10, 20, and 30 mL/min feedstock flow rates are shown in Figure 5. Also, it was found that the coatings deposited at 100 mm SOD using a 30 mL/min feedstock flow rate exhibited high carbon content which could be possible because of the presence of surfactant (0.1%–1.0%) in the TiO_2 suspension. When more feedstock is injected into the plasma, due to the reduction in enthalpy, poor melting takes place (which is also evident from SEM data), implying the presence of the surfactant in the Ti-100-30 sample.

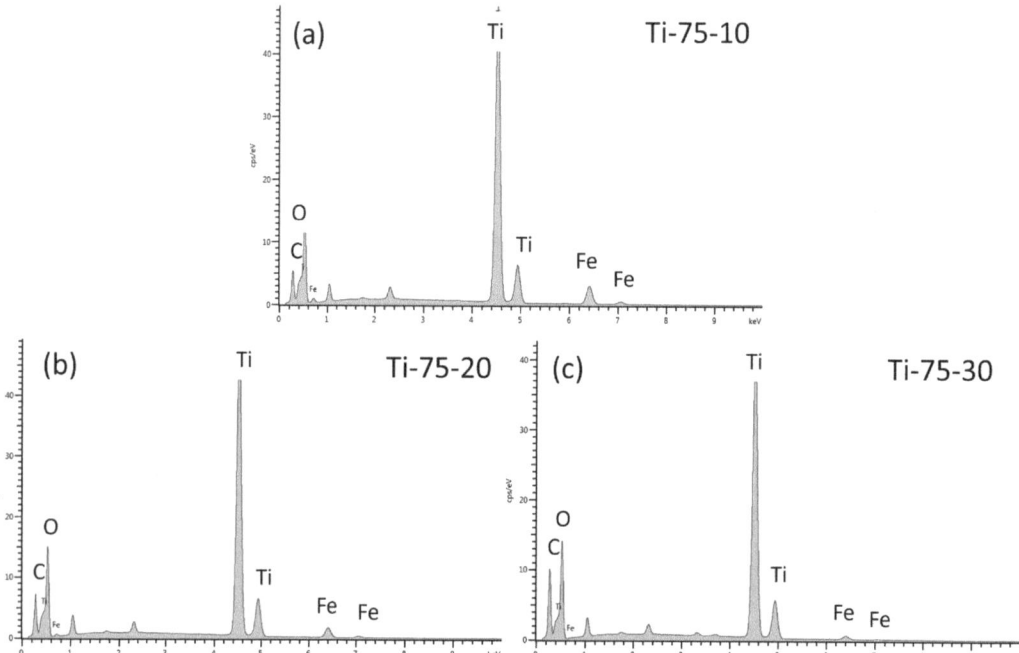

Figure 5. EDX pattern of TiO_2 coatings (a) Ti-75-10, (b) Ti-75-20, and (c) Ti-75-30.

Cross-sectional back-scattered SEM images of developed coatings are shown in Figure 6. Since the substrate was pre-treated with alumina grit blasting before coating deposition for better adhesion between the coating and the substrate, the formation of columnar structures is more favourable than vertical cracks or smooth coatings [28]. This happens because when particle droplets travel toward the substrate perpendicularly, only big particles possessing a high moment of inertia hit the substrate perpendicularly. Smaller particles easily be influenced by the drag force of the plasma trajectory and deviate from their original direction, impacting the substrate at a shallow angle, and deposit on the nearest asperities, forming columnar features (shadowing effect). After subsequent deposition and vertical and horizontal growth, cone-shaped structures are developed, where inter-columnar voids separate these cones/columns, and the heads of the columns give rise to a typical cauliflower-like morphology [29].

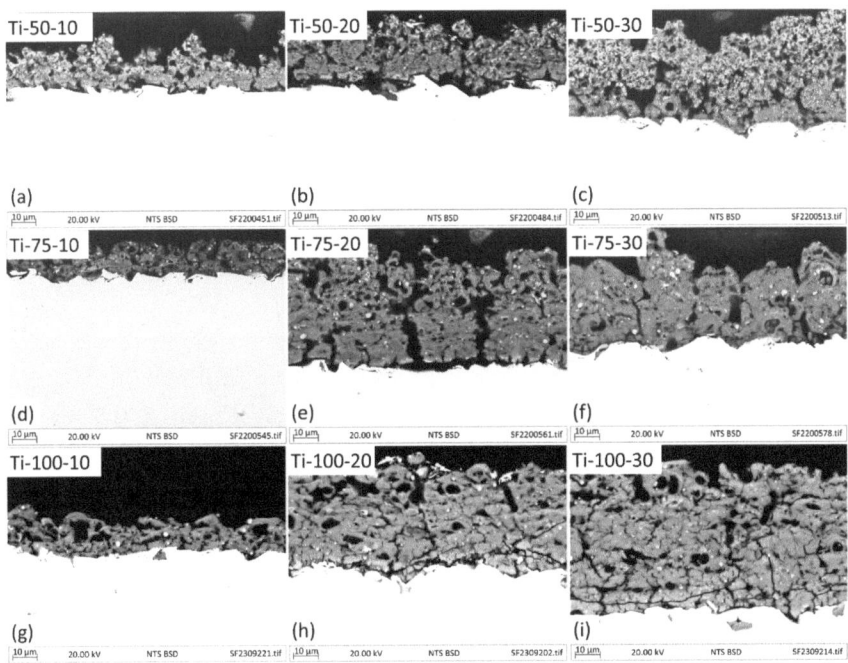

Figure 6. Cross-sectional back-scattered SEM images of different TiO_2 coatings.

Coatings deposited using 10 mL/min do not exhibit column formation prominently due to the inefficient coverage of the substrate (thinner coatings). Simultaneously, coatings deposited at 50 mm SOD show sharper peaks in cross-sections because of the high impact and kinetic energy of the particles, which hit the substrate with high accelerated force, depositing well-spread splats over the substrate's asperities. As the coating becomes thicker, column formation and their separation (inter-columnar voids) can be observed clearly (Figure 6e,f) [29]. At longer SODs, due to inefficient kinetic and thermal energy exchange, thicker splats are formed that give rise to wide and comparatively round-shaped peaks. Also, it was found that with the increased feedstock flow rate, partially melted, unmelted, and agglomerated/sintered feedstock particles were observed more prominently, leading to more porous coating with vertical crack formation (Figure 6c). The coatings at 50 mm SOD, especially deposited using 20 and 30 mL/min feedstock flow rates, exhibited the presence of more partially melted or unmelted feedstock particles and more porosity as compared to the coatings developed at 75 mm SOD and 20 and 30 mL/min feedstock flow rates (Figure 6e,f). This can be explained by the fact that at a shorter distance, feedstock droplets do not have sufficient time for solvent evaporation, agglomeration, sintering, and melting, consequently giving rise to more porous coatings [30]. Meanwhile, at longer SODs, melted material sometimes cools down even before hitting the substrate, and rather than forming a splat, adheres to the previously deposited splats.

The porosity data are presented in Figure 7, and it was found that the total porosity area percentage for all the coatings was in a range from 23%–37%. There was no discernible trend observed based on the variations in spray parameters. In terms of different pore sizes, the percentage of fine pores was almost similar for all coatings. For coatings deposited at 75 mm, the area percentage of large pores was higher than coatings deposited at 100 mm SOD. The possible reason behind this could be that the column formation in 75 mm SOD coatings gave rise to vertical cracks which acted as large pores during image analysis. However, for 100 mm SOD coatings, the major contribution in area percentage of large pores was from the voids that developed because of poorly-/semi-molten or unmelted coating particles.

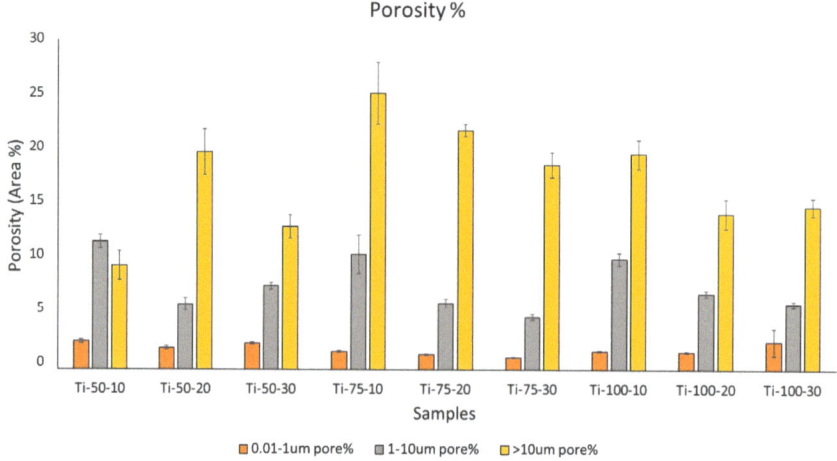

Figure 7. Comparison of relative pore range contribution to the total porosity of various TiO_2 coatings.

The spray parameters are critical in influencing the particle impact energy, coating microstructures, and coating roughness. The R_a (average roughness), R_q (root mean square roughness), R_z (maximum height roughness), R_{ku} (kurtosis: sharpness of the surface's peaks and valleys), and R_{sk} (skewness: asymmetry of the surface) values are shown in Figure 8. It was found that with an increased feedstock flow rate, the R_a, R_q, R_z, and R_{sk} increase because at a higher feedstock flow rate, particle accumulation, over-deposition, and insufficient particle flattening take place, resulting in a coating with higher porosity and rougher surface texture. With the feedstock flow rate, R_{ku} was increased but only for 50 mm SOD. In comparison, for 75 mm and 100 mm SODs, R_{ku} was almost similar, which means that at 50 mm SOD, due to the comparatively high impact of particles, sharper peaks and valleys are formed on the substrate's asperities. For Ti-50-20, the R_{sk} value was negative, indicating that there were more valleys or depressions on the surface than peaks.

The roughness values were almost similar for the coatings deposited at different SODs using a low feedstock flow rate, i.e., 10 mL/min. In comparison, this trend changes at higher 20 and 30 mL/min, resulting in a decreased roughness with the increased SOD. This can be elucidated by the fact that, at lower feedstock flow rates, the formation of flattened splats with extensive substrate coverage occurs. This results in a well-adhered coating due to the adequate melting of the coating material. Also, it can be concluded that at higher SODs, the sharpness of peaks/columns decreases because the insufficient kinetic and thermal energy exchange gives rise to thicker (disc-like) splats, forming columns with globular heads after subsequent deposition [31,32].

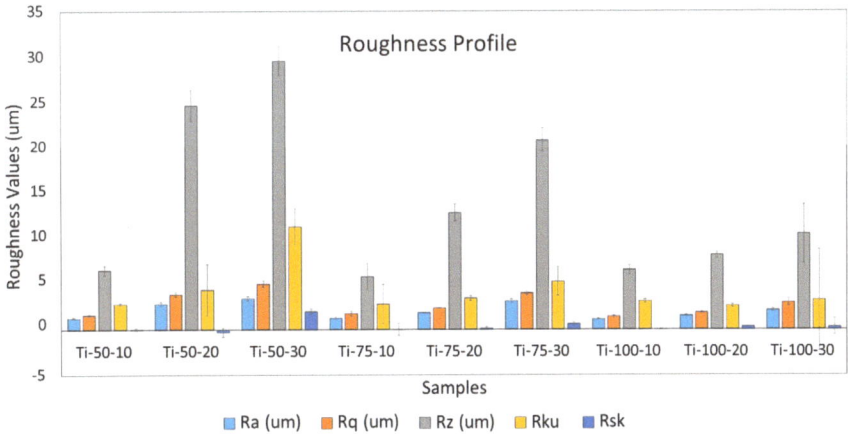

Figure 8. Roughness profile of different TiO$_2$ coatings.

The crystalline structure of TiO$_2$ coatings deposited at different spray conditions was analysed by comparing XRD data, shown in Figure 9. All coatings were found to show two main phases of TiO$_2$, i.e., anatase (PDF ID: 03-065-5714) and rutile (PDF ID: 01-089-0555), and characteristic peaks related to the substrate. The peak intensity was highest for the coatings developed using a low feedstock flow rate because of low coating thickness. As the feedstock flow rate increases, the peak intensity of the substrate's characteristic peaks decreases. Various spray parameters, including SOD and feedstock flow rate, affect the phase formation and crystallite size of SPS coatings [18–20]. The crystallite size of coatings was calculated using the Scherrer equation applied to anatase (101) and rutile (110) peaks (Table 2). Regarding the anatase crystallite size, a reduction was noted with higher feedstock flow rates and a decrease in stand-off distance (SOD). This can be clarified by understanding that at longer spray distances, there is more time in-flight before reaching the substrate. This extended travel favours the formation of larger agglomerates, allowing for sufficient sintering of the material and leading to an increase in crystallite size [33]. At a higher feedstock flow rate, insufficient sintering or re-solidification due to the insertion of more solvent and reduced flame enthalpy leads to reduced crystallite size. For the rutile phase, it was observed that coatings deposited using 20 mL/min exhibited larger crystallites. The possible explanation behind this could be that at 20 mL/min, more solid content is present in the droplets, which, after efficient melting, agglomeration, and sintering, form larger crystallites as compared to the coatings deposited using 30 mL/min, where comparatively insufficient melting occurs. Although no significant difference was observed in the anatase content of all deposited coatings, a slight increase was found in the anatase content with the increased SOD (Table 2). With the increased SOD, agglomerates have more time to re-solidify during flight, forming anatase phase proportions due to homogeneous nucleation. At the same time, agglomerates that solidify on the substrate (in the case of shorter SODs) can form anatase or rutile phases depending on various factors, including agglomerate size. Usually, for agglomerates solidifying on the substrate, rutile phase formation is favoured due to the lower Gibbs free energy. However, if the agglomerate size is in the submicron range, metastable anatase phase formation takes place due to the fast solidification, suppressing the heterogeneous nucleation [33–35].

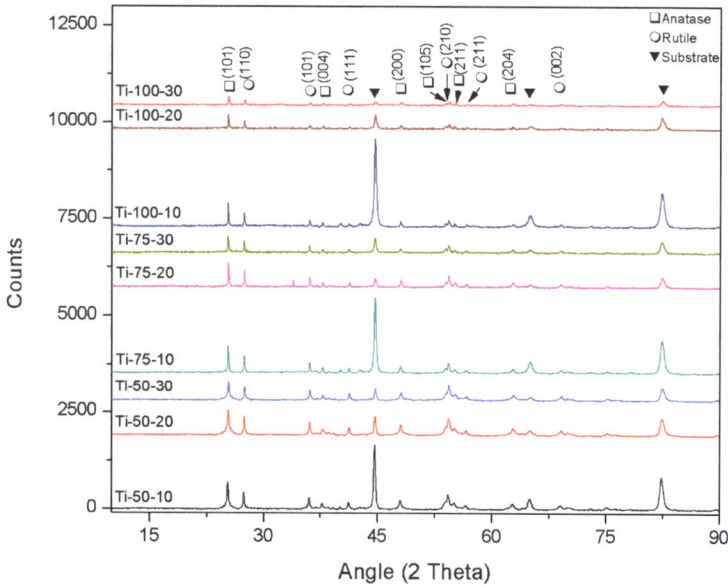

Figure 9. Comparison of XRD pattern of different TiO$_2$ coatings developed at varied SODs and feedstock flow rates.

Table 2. Comparison of volume percentage of anatase phase and anatase and rutile phase crystallite size of different TiO$_2$ coatings developed at varied SOD and feedstock flow rate.

Sample	Anatase Content (C$_A$); %	Crystallite Size (nm)	
		Anatase	Rutile
Ti-50-10	48	24.15	31.94
Ti-50-20	41	17.2	36.57
Ti-50-30	42	18.03	29.25
Ti-75-10	47	44.99	45.29
Ti-75-20	45	37.62	91.33
Ti-75-30	44	33.96	48.19
Ti-100-10	50	67.06	52.83
Ti-100-20	45	50.99	136.95
Ti-100-30	43	36.86	71.28

The macrographs of different coatings after performing tape adhesion tests (shown in Figure 10) show that the coatings deposited at a lower SOD, i.e., 50 mm, were well adhered to the substrate due to high impact and comparatively better thermal energy transfer at low SODs between the torch and feedstock particles. Also, coatings deposited using a low feedstock flow rate, i.e., 10 mL/min, showed good adhesion. When higher feedstock flow rates are used, along with more material, more solvent (water) is introduced, hence improper melting of particles due to cooling down of the plasma flame [14]. Therefore, Ti-75-20, Ti-75-30, Ti-100-20, and Ti-100-30 showed a high degree of delamination compared to others.

Figure 10. Macrographs of different TiO$_2$ coatings after tape adhesion tests.

4. Summary and Conclusions

Coatings with superior performance help extend the lifetime of geothermal heat exchangers and improve the overall efficiency of geothermal power plants. Thermal spray TiO$_2$ coating could be an option to protect the components of geothermal heat exchangers from corrosion and scaling. In this study, TiO$_2$ coatings were deposited through suspension plasma spray using a uniform aqueous suspension of 5 wt.% TiO$_2$ nanoparticles on carbon steel (S275JR, EN 10025-2) substrates. To understand the role of different spray parameters in coating microstructure formation, the coatings were deposited at different stand-off distances using different feedstock flow rates. It was found that at a small SOD, due to the proximity with the plasma torch, coating particles possessed high thermal and kinetic energy, which led to well-melted and smooth splat formation, hence, compact coatings. However, as the feedstock flow rate increased at a small SOD, poor melting of particles due to cooling of plasma plume, excessive material insertion, and insufficient time took place, and hence porous coatings were formed. At larger SODs, due to the poor thermal and kinetic energy exchange between the torch and feedstock particles, resolidification of feedstock particles occurred and gave rise to porous coatings. The tape adhesion tests showed that coatings deposited at 50 mm SOD (Ti-50-10, Ti-50-20 and Ti-50-30) and coatings deposited using a 10 mL/min feedstock flow rate (Ti-75-10 and Ti-100-10) were well adhered to the substrate in comparison to the coatings deposited using higher feedstock flow rates and at longer SODs (Ti-75-20, Ti-75-30, Ti-100-20, and Ti-100-30). The outcomes of this work offer a platform for further optimization and development of protective coatings for geothermal heat exchangers.

Author Contributions: Conceptualization, S.P.; methodology, S.P. and G.M.; investigation, G.M. and N.G.M.; formal analysis, G.M.; writing—original draft preparation, G.M. and A.B.N.J.; writing—review and editing, S.P.; supervision, S.P. and D.M.; project administration, D.M. and N.K. All authors have read and agreed to the published version of the manuscript.

Funding: This project has received funding from the European Union's Horizon 2020 research and innovation programme under the project: GeoHex-advanced material for cost-efficient and enhanced heat exchange performance for geothermal application (Grant agreement 851917).

Institutional Review Board Statement: Not applicable.

Informed Consent Statement: Not applicable.

Data Availability Statement: Data are contained within the article.

Acknowledgments: We want to acknowledge the contribution of the GeoHex consortium to this project.

Conflicts of Interest: The authors declare no conflict of interest.

References

1. Nogara, J.; Zarrouk, S.J. Corrosion in geothermal environment: Part 1: Fluids and their impact. *Renew. Sustain. Energy Rev.* **2018**, *82*, 1333–1346. [CrossRef]
2. Benea, L.; Bounegru, I.; Forray, A.; Axente, E.R.; Buruiana, D.L. Preclinical EIS Study of the Inflammatory Response Evolution of Pure Titanium Implant in Hank's Biological Solution. *Molecules* **2023**, *28*, 4837. [CrossRef] [PubMed]
3. Nakashima, Y.; Umehara, N.; Kousaka, H.; Tokoroyama, T.; Murashima, M.; Mori, D. Carbon-based coatings for suppression of silica adhesion in geothermal power generation. *Tribol. Int.* **2023**, *177*, 107956. [CrossRef]
4. Cadelano, G.; Bortolin, A.; Ferrarini, G.; Bison, P.; Dalla Santa, G.; Di Sipio, E.; Bernardi, A.; Galgaro, A. Evaluation of the Effect of Anti-Corrosion Coatings on the Thermal Resistance of Ground Heat Exchangers for Shallow Geothermal Applications. *Energies* **2021**, *14*, 2586. [CrossRef]
5. Zhang, F.; Tabecki, A.; Bennett, M.; Begg, H.; Lionetti, S.; Paul, S. Feasibility Study of High-Velocity Oxy-fuel (HVOF) Sprayed Cermet and Alloy Coatings for Geothermal Applications. *J. Therm. Spray Technol.* **2023**, *32*, 339–351. [CrossRef]
6. Kim, D.E.; Yu, D.I.; Jerng, D.W.; Kim, M.H.; Ahn, H.S. Review of boiling heat transfer enhancement on micro/nanostructured surfaces. *Exp. Therm. Fluid Sci.* **2015**, *66*, 173–196. [CrossRef]
7. Fanicchia, F.; Karlsdottir, S.N. Research and Development on Coatings and Paints for Geothermal Environments: A Review. *Adv. Mater. Technol.* **2023**, *8*, 2202031. [CrossRef]
8. Oppong Boakye, G.; Straume, E.O.; Gunnarsson, B.G.; Kovalov, D.; Karlsdottir, S.N. Corrosion behaviour of HVOF developed Mo-based high entropy alloy coating and selected hard coatings for high temperature geothermal applications. *Mater. Des.* **2023**, *235*, 112431. [CrossRef]
9. Kumar, M.; Upadhyaya, R. Corrosion and wear analysis of HVOLF sprayed WC-10CO-4Cr coating on geothermal turbine blade. *World J. Eng.* **2019**, *16*, 768–774. [CrossRef]
10. Thorhallsson, A.I.; Fanicchia, F.; Davison, E.; Paul, S.; Davidsdottir, S.; Olafsson, D.I. Erosion and Corrosion Resistance Performance of Laser Metal Deposited High-Entropy Alloy Coatings at Hellisheidi Geothermal Site. *Materials* **2021**, *14*, 3071. [CrossRef]
11. Jayakumari, A.B.N.; Malik, N.G.; Mittal, G.; Martelo, D.; Kale, N.; Paul, S. Coatings for Geothermal Heat Exchangers. *Preprints* **2023**. [CrossRef]
12. Cai, Y.; Quan, X.; Li, G.; Gao, N. Anticorrosion and Scale Behaviors of Nanostructured ZrO_2–TiO_2 Coatings in Simulated Geothermal Water. *Ind. Eng. Chem. Res.* **2016**, *55*, 11480–11494. [CrossRef]
13. Mahu, G.; Munteanu, C.; Istrate, B.; Benchea, M.; Lupescu, S. Influence of Al_2O_3-13TiO_2 powder on a C45 steel using atmospheric plasma spray process. *IOP Conf. Ser. Mater. Sci. Eng.* **2018**, *444*, 032010. [CrossRef]
14. Mittal, G.; Paul, S. Suspension and Solution Precursor Plasma and HVOF Spray: A Review. *J. Therm. Spray Tech.* **2022**, *31*, 1443–1475. [CrossRef]
15. Fauchais, P.; Vardelle, M.; Vardelle, A.; Goutier, S. What Do We Know, What are the Current Limitations of Suspension Plasma Spraying? *J. Therm. Spray Technol.* **2015**, *24*, 1120–1129. [CrossRef]
16. Tejero-Martin, D.; Rezvani Rad, M.; McDonald, A.; Hussain, T. Beyond Traditional Coatings: A Review on Thermal-Sprayed Functional and Smart Coatings. *J. Therm. Spray Technol.* **2019**, *28*, 598–644. [CrossRef]
17. Chen, X.; Kuroda, S.; Ohnuki, T.; Araki, H.; Watanabe, M.; Sakka, Y. Effects of Processing Parameters on the Deposition of Yttria Partially Stabilized Zirconia Coating During Suspension Plasma Spray. *J. Am. Ceram. Soc.* **2016**, *99*, 3546–3555. [CrossRef]
18. Zhang, X.; Wang, Y.; Zhao, D.; Guo, J. Improved thermal performance of heat exchanger with TiO_2 nanoparticles coated on the surfaces. *Appl. Therm. Eng.* **2017**, *112*, 1153–1162. [CrossRef]
19. Yan, W.; Lin-lin, W.; Ming-yan, L. Antifouling and enhancing pool boiling by TiO_2 coating surface in nanometer scale thickness. *AIChE J.* **2007**, *53*, 3062–3076. [CrossRef]
20. Wang, L.L.; Liu, M.Y. Pool boiling fouling and corrosion properties on liquid-phase-deposition TiO_2 coatings with copper substrate. *AIChE J.* **2011**, *57*, 1710–1718. [CrossRef]
21. ISO 17836:2017; Thermal Spraying Determination of the Deposition Efficiency for Thermal Spraying. ISO: Geneva, Switzerland, 2017. Available online: https://www.iso.org/standard/69754.html (accessed on 30 August 2023).
22. Berger-Keller, N.; Bertrand, G.; Filiatre, C.; Meunier, C.; Coddet, C. Microstructure of plasma-sprayed titania coatings deposited from spray-dried powder. *Surf. Coat. Technol.* **2003**, *168*, 281–290. [CrossRef]
23. Toma, F.-L.; Sokolov, D.; Bertrand, G.; Klein, D.; Coddet, C.; Meunier, C. Comparison of the photocatalytic behavior of TiO_2 coatings elaborated by different thermal spraying processes. *J. Therm. Spray Technol.* **2006**, *15*, 576–581. [CrossRef]
24. Standard Test Methods for Rating Adhesion by Tape Test. Available online: https://www.astm.org/d3359-17.html (accessed on 30 August 2023).
25. Mauer, G.; Vaßen, R. Coatings with Columnar Microstructures for Thermal Barrier Applications. *Adv. Eng. Mater.* **2020**, *22*, 1900988. [CrossRef]

26. Du, L.; Coyle, T.W.; Chien, K.; Pershin, L.; Li, T.; Golozar, M. Titanium Dioxide Coating Prepared by Use of a Suspension-Solution Plasma-Spray Process. *J. Therm. Spray Technol.* **2015**, *24*, 915–924. [CrossRef]
27. Rampon, R.; Marchand, O.; Filiatre, C.; Bertrand, G. Influence of suspension characteristics on coatings microstructure obtained by suspension plasma spraying. *Surf. Coat. Technol.* **2008**, *202*, 4337–4342. [CrossRef]
28. Caio, F.; Moreau, C. Influence of Substrate Shape and Roughness on Coating Microstructure in Suspension Plasma Spray. *Coatings* **2019**, *9*, 746. [CrossRef]
29. Bernard, B.; Bianchi, L.; Malié, A.; Joulia, A.; Rémy, B. Columnar suspension plasma sprayed coating microstructural control for thermal barrier coating application. *J. Eur. Ceram. Soc.* **2016**, *36*, 1081–1089. [CrossRef]
30. Joulia, A.; Duarte, W.; Goutier, S.; Vardelle, M.; Vardelle, A.; Rossignol, S. Tailoring the Spray Conditions for Suspension Plasma Spraying. *J. Therm. Spray Technol.* **2015**, *24*, 24–29. [CrossRef]
31. Sokołowski, P.; Kozerski, S.; Pawłowski, L.; Ambroziak, A. The key process parameters influencing formation of columnar microstructure in suspension plasma sprayed zirconia coatings. *Surf. Coat. Technol.* **2014**, *260*, 97–106. [CrossRef]
32. Sokołowski, P.; Pawłowski, L.; Dietrich, D.; Lampke, T.; Jech, D. Advanced Microscopic Study of Suspension Plasma-Sprayed Zirconia Coatings with Different Microstructures. *J. Therm. Spray Technol.* **2016**, *25*, 94–104. [CrossRef]
33. Robinson, B.W.; Tighe, C.J.; Gruar, R.I.; Mills, A.; Parkin, I.P.; Tabecki, A.K.; de Villiers Lovelock, H.L.; Darr, J.A. Suspension plasma sprayed coatings using dilute hydrothermally produced titania feedstocks for photocatalytic applications. *J. Mater. Chem. A* **2015**, *3*, 12680–12689. [CrossRef]
34. Alebrahim, E.; Tarasi, F.; Rahaman, M.S.; Dolatabadi, A.; Moreau, C. Fabrication of titanium dioxide filtration membrane using suspension plasma spray process. *Surf. Coat. Technol.* **2019**, *378*, 124927. [CrossRef]
35. Bemporad, E.; Bolelli, G.; Cannillo, V.; De Felicis, D.; Gadow, R.; Killinger, A.; Lusvarghi, L.; Rauch, J.; Sebastiani, M. Structural characterisation of High Velocity Suspension Flame Sprayed (HVSFS) TiO_2 coatings. *Surf. Coat. Technol.* **2010**, *204*, 3902–3910. [CrossRef]

Disclaimer/Publisher's Note: The statements, opinions and data contained in all publications are solely those of the individual author(s) and contributor(s) and not of MDPI and/or the editor(s). MDPI and/or the editor(s) disclaim responsibility for any injury to people or property resulting from any ideas, methods, instructions or products referred to in the content.

Article

Fe/TaC Coatings Produced on 145Cr6 Steel by Laser Alloying—Manufacturing Parameters and Material Characterization

Dariusz Bartkowski [1,*] and Aneta Bartkowska [2]

[1] Faculty of Mechanical Engineering, Poznan University of Technology, 60-965 Poznań, Poland
[2] Faculty of Materials Engineering and Technical Physics, Poznan University of Technology, 60-965 Poznań, Poland; aneta.bartkowska@put.poznan.pl
* Correspondence: dariusz.bartkowski@put.poznan.pl

Abstract: This paper focuses on Fe/TaC composite coatings produced on 145Cr6 steel by laser alloying a TaC precoat in paste form. Fe/TaC coatings were produced in two consecutive steps. The first stage was the application of a precoat in paste form made from tantalum carbide and water glass on a steel substrate. Three TaC precoat thicknesses were produced: 30 μm, 60 μm and 90 μm. In the second step, the TaC precoat was remelted on a steel substrate using a 3 kW rated diode laser beam. A constant laser beam scanning speed of 3 m/min and three laser beam powers were used: 500 W, 800 W and 1100 W. In the study, microstructure, microhardness, chemical and phase composition and wear resistance were tested. The aim of the research was to check the possibility of producing composite coatings in which the reinforcing phase will be TaC, and the role of the matrix will be played by the material from the substrate. It was found that it is possible to produce the continuous composite coatings by remelting the TaC precoat with steel substrate. As microhardness increased, so did wear resistance. The coating microhardness obtained ranged from about 750 to 850 HV0.05 depending on the parameters used.

Keywords: laser processing; tantalum carbide; Fe/TaC coating; metal matrix composite coating; microhardness; wear resistance

1. Introduction

The construction and tooling materials currently used for specialist products often have a fairly high price. An alternative way to increase their durability is to apply an appropriate surface treatment that will increase hardness and ensure good friction wear resistance in the subsurface zone while maintaining a ductile core. The most commonly used surface treatments are carburizing and nitriding processes. Despite their many advantages, a significant drawback is their high energy intensity. Therefore, special attention should be paid to technologies that use high-energy heat sources such as a laser beam or plasma [1–3]. This type of heat source is used in laser [1,2] or plasma remelting [3], laser hardening [1,2], laser alloying [4,5] or laser cladding [6–12]. These technologies are used to modify tool surfaces or machine parts [1,13]. A laser beam is used on various metal alloys (Fe, Ni, Ti and Al), out of which iron-based alloys are the most common [5,8,9]. Laser processing allows for the modification of selected fragments of the product, often in hard-to-reach places. In addition, these processes can be easily automated, and thus changes in the material can be controlled. A laser beam is successfully used to produce hard carbide-containing coatings on products used in mining and agriculture [8]. Remelting substrate material with carbides produces a composite layer. The most commonly used reinforcing phases are tungsten carbides such as WC and W_2C [8,11,12], silicon carbides [6], boron carbides [7], titanium carbides [9] or zirconium carbides [10]. In scientific publications, much less attention is paid to tantalum carbides, which also have many important properties. Tantalum carbide has one of the highest melting points of any known chemical compound. It is used as a material for the production of tools as well as a component of casting molds, where it has a positive effect by reducing friction between meld surfaces and the object cast, and is an important component of cermets. One of the few dated research papers in which the authors studied topics related to the production of the TaC layer was the work of Teghil et al. [14]. The authors used the pulsed laser deposition technique to obtain TaC layers deposited on graphite substrates. Based on the obtained results, they only showed that the laser technique may be suitable for the production of TaC films with noticeable homogeneity and adhesion. There are studies available in which researchers focus on the surface treatment of sintered carbide tools containing carbides WC, TiC and TaC [13,15]. In paper [15], the authors carried out laser heating of the surfaces of sintered carbide tools. They found that Co and TaC melt in the early heating stage due to their relatively lower thermal conductivity than WC. On the other hand, in paper [16] the authors deposited thin layers of Ti, Zr, Hf and Ta carbides on a titanium substrate using the pulsed laser ablation method. The layers obtained were smooth and compact. A majority of available studies focus on the production of a layer containing tantalum carbides by in situ synthesis [17–23]. In study [17], an in situ synthesized TaC composite coating reinforced with Ni-based solid particles was produced on steel in the process of laser cladding a mixture of Ni60 alloy powder with admixture (Ta_2O_5 + C). It was shown that the coating is metallurgically bonded to the substrate and has a homogeneous, fine-grained microstructure. Compared to the Ni60 coating, the hardness of the TaC/Ni60 composite coating was increased. In paper [19], the authors aiming to improve the wear resistance of copper strengthened it by adding TaC particles, which were synthesized in situ by laser surface modification with a mixture of NiCrBSi + (Ta_2O_5 + C) powders. TaC particles synthesized in situ were uniformly dispersed in the solid solution matrix. The modified layer showed higher hardness and better wear resistance. In [20], the authors added tantalum to the laser-clad NiCrBSi coating in order to improve the wear resistance. The in situ synthesized TaC particles of a nearly equiaxed shape were uniformly dispersed in the coating. TaC molecules had good bonding with the matrix and tended to crumble and squeeze into the matrix instead of being pulled out of the wear surface in the abrasion process. NiCrBSi coating with Ta showed higher crack resistance and higher abrasive and adhesive wear resistance than the NiCrBSi coating. In paper [22], the authors produced Ta-reinforced, cobalt-based composite coatings using direct laser deposition technology on the surface of martensitic stainless steel. The results indicate that the in situ synthesized TaC phases can effectively mitigate the impact of particulate matter in the erosion process and improve the erosion resistance of cobalt-based composite coating. The authors in study [23] examined the problem of cracking and the porosity of high-strength composite Ni60A/WC laser clad coatings. The influence of varying contents of Ta powder on phase composition, microstructure, microhardness and friction properties of the Ni60A/WC composite coating was investigated. The Ta powder was added to synthesize TaC in situ with carbon in a melt pool. The results show that Ta combines more easily with carbon. Fine TaC particles synthesized in situ are used as nucleation sites to significantly fragment the grains and improve microstructure homogeneity. As the content of Ta powder increases, the microhardness and wear resistance of the composite coating first increases and then decreases. This finding is closely related to the fragmentation of the microstructure, uniform distribution of the hard phase and dissolution of WC particles. The authors of study [24] investigated the properties of laser-clad nickel-based powders with the addition of TaC nanopowders. They showed that the microhardness of the coating with the addition of tantalum carbide nanopowder differs from the microhardness of the cladding of a standard nickel-based powder. A tendency to increase microhardness along with a reduced concentration of TaC nanoparticles was also shown. In paper [25], the authors produced laser-clad composite coatings deposited on a Ti_6Al_4V substrate using mixed Ni-based alloy powders with a varying TaC content (0, 5, 10, 15, 20, 30 and 40% by weight). They found that with an increase in TaC content, the coefficient of friction of these coatings showed a decreasing trend due to the formation of Ta_2O_5. In paper [26], the authors added tantalum during laser cladding of the NiCrBSi coating in order to improve high temperature wear resistance. At 700 °C, the wear of the original NiCrBSi coating is 1.5 times greater than that of the Ta-reinforced coating. This is attributed to the Ni matrix reinforced with Ta, TaC of high hardness and thermal stability. In paper [27], the authors examined composite coatings based on TaC/Stellite X-40 Co produced on a nickel–aluminum bronze substrate by laser surface plating. The microhardness, wear resistance and corrosion resistance of the TaC MMC coatings were significantly improved compared to the substrate.

A review of the literature on the subject clearly indicates that available publications mainly focus on the in situ production of tantalum carbide layers, and the majority of studies focused on the analysis of the impact of tantalum addition on tungsten carbide. In this paper, only tantalum carbide particles without the addition of other particles were used. This study presents a continuation of preliminary studies where in the first stage of the research, mainly microstructure changes in single tracks were analyzed. The purpose of these presented studies is to determine the impact of tantalum carbide on the friction wear resistance of multiple tracks produced in terms of their application.

2. Materials and Methods of Research

2.1. Materials

Fe/TaC composite coatings were produced on 145Cr6 tool steel. The chemical composition of the steel substrate is presented in Table 1 and it is in accordance with the manufacturer's attestation. Tantalum carbide powder (TaC) with a density of 14.3 g/cm^3 was used to make the primer. The morphology of TaC powder particles was observed by scanning electron microscopy (SEM) and is shown in Figure 1. TaC molecules were characterized by an irregular shape and the average particle size (APS) was less than 6 µm. The purity of the TaC powder used in the tests was 99.9%.

Table 1. Chemical composition of steel used [wt.%].

C	Mn	Si	P	S	Cr	V	Fe
1.35	0.60	0.30	0.02	0.02	1.45	0.20	Balance

Figure 1. Morphology of TaC powder particles (SEM image).

2.2. Precoat Production and Laser Processing Parameters

Before testing, the surface of the steel samples was ground to obtain uniform surface roughness. The samples were then purified with alcohol and finally degreased with acetone.

In the first step of producing the composite coatings, paste-form TaC precoats were applied to the steel surfaces. This paste consisted of TaC powder as a base material and a solution of sodium water glass with distilled water as a binder. The consistency of the prepared paste was crucial due to the ease and possibility of applying it with a brush. A paste that is too liquid does not allow for the formation of a precoat of a suitable thickness. Too dense paste, on the other hand, does not allow for maintaining a homogeneous thickness. The paste was therefore prepared in such a way that for every 10 g of tantalum carbide, there were 3 mL of water glass and 3 mL of distilled water. Exactly the same proportions were used in the preliminary studies [28]. Precoat thicknesses were measured using the ultrasonic thickness sensor PosiTector® 6000 Advance (DeFelsko Corporation, Ogdensburg, NY, USA) with an accuracy of ±2 µm. Only precoats characterized by the same thickness over the entire surface of the sample were subjected to laser treatment. The second stage was the laser remelting process. A TruDiode 3006 (TRUMPF, Ditzingen, Germany) diode laser with a rated power of 3 kW was used. The laser head was controlled by a 5-axis KR16-2 robotic arm (KUKA, Augsburg, Germany). Three continuous laser beam powers of 500 W, 800 W and 1100 W were used. The laser beam scanning speed was constant during all processes and was 3 m/min. The laser beam diameter was 1 mm while its wavelength was 1040 nm. In contrast to preliminary tests, in which only single tracks were performed, in the tests described in this article we focused on multiple tracks that allow for the production of full-size coatings. This process involved shifting the laser beam from one edge of the sample to the other and turning off the laser. Then, the laser head was returned to the starting point, after which it was moved transversely by 0.5 mm. The laser was turned on and the first move was repeated. This was repeated until the entire surface of the sample was laser remelted. The laser track overlap was 50%. The laser remelting scheme is shown in Figure 2.

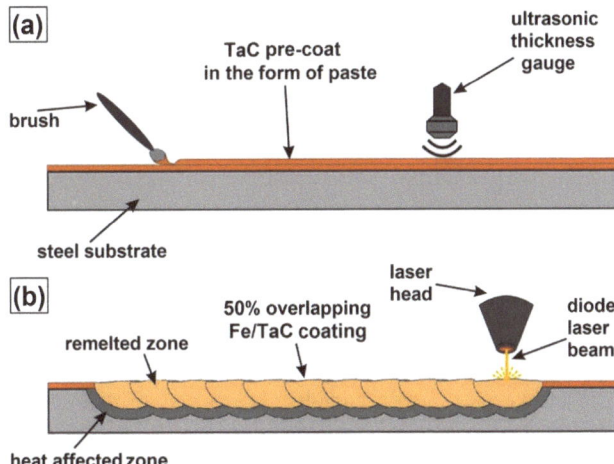

Figure 2. Scheme of Fe/TaC coatings production using laser remelting: pre-coat preparation (**a**), laser processing (**b**).

2.3. Microstructure Investigation

Microstructure observations were made on cross-sections of samples perpendicular to the produced coatings on a scanning electron microscope with the Schottky emission gun (FEG SEM) MIRA3 (TESCAN, Brno, Czech Republic) and Eclipse MA200 light microscope (Nikon, Tokyo, Japan). Cross-sections of steel samples with Fe/TaC composite coatings produced were ground and polished using the Mecatech 250 device from PRESI (Eybens, France). Grinding and polishing discs were used for hard materials in accordance with the manufacturer's recommendations. In order to reveal the microstructure of the produced coatings, the cross-sections were digested in a 5% solution of HNO_3 for 45 s.

2.4. Chemical and Phase Composition Examination

In order to determine the chemical composition of Fe/TaC composite coatings, energy dispersive X-ray spectroscopy (EDS) was used. The Ultim Max 65 spectrometer (Oxford Instruments, High Wycombe, UK) and Aztec Energy Live Standard software (Oxford Instruments, Abingdon, UK) were applied. The results of the study are presented as a map of the chemical elements. During this study, a working distance equal to 15.0 mm with accelerating voltage 10.0 kV was used. Phase composition studies were also carried out using the X-ray diffraction (XRD) method. An EMPYREAN (Malvern PANalytical, Malvern, UK) X-ray diffractometer equipped with a ceramic X-ray tube with a Cu anode was used to identify the phases. In the tests, a voltage of 45 kV and a current of 40 mA were used, and the temperature during the tests was 25 °C. The step size was 0.0330 °2Θ and scan step time was 596,900 s.

2.5. Microhardness Measurements

Microhardness tests of the Fe/TaC coatings were carried out on cross-sections perpendicular to the coating from the surface to the steel core using Future-Tech's FM-810 (Kawasaki, Japan) microhardness meter equipped with an FT-Zero (Future-Tech, Kawasaki, Japan) automatic measurement program. The Vickers method was used. Each recess was made at a load of 50 g and a loading time of 15 s. The results obtained allowed for the execution of microhardness profiles. The TaC hardness is known (reaching even 2000 HV); therefore, attempts were made to perform hardness measurements not directly in the place of the reinforcing phase but between them—in the matrix.

2.6. Wear Resistance Tests

Wear resistance tests were carried out on an MBT-01 Amsler type (Poznan, Poland) tribometer. The test sample was a steel sample with an Fe/TaC composite coating, while the counter-sample was a ring made of Hardox steel with a hardness of 35 HRC. Dry friction conditions were used with the following parameters: counter-sample rotational speed of 250 rpm, pressure force on the sample of 147N and friction time of 60 min. The weight loss of the sample was checked every 10 min by measuring on the AS220 analytical scale R2 from RADWAG (Poznan, Poland). The aim of abrasion resistance tests was to find a correlation between laser beam power, precoat thickness, microhardness and friction effects on the produced Fe/TaC composite coatings.

3. Results and Discussion

3.1. Macroscopic Observation and Microstructure Analysis

Figure 3 shows macroscopic images of the Fe/TaC coatings produced. It can be seen that track geometry is influenced by both laser beam power and precoat thickness. Increasing the laser beam power from 500 W to 1100 W clearly resulted in the formation of irregular tracks, regardless of the precoat thickness. For example, comparing two Fe/TaC coatings produced using a 30 μm precoat, but one using a 500 W laser beam power (Figure 3a), and the other at 1100 W (Figure 3g), the widening of the tracks of the latter and its irregular shape are clearly visible. A higher laser beam power causes a very large remelting and mixing of the material on the surface. It should be noted, however, that the thicker the precoat, the greater the irregularity. This is related to the consistency of the applied paste and difficulties in its application in thicker precoats.

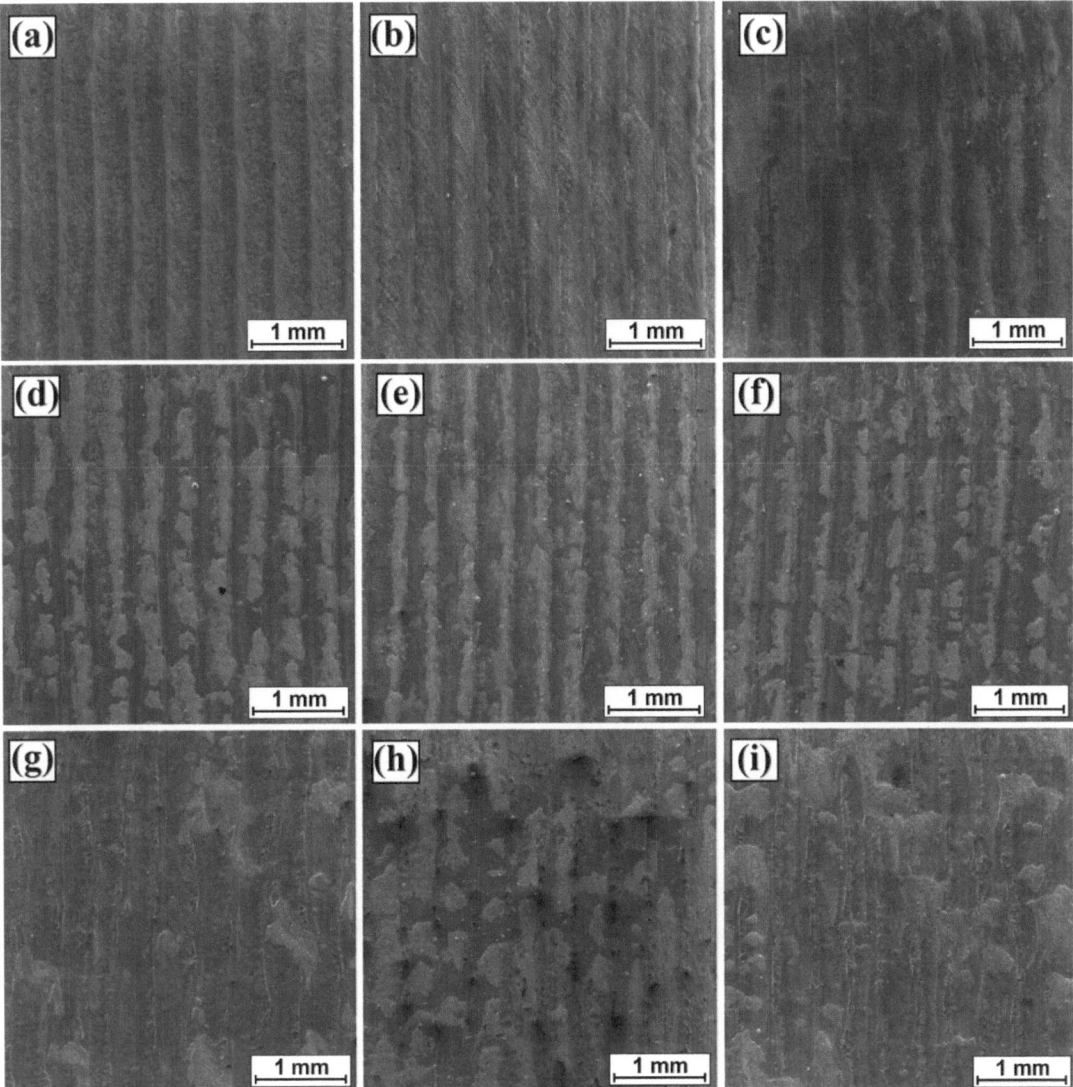

Figure 3. Macroscopic images of Fe/TaC coatings produced using: (**a**) 500 W and 30 μm precoat, (**b**) 500 W and 60 μm precoat, (**c**) 500 W and 90 μm precoat, (**d**) 800 W and 30 μm precoat, (**e**) 800 W and 60 μm precoat, (**f**) 800 W and 90 μm precoat, (**g**) 1100 W and 30 μm precoat, (**h**) 1100 W and 60 μm precoat and (**i**) 1100 W and 90 μm precoat.

Full-size coatings produced using the parameters developed as part of the preliminary study were described in the work [28]. Figure 4 shows all the coatings along their entire cross-sections. It can be seen that as laser beam power increases, a fairly thick heat-affected zone is produced. Due to the fact that the sample heats up in the process, this zone increases in the direction of coating production. Bright spots on the sample cross-section are the sites of TaC powder agglomerates, which, due to the high melting point, were not remelted by the laser beam. These agglomerates formed mainly when the thickest precoat was produced and are most visible in Figure 4c,i.

Figure 4. Cross-section of Fe/TaC coatings produced using: (**a**) 500 W and 30 μm precoat, (**b**) 500 W and 60 μm precoat, (**c**) 500 W and 90 μm precoat, (**d**) 800 W and 30 μm precoat, (**e**) 800 W and 60 μm precoat, (**f**) 800 W and 90 μm precoat, (**g**) 1100 W and 30 μm precoat, (**h**) 1100 W and 60 μm precoat and (**i**) 1100 W and 90 μm precoat.

In order to take a closer look at the microstructure of individual Fe/TaC coatings, images were taken in scanning microscopy (Figures 5–7). Figure 5 shows the coatings produced at a laser beam power of 500 W and the thickness of the TaC precoat coating, respectively: 30 μm (Figure 5a,d,g), 60 μm (Figure 5b,e,h) and 90 μm (Figure 5c,f,i). It was found that as the thickness of the TaC precoat increased, the tendency to form TaC powder agglomerates increased. These agglomerates are clearly visible in Figure 5c. An increased number of TaC particles makes it difficult to produce a homogeneous paste. TaC powder particles with a melting point of over 3700 °C melt in the laser processing and by combining bring about agglomerate formation. This is a problem that was partially solved by increasing the laser beam power. An analysis of the microstructure at higher magnification can lead to a conclusion that the coatings are made of matrix (dark area) and carbide mesh (Figure 5d), in which eutectic can be observed (Figure 5g). An increase in the amount of TaC carbides caused by the use of a thicker precoat gradually gives rise to the appearance of only partially remelted (Figure 5h) or unmelted TaC particles (Figure 5f). However, it can be stated that these particles also have the form of two or more tantalum carbide particles joined together (Figure 5i). Such a structure proves the composite nature of some of the produced Fe/TaC coatings.

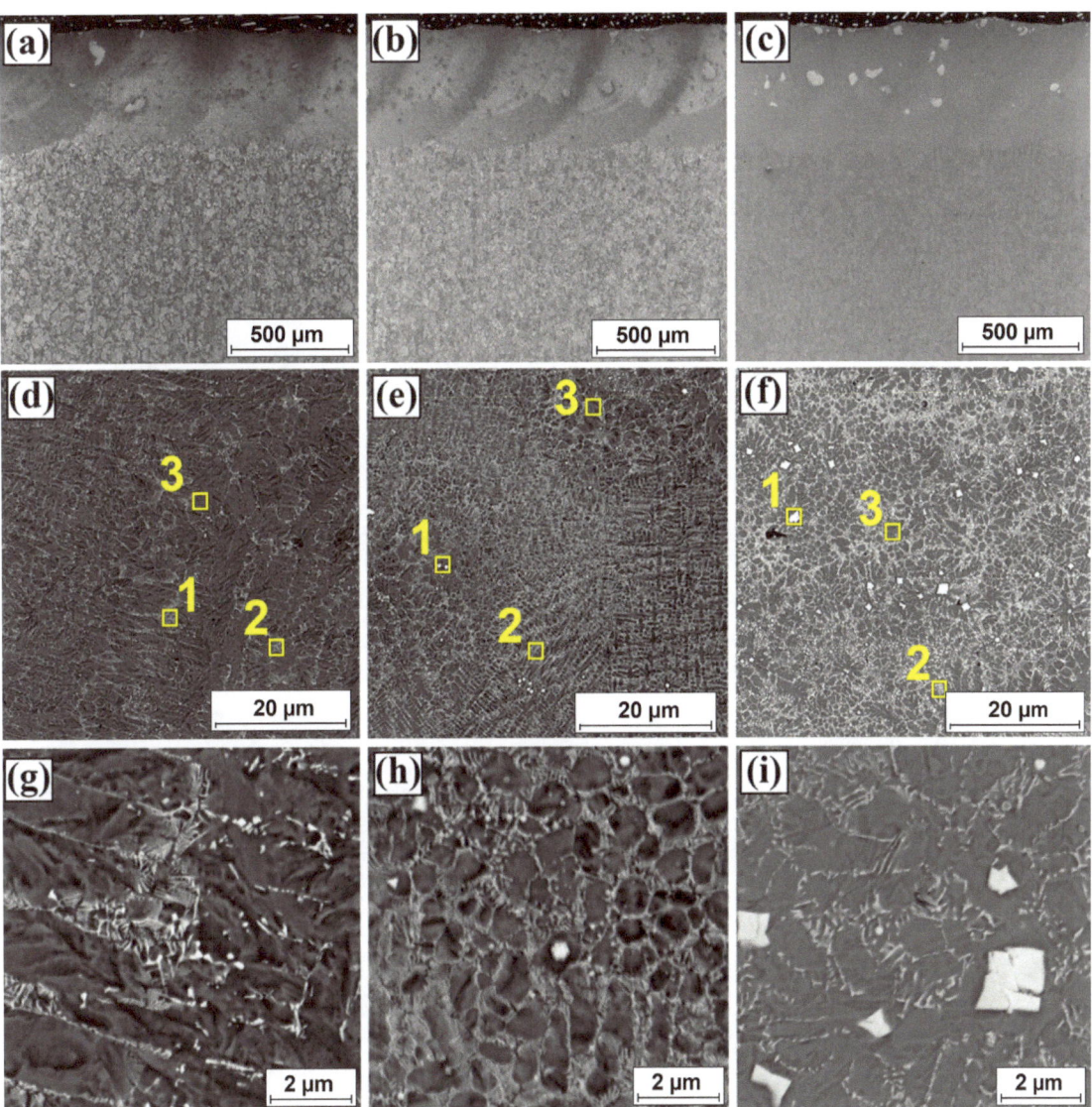

Figure 5. Microstructure of Fe/TaC coatings produced using: (**a,d,g**) 500 W and 30 μm precoat, (**b,e,h**) 500 W and 60 μm precoat and (**c,f,i**) 500 W and 90 μm precoat.

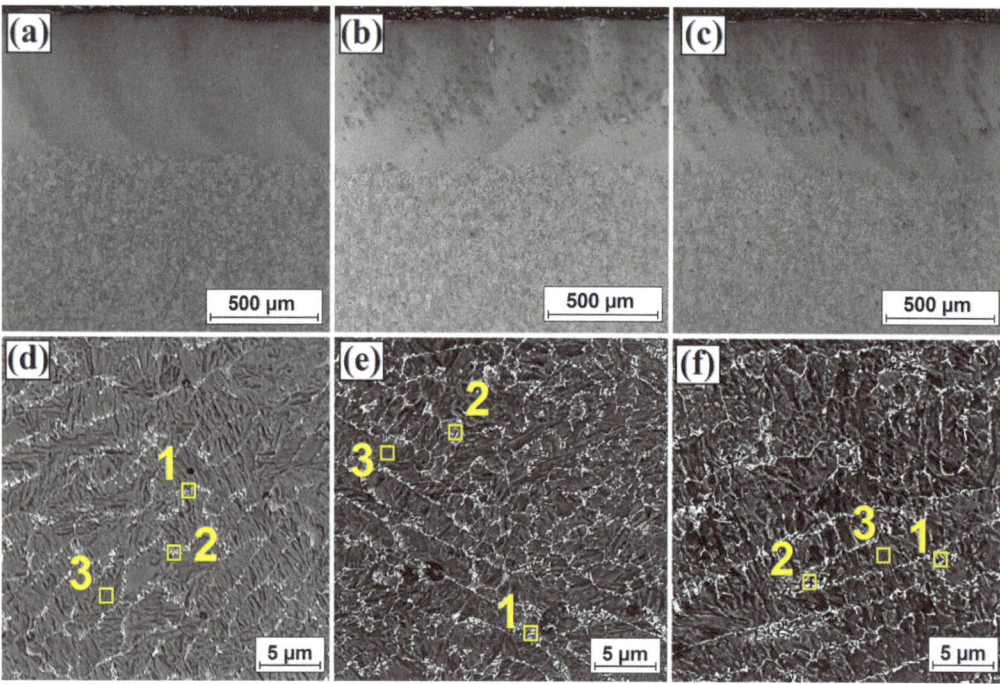

Figure 6. Microstructure of Fe/TaC coatings produced using: (**a**,**d**) 800 W and 30 μm precoat, (**b**,**e**) 800 W and 60 μm precoat and (**c**,**f**) 800 W and 90 μm precoat.

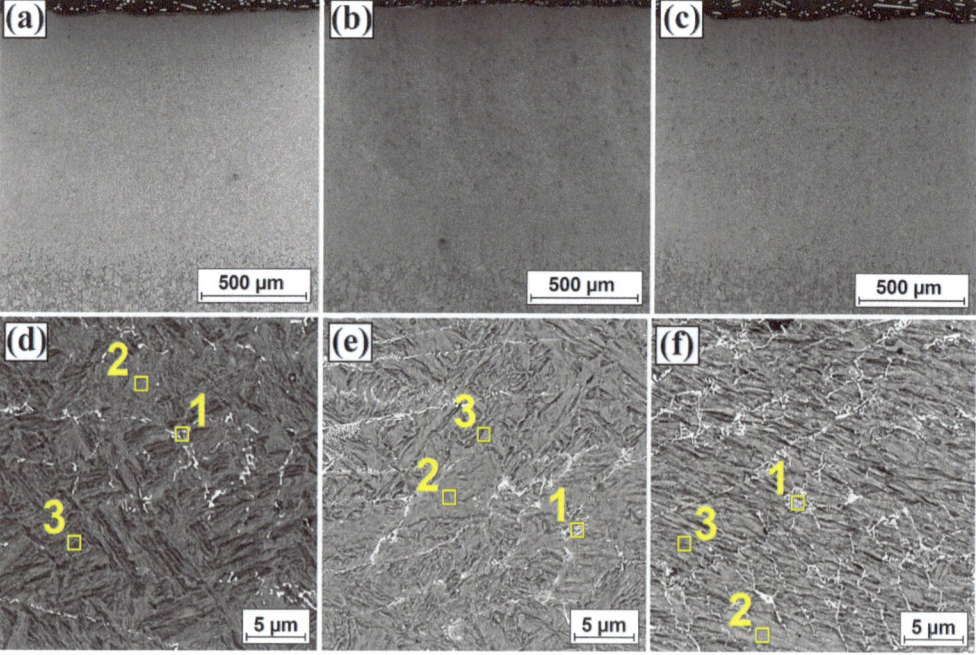

Figure 7. Microstructure of Fe/TaC coatings produced using: (**a**,**d**) 1100 W and 30 μm precoat, (**b**,**e**) 1100 W and 60 μm precoat and (**c**,**f**) 1100 W and 90 μm precoat.

Figure 6 shows the coatings produced at a laser beam power of 500 W and the thickness of the TaC precoat, respectively, as 30 μm (Figure 6a,d), 60 μm (Figure 6b,e) and 90 μm (Figure 6c,f). An increase in the thickness of the remelted zone was found in comparison to the coatings produced at a laser beam power of 500 W. The increase in the laser beam power resulted in a complete remelting of both TaC particles and possibly agglomerates formed of those particles that may have been present in the precoat. It was found that the most significant change in the microstructure was the increased intensity of the occurrence of the secondary carbide mesh. It can be seen that when using a 30 μm precoat (Figure 6d), white precipitates in mesh form are much fewer than in a coating produced using a precoat with a thickness of 90 μm (Figure 6f). The same relationship can be observed in Figure 7, where the coatings produced at a laser beam power of 1100 W and the thickness of the TaC precoat of 30 μm (Figure 7a,d), 60 μm (Figure 7b,e) and 90 μm (Figure 7c,f) are shown, respectively. An increase in the precoat thickness resulted in an increase in the number of secondary carbide precipitates in mesh form. However, the use of a laser beam power of 1100 W significantly contributed to the overall reduction of the presence of the mesh in the structure. This is due to an increase in the amount of iron in the produced coating which comes from the steel substrate.

3.2. Chemical and Phase Composition Results

The results of the EDS point analysis for the sites marked with yellow squares in Figures 5–7 are presented in Table 2. Characteristic areas of the obtained microstructure were analyzed, i.e., light areas in the form of precipitates and mesh, as well as dark areas forming the coating matrix. Three chemical elements were taken into account: tantalum, carbon and iron, all derived from the steel substrate. It can be seen that increased tantalum content occurs in bright areas. Bright sharp precipitates are most likely tantalum carbides, as indicated by their high content at the level of approximately 71 wt.% (point 1 of Figure 5, Table 2). Increased tantalum content was also found in bright areas forming the eutectic. Here, its percentage content is about 20–30 wt.%, depending on the pre-applied paste thickness and laser beam power. It can be clearly seen that with the increase in the thickness of the TiC precoat, the share of tantalum increases both in the precipitates in mesh form and in the matrix of the coating produced. A similar relationship also occurs when the laser beam power is changed. Here, the greater the laser beam power, the lesser the tantalum amount. On the other hand, the iron content increases significantly. An analysis of the effect of laser beam power on the chemical composition of the Fe/TaC coating produced using the thickest TaC precoat showed that tantalum content in the substrate area decreased from 16.7 wt.% at 500 W to 1.9 wt.% at 1100 W. A similar relationship was also observed in paper [23], where the authors analyzed the effect of the amount of Ta introduced on the properties of the Ni60A/WC laser-alloyed composite coatings.

In order to check what kind of phases were present in the produced surface layers, XRD phase composition analysis tests were carried out (Figure 8). The presence of peaks derived from the TaC and Ta_2C carbide phases as well as from the $TaFe_2$ and Fe phases was found. The influence of laser beam power on phase composition for the thickest TaC precoat was analyzed. It can be seen that at a laser beam power of 500 W, there were distinct peaks derived from TaC and Ta_2C carbides as well as from the tantalum-rich phase $TaFe_2$. An increase in laser beam power caused a decrease in the intensity of the peaks or even their complete disappearance with a simultaneous increase in the intensity of the peak derived from iron.

Table 2. Chemical composition of the areas marked with a squares in Figures 5–7 [wt.%].

Coating Type/Test Zone	No	Fe	Ta	C
Figure 5. 500 W 30 µm	1	81.6	12.6	5.8
	2	76.8	16.7	6.5
	3	86.2	7.7	6.1
Figure 5. 500 W 60 µm	1	42.2	51.9	5.9
	2	72.4	21.7	5.8
	3	79.2	14.3	6.5
Figure 5. 500 90 µm	1	22.8	71.6	5.6
	2	64.9	30.8	4.3
	3	78.4	16.7	4.8
Figure 6. 800 W 30 µm	1	83.2	12.5	4.3
	2	78.9	15.2	5.9
	3	92.5	3.5	4.1
Figure 6. 800 W 60 µm	1	75.6	17.8	6.7
	2	75.5	18.3	6.2
	3	90.1	3.9	6.0
Figure 6. 800 W 90 µm	1	67.9	24.4	7.7
	2	77.0	16.3	6.7
	3	89.0	3.8	7.3
Figure 7. 1100 W 30 µm	1	76.6	17.5	5.9
	2	93.5	1.4	5.1
	3	94.4	1.4	4.2
Figure 7. 1100 W 60 µm	1	75.0	18.9	6.1
	2	90.9	3.0	6.1
	3	91.7	2.6	5.7
Figure 7. 1100 W 90 µm	1	71.3	19.9	8.8
	2	91.8	2.9	5.4
	3	91.9	1.9	6.2

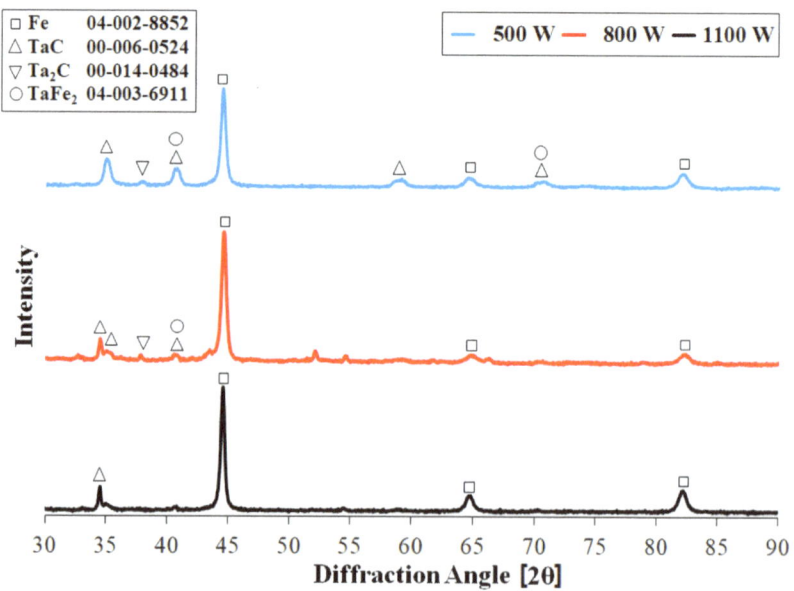

Figure 8. Phase composition (XRD) of Fe/TaC coatings produced using a 90 µm precoat and different types of laser beam power.

3.3. Microhardness Observations

Figures 9–11 show the results of the measurements of the microhardness of tantalum carbide laser-alloyed coatings for the precoat thickness of 30 µm, 60 µm, and 90 µm, respectively. In all the analyzed samples, microhardness gently decreased from the remelted zone enriched in tantalum, through the heat-affected zone, to the substrate, reaching a hardness of approximately 300 HV0.05. Despite the intense movements caused by Marangoni convection forces, the obtained microhardness profiles for the analyzed samples are characterized by a very uniform distribution of microhardness over the entire depth of the remelted zone. A gradual reduction of hardness from the surface towards the substrate of the material is advantageous for potential applications of the coating produced. Figure 9 shows the microhardness profiles for a 30 µm TaC precoat thickness. The highest hardness was obtained by remelting the coating at a laser beam power of 500 W. Here, the hardness in the remelted zone was approximately 750 HV0.05. An increase in laser beam power to 800 W and 1100 W contributed to a reduction in hardness and its value in the remelted zone ranged from 700 HV0.05 to 620 HV0.05. Figure 10 shows the microhardness profiles for a TaC precoat thickness of 60 µm. Here, an increase in hardness for the laser beam powers of 500 W and 800 W by approximately 50–80 units can be seen in comparison to the coating produced at 30 µm. The use of a higher laser beam power does not favorably affect the microhardness. It can be seen, however, that a higher laser beam power results in a lower hardness in the SWC. This is due to the intense heating of one track from the other, and thus slower cooling in this zone. The best results of microhardness measurements were obtained for the precoat thickness of TaC 90 µm (Figure 11). As a result of remelting such a coating thickness with a laser beam power of 500 W, a microhardness of about 850 HV0.05 was obtained. The increase in laser beam power resulted in lower microhardness values at the level of approximately 750 HV0.05 for 800 W and 650 HV0.05 for 1100 W. Analyzing the microhardness graphs, it can be concluded that the use of a low laser beam power and the thickest precoat results in the production of a coating with a favorable hardness. This is also influenced by the rate of heating and cooling, which at a lower power results in a finer microstructure. The authors of paper [27] laser-alloyed a TaC/Stellite X-40 composite coating with nickel–aluminum bronze. Here, the Stellite X40 alloy contained a large amount of Cr element, which also strengthens the solution and contributed to increasing the coating hardness. As in this paper, the authors found that coating hardness increased significantly with an increased TaC content.

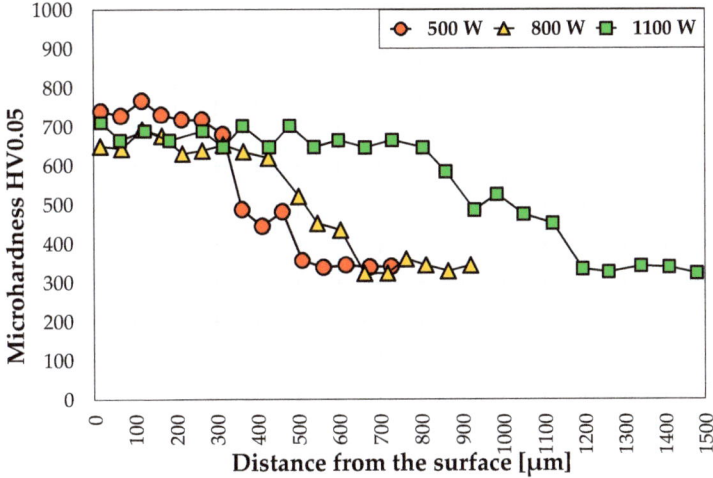

Figure 9. Microhardness of Fe/TaC coatings produced using a 30 µm precoat and different types of laser beam power.

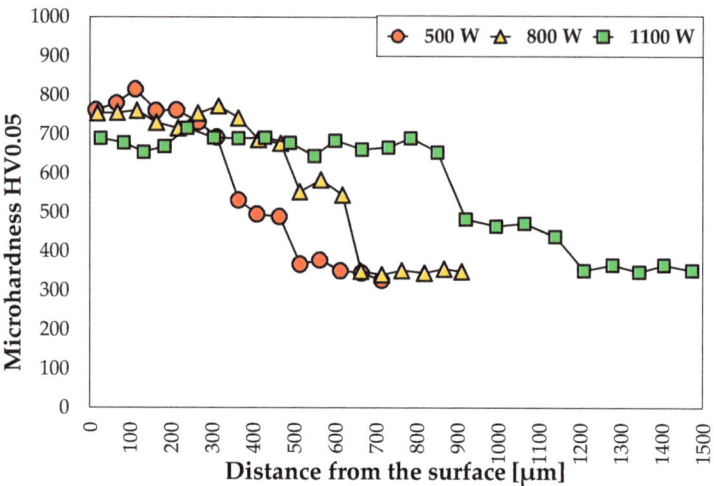

Figure 10. Microhardness of Fe/TaC coatings produced using a 60 μm precoat and different types of laser beam power.

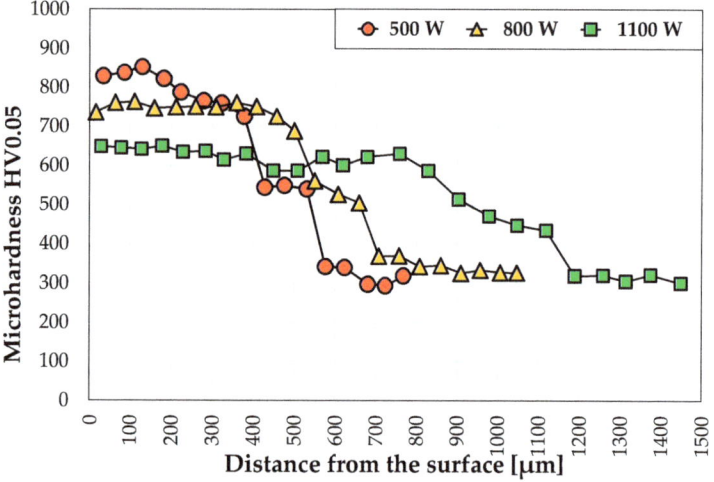

Figure 11. Microhardness of Fe/TaC coatings produced using a 90 μm precoat and different types of laser beam power.

3.4. Wear Resistance

The produced Fe/TaC composite coatings were subjected to a dry friction wear test. The results of the weight loss measurements of the samples are shown in Figure 12. Figure 12a shows the effect of the thickness of the precoat TaC and the laser beam power of 500 W on the wear resistance of the finally produced Fe/TaC coating. The test results obtained for the coatings were compared with the results obtained for the hardened and tempered steel sample. It can be seen that the production of the Fe/TaC composite coating positively increases the wear resistance of the substrate compared to hardened steel. The greatest resistance to friction wear was found for coatings produced with a TaC precoat thickness of 90 μm. The weight loss of all samples in a time unit is directly proportional. Analyzing the impact of laser beam power at a constant coating thickness of 90 μm, it was found that the best friction wear resistance Is shown by a sample affected by a 500 W laser beam. The use of a higher laser beam power contributed to an increase in weight loss by approximately 3 mg after 60 min in relation to the best coating. Wear resistance is closely correlated to microhardness results—the higher the microhardness, the better the wear resistance. The conducted tests showed that the wear resistance of the Fe/TaC coatings produced is very good compared to the hardened substrate. All produced coatings have about 2.5 times higher wear resistance, which is very important from the application point of view.

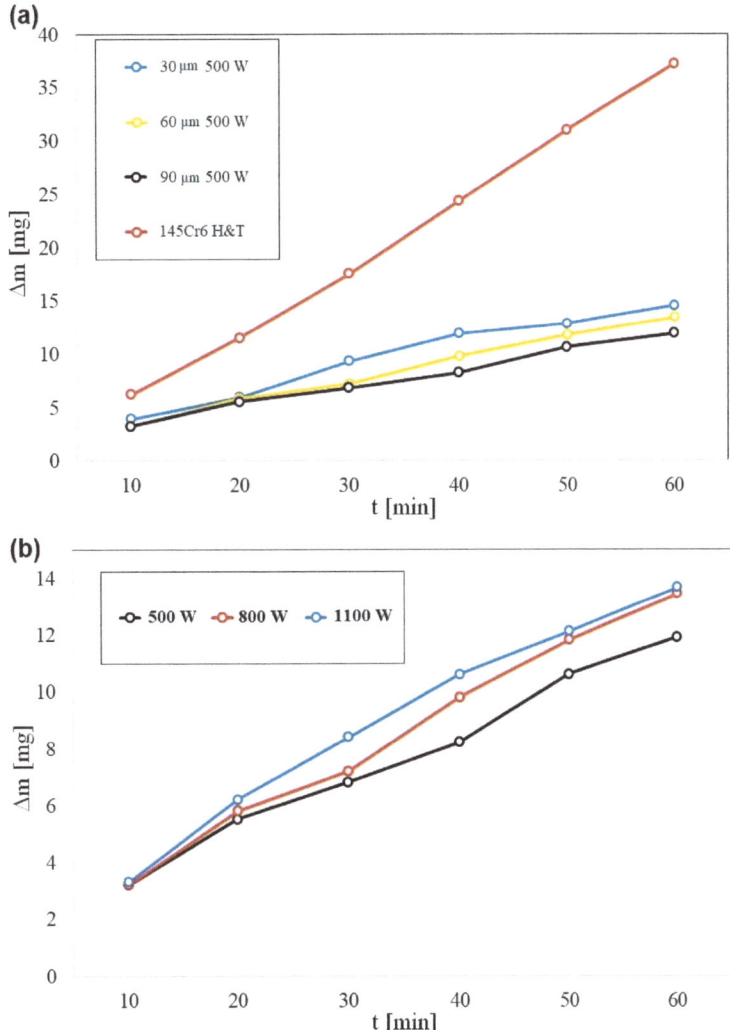

Figure 12. Wear resistance of Fe/TaC coatings: (**a**) influence of precoat thickness on wear resistance of coatings produced using 500 W laser beam power and (**b**) influence of laser beam power on wear resistance of coatings produced using a precoat of 90 μm.

Figures 13–15 show sample surfaces following wear resistance tests. Figure 13 shows the surface after a wear test at a constant laser beam power of 500 W and a variable thickness of the TaC precoat. The image in the upper figure shows a view of the wear track with fragments of the laser tracks produced. White areas in the wear track zone are tantalum carbides whose amount is greater with an increase in the thickness of the TaC precoat. The figures beneath show an enlarged fragment from the wear track in SE and BSE contrasts.

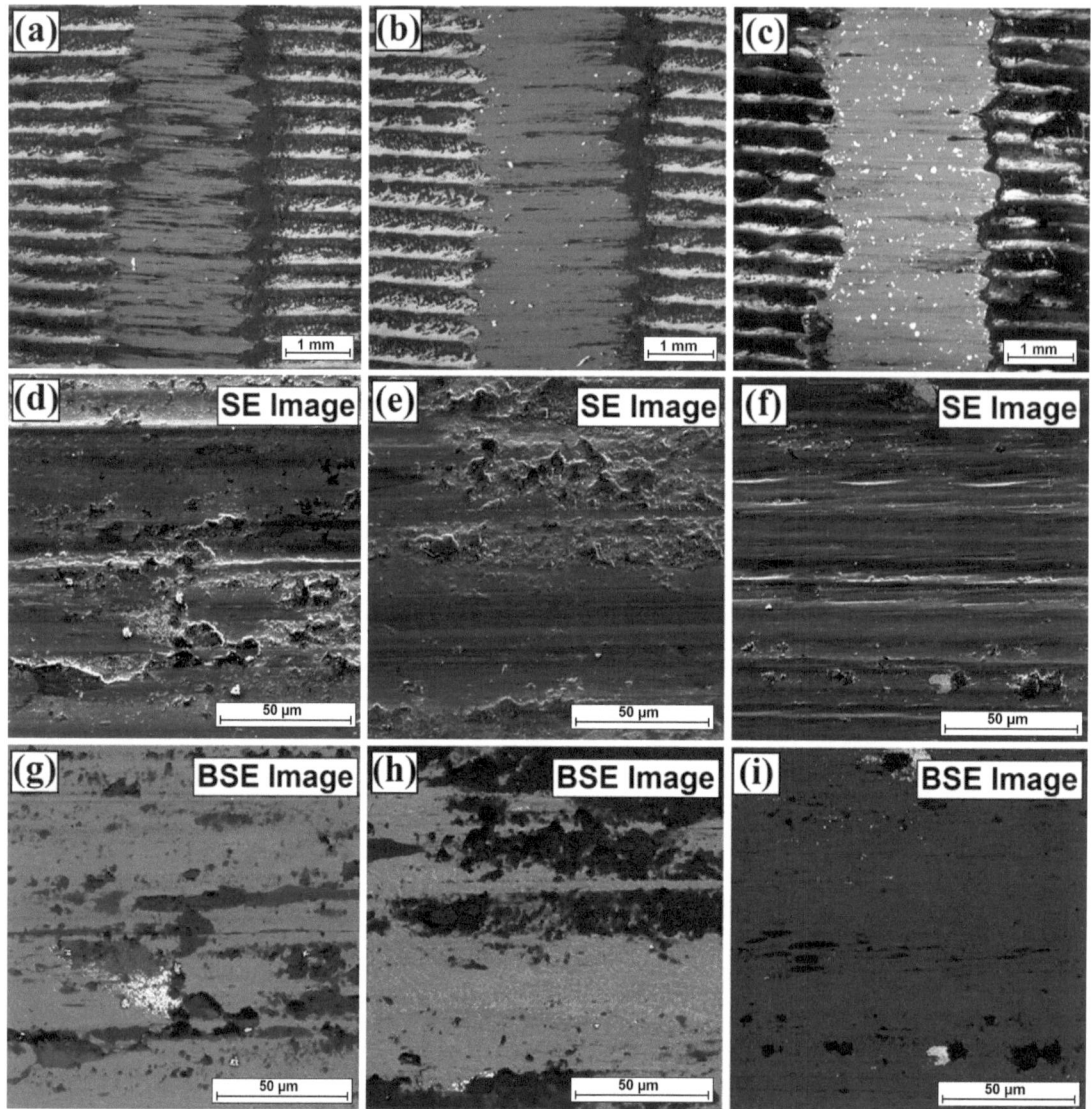

Figure 13. Surface condition after wear tests of Fe/TaC coatings produced using: (**a,d,g**) 500 W and 30 µm precoat, (**b,e,h**) 500 W and 60 µm precoat and (**c,f,i**) 500 W and 90 µm precoat.

On the surface of the sample, tears from carbide particles are visible. Some carbides that remained in the agglomerates crumble under friction and break away from the produced coating. Such particles then move between the sliding surface causing destruction of the coating and of the counter-sample.

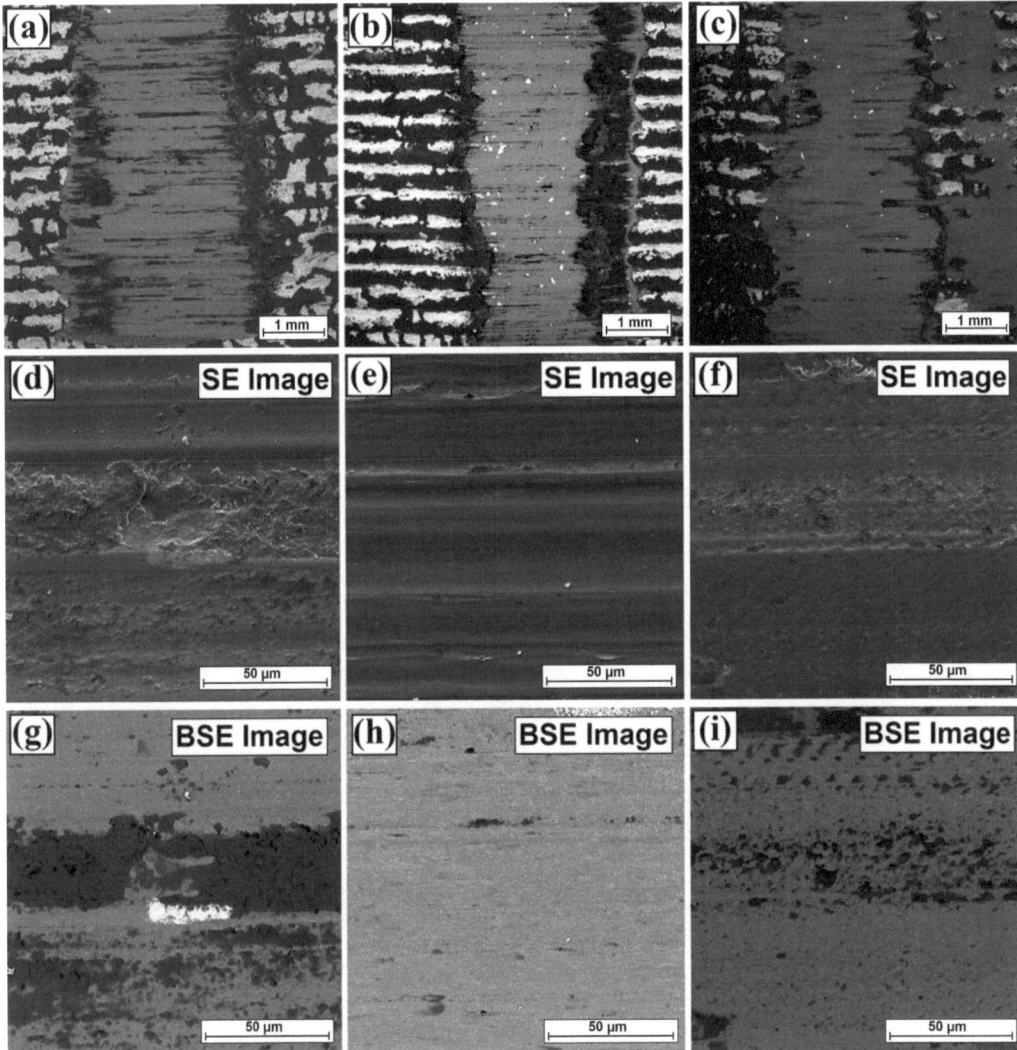

Figure 14. Surface condition after wear tests of Fe/TaC coatings produced using: (**a,d,g**) 800 W and 30 μm precoat, (**b,e,h**) 800 W and 60 μm precoat and (**c,f,i**) 800 W and 90 μm precoat.

Therefore, the mechanism of abrasion and microflaking can be mentioned here. In addition, a high oxygen content was found in the analyzed samples, which indicates that surface oxidation takes place in the wear process (Figure 16). It can be assumed that in the wear test, the worn surface of the Fe/TaC composite coating is covered with an oxide layer, which may form faster than the resulting wear tracks; hence, there are clearly visible areas in which the tested coating is oxidized. On the basis of the analyzed samples, it can be assumed that one of the leading wear mechanisms of the composite coating is abrasion combined with the oxidation process. Similar relationships were observed by the authors in paper [20], where tantalum was added to the NiCrBSi coating in the laser alloying process. The authors found that the main wear mechanism of the composite coating containing in situ synthesized TaC particles in the dry sliding wear process is abrasion combined with oxidation. In their studies, the authors of paper [20] used 30 N and 75 N loads. Some carbide particles are resistant to pulling out from the matrix in the friction process.

Oxides accumulate around them which can protect the coating from direct contact with the counters ample. In the conducted studies, a load of 147 N was applied; therefore, TaC particles may show some tendency to crumble and then squeeze into the matrix instead of pulling out of the wear surface (Figure 13c). Thus, oxide residues and TaC residues inserted into the worn surface will have a positive effect on wear resistance. In all these cases, no cracks are observed at the matrix/reinforcing phase interface. This indicates a reduction in crack initiation and propagation. The structure of the carbide mesh found in Fe/TaC coatings (the thickest precoat) is characterized by good resistance to micro-ploughing, micro-cutting, and thus provides the composite coating with high abrasive and adhesive wear resistance.

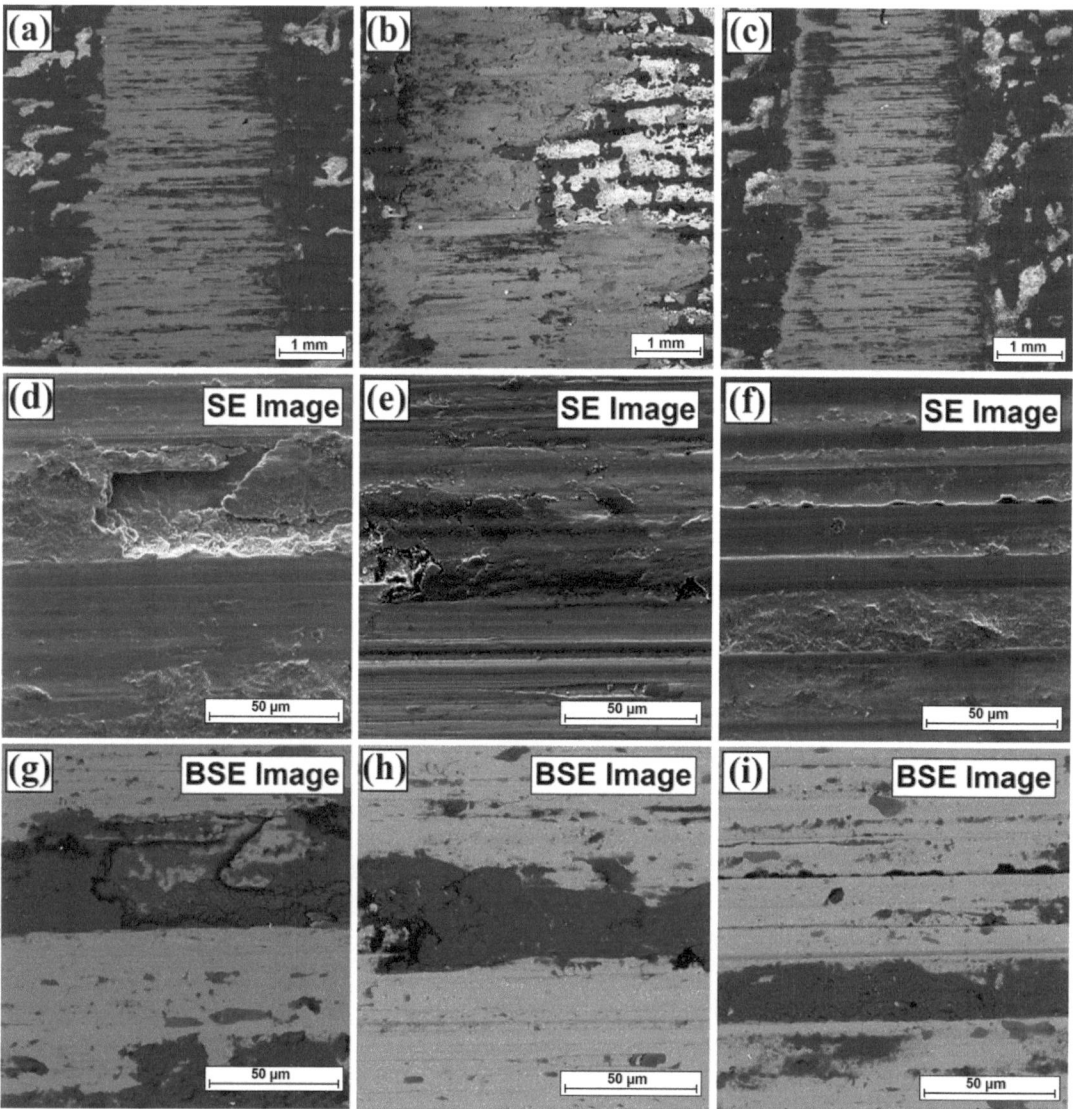

Figure 15. Surface condition after wear tests of Fe/TaC coatings produced using: (a,d,g) 1100 W and 30 µm precoat, (b,e,h) 1100 W and 60 µm precoat and (c,f,i) 1100 W and 90 µm precoat.

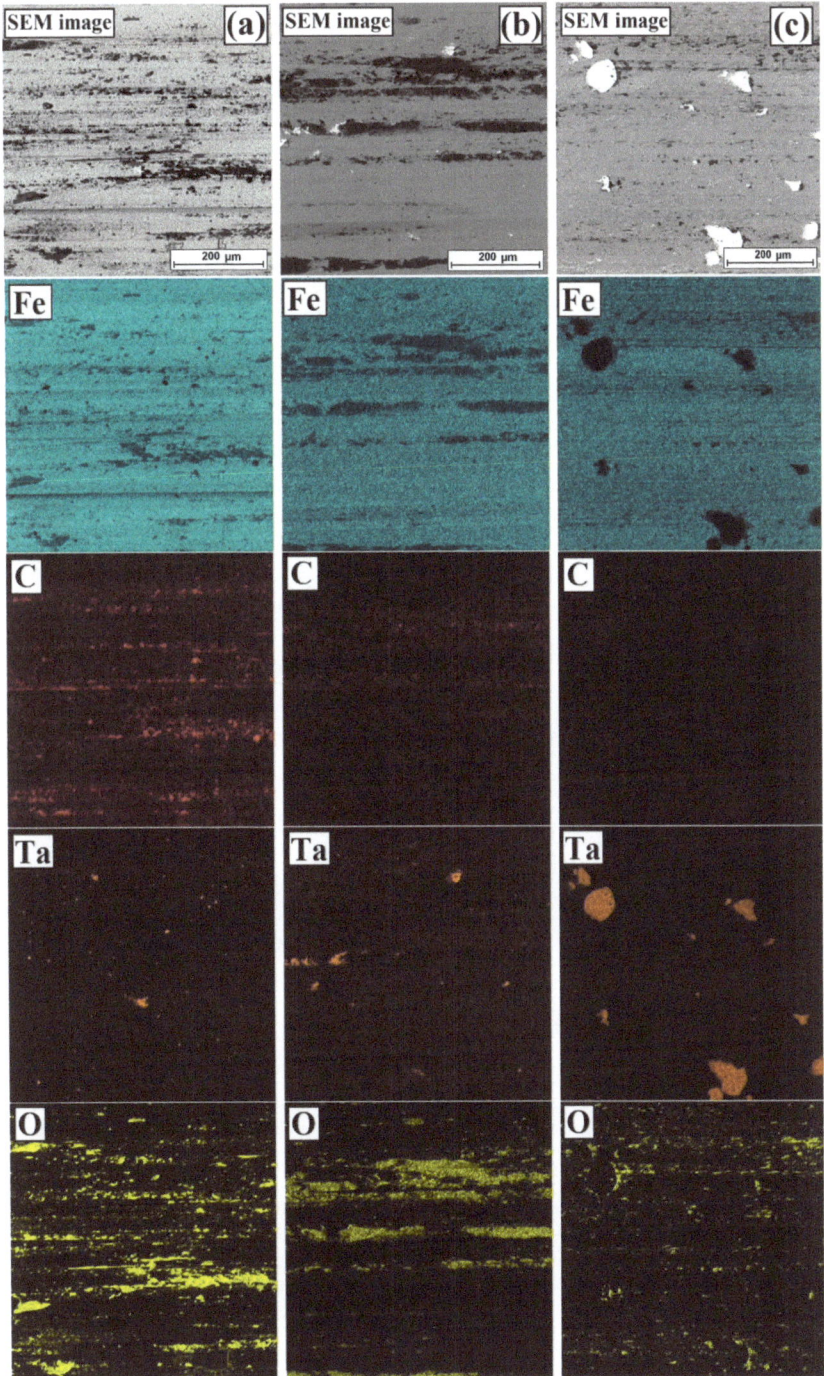

Figure 16. Surface condition and EDS mapping after wear tests of Fe/TaC coatings produced using: (**a**) 500 W and 30 μm precoat, (**b**) 500 W and 60 μm precoat and (**c**) 500 W and 90 μm precoat.

Figure 14 shows the surface after wear at a constant laser beam power of 800 W and a varying thickness of the TaC precoat. Here, too, a similar mechanism can be observed as for the lower laser beam power. On the surface of the coating grooves are also visible, and the hard reinforcing phase in the form of TaC mesh and carbides occurring in the microstructure makes surface abrasion difficult (clearly visible in Figure 14b). Under the applied load, the large hard phase easily cracks and peels, which leads to the formation of craters (Figure 14a). In the coatings produced at the highest laser beam power (Figure 15), the main wear mechanism was abrasion and adhesion combined with oxidation. Irregular pits prove adhesive wear behavior, while longitudinal grooves indicate the micro-cutting and micro-ploughing effect. The interaction between contacting inequalities of the resulting oxides leads to increased coating wear. It should be noted that the most uniform wear surface track was observed for the Fe/TaC composite coatings produced at the lowest laser beam power. The increase in laser beam power resulted in uneven abrasion of the samples, which can be seen in the wear tracks.

4. Conclusions

Based on the studies carried out for Fe/TaC composite coatings, the following conclusions can be drawn:

1. It is possible to produce full-size composite coatings in which the reinforcing phase is TaC and the matrix is iron from the steel substrate.
2. An increase in laser beam power reduces the number of TaC particles, simultaneously affecting the composite character of the coatings. In the microstructure, the eutectic of tantalum is formed as a mesh, which creates a reinforcing phase.
3. The use of a low laser beam power results in a higher tantalum content in the Fe/TaC coating. On the other hand, an increase in power contributes to the reduction of its share and thus to a reduction in microhardness.
4. In terms of wear resistance, it is the most favorable to produce an Fe/TaC coating using a laser beam power of 500 W and a 90 µm precoat.

Author Contributions: Conceptualization, D.B.; investigation, D.B. and A.B.; methodology, D.B. and A.B.; validation, D.B.; visualization, D.B.; writing—original draft, D.B. and A.B.; writing—review and editing, D.B. All authors have read and agreed to the published version of the manuscript.

Funding: The presented research results were funded by grants for education allocated by the Ministry of Science and Higher Education in Poland.

Institutional Review Board Statement: Not applicable.

Informed Consent Statement: Not applicable.

Data Availability Statement: The data presented in this study are available on request from the corresponding author.

Conflicts of Interest: The authors declare no conflict of interest.

References

1. Lawrence, J.R.; Waugh, D. Laser surface engineering: Processes and applications. In *Woodhead Publishing Series in Metals and Surface Engineering Book*, 1st ed.; Kindle Edition: Chester, UK, 2014.
2. Dowden, J.; Schulz, W. *The Theory of Laser Materials Processing: Heat and Mass Transfer in Modern Technology*, 2nd ed.; Springer Series in Materials Science; Springer Dordrecht: Bristol, UK, 2017; ISBN-13: 978-3319567105.
3. Fan, L.I.; Dong, Y.; Chen, H.; Dong, L.; Yin, Y. Wear Properties of Plasma Transferred Arc Fe-based Coatings Reinforced by Spherical WC Particles. *J. Wuhan Univ. Technol.-Mater. Sci. Edit.* **2019**, *34*, 433–439. [CrossRef]
4. Kukliński, M.; Bartkowska, A.; Przestacki, D. Microstructure and selected properties of Monel 400 alloy after laser heat treatment and laser boriding using diode laser. *Int. J. Adv. Manuf. Technol.* **2018**, *98*, 3005–3017. [CrossRef]
5. Dobrzański, L.A.; Labisz, K.; Piec, M.; Klimpel, A. Modelling of surface layer of the 31CrMoV12-18 tool steel using HPDL laser for alloying with TiC powder. *J. Achiev. Mater. Manuf. Eng.* **2007**, *24*, 27–34.
6. Abbas, G.; West, D.R.F. Laser surface cladding of Stellite and Stellite-SiC composite deposits for enhanced hardness and wear. *Wear* **1991**, *143*, 353–363. [CrossRef]

7. Chao, M.-J.; Niu, X.; Yuan, B.; Liang, E.-J.; Wang, D.-S. Preparation and characterization of in situ synthesized B_4C particulate reinforced nickel composite coatings by laser cladding. *Surf. Coat. Technol.* **2006**, *201*, 1102–1108. [CrossRef]
8. Bartkowski, D.; Bartkowska, A. Wear resistance in the soil of Stellite-6/WC coatings produced using laser cladding method. *Int. J. Refract. Met. Hard Mater.* **2017**, *64*, 20–26. [CrossRef]
9. Ertugrul, O.; Enrici, T.M.; Paydas, H.; Saggionetto, E.; Boschini, F.; Mertens, A. Laser cladding of TiC reinforced 316L stainless steel composites: Feedstock powder preparation and microstructural evaluation. *Powder Technol.* **2020**, *375*, 384–396. [CrossRef]
10. Lv, X.; Zhan, Z.; Cao, H.; Guo, C. Microstructure and properties of the laser cladded in-situ ZrB_2-ZrC/Cu composite coatings on copper substrate. *Surf. Coat. Technol.* **2020**, *396*, 125937. [CrossRef]
11. Bartkowski, D.; Bartkowska, A.; Jurči, P. Laser cladding process of Fe/WC metal matrix composite coatings on low carbon steel using Yb: YAG disk laser. *Opt. Laser Technol.* **2021**, *136*, 106784. [CrossRef]
12. Davoren, B.; Sacks, N.; ·Theron, M. Microstructure characterization of WC-9.2wt%Monel 400 fabricated using laser engineered net shaping. *Prog. Addit. Manuf.* **2021**, *6*, 431–443. [CrossRef]
13. Yilbas, B.S.; Arif, A.F.M.; Karatas, C.; Ahsan, M. Cemented carbide cutting tool: Laser processing and thermal stress analysis. *Appl. Surf. Sci.* **2007**, *253*, 5544–5552. [CrossRef]
14. Teghil, R.; D'Alessio, L.; De Maria, G.; Ferro, D. Pulsed-laser deposition and characterization of TaC films. *Appl. Surf. Sci.* **1995**, *86*, 190–195. [CrossRef]
15. Karatas, C.; Yilbas, B.S.; Aleem, A.; Ahsan, M. Laser treatment of cemented carbide cutting tool. *J. Mater. Process. Technol.* **2007**, *183*, 234–240. [CrossRef]
16. Ferro, D.; Rau, J.V.; Rossi Albertini, V.; Generosi, A.; Teghil, R.; Barinov, S.M. Pulsed laser deposited hard TiC, ZrC, HfC and TaC films on titanium: Hardness and an energy-dispersive X-ray diffraction study. *Surf. Coat. Technol.* **2008**, *202*, 1455–1461. [CrossRef]
17. Chao, M.; Wang, W.; Liang, E.; Ouyang, D. Microstructure and wear resistance of TaC reinforced Ni-based coating by laser cladding. *Surf. Coat. Technol.* **2008**, *202*, 1918–1922. [CrossRef]
18. Yu, T.; Deng, Q.; Dong, G.; Yang, J. Effects of Ta on microstructure and microhardness of Ni based laser clad coating. *Appl. Surf. Sci.* **2011**, *257*, 5098–5103. [CrossRef]
19. Li, M.Y.; Chao, M.J.; Liang, E.J.; Li, D.C.; Yu, J.M.; Zhang, J.J. Laser synthesised TaC for improving copper tribological property. *Surf. Eng.* **2013**, *29*, 616–6219. [CrossRef]
20. Yu, T.; Deng, Q.L.; Zheng, J.F.; Dong, G.; Yang, J.G. Microstructure and wear behaviour of laser clad NiCrBSi+Ta composite coating. *Surf. Eng.* **2012**, *28*, 357–363. [CrossRef]
21. Hu, D.; Liu, Y.; Chen, H.; Wang, M.; Liu, J. Microstructure and properties of in-situ synthesized Ni_3Ta-TaC reinforced Ni-based coatings by laser cladding. *Surf. Coat. Technol.* **2021**, *405*, 126599. [CrossRef]
22. Liu, Y.; Ding, T.; Lv, H.; Hu, D.; Zhang, Y.; Chen, H.; Chen, Y.; She, J. Microstructure and properties of Ta-reinforced cobalt based composite coatings processed by direct laser deposition. *Surf. Coat. Technol.* **2022**, *447*, 128874. [CrossRef]
23. Huang, X.; Yu, J.; Jiang, J.; Lian, G.; Chen, C.; Zhou, M.; Xu, W.; Hu, X. Effect of Ta Content on the Microstructure and Properties of Laser Cladding Ni60A/WC Composite Coatings. *JOM* **2023**, *75*, 97–108. [CrossRef]
24. Murzakov, M.A.; Petrovskiy, V.N.; Polski, V.I.; Mironov, V.D.; Prokopova, N.M.; Tret'yakov, E.V. Influence of additions of nanoparticles TaC on a microstructure laser cladding. *J. Phys. Conf. Ser.* **2015**, *594*, 012032. [CrossRef]
25. Lv, Y.H.; Li, J.; Tao, Y.F.; Hu, L.F. High-temperature wear and oxidation behaviors of TiNi/Ti2Ni matrix composite coatings with TaC addition prepared on Ti6Al4V by laser cladding. *Appl. Surf. Sci.* **2017**, *402*, 478–494. [CrossRef]
26. Yu, T.; Chen, J.; Wen, J.; Deng, Q. High temperature phase stability and wear behavior of laser clad Ta reinforced NiCrBSi coating. *Appl. Surf. Sci.* **2021**, *547*, 149171. [CrossRef]
27. Li, Z.; Yan, H.; Zhang, P.; Guo, J.; Yu, Z.; Ringsberg, J.W. Improving surface resistance to wear and corrosion of nickel-aluminum bronze by laser-clad TaC/Co-based alloy composite coatings. *Surf. Coat. Technol.* **2021**, *405*, 126592. [CrossRef]
28. Bartkowski, D. Manufacturing Technology and Properties of Fe/TaC Metal Matrix Composite Coatings Produced on Medium Carbon Steel Using Laser Processing—Preliminary Study on the Single Laser Tracks. *Materials* **2021**, *14*, 5367. [CrossRef]

Disclaimer/Publisher's Note: The statements, opinions and data contained in all publications are solely those of the individual author(s) and contributor(s) and not of MDPI and/or the editor(s). MDPI and/or the editor(s) disclaim responsibility for any injury to people or property resulting from any ideas, methods, instructions or products referred to in the content.

Article

Microstructure Refinement of EB-PVD Gadolinium Zirconate Thermal Barrier Coatings to Improve Their CMAS Resistance

Christoph Mikulla [1,*], Lars Steinberg [2,†], Philipp Niemeyer [1], Uwe Schulz [1] and Ravisankar Naraparaju [1]

1. German Aerospace Center (DLR), Institute of Materials Research, Cologne, Linder Hoehe, D-51147 Cologne, Germany
2. TU Dresden, Institute of Materials Science (IfWW), D-01062 Dresden, Germany
* Correspondence: christoph.mikulla@dlr.de
† Current address: Engel Austria GmbH, 4311 Schwertberg, Austria.

Abstract: Rare-earth zirconates are proven to be very effective in restricting the CMAS attack against thermal barrier coatings (TBCs) by forming quick crystalline reaction products that seal the porosity against infiltration. The microstructural effects on the efficacy of Electron Beam-Physical Vapor Deposition gadolinium zirconate (EB-PVD GZO) against CMAS attack are explored in this study. Four distinct GZO microstructures were manufactured and the response of two selected GZO variants to different CMAS and volcanic ash melts was studied for annealing times between 10 min and 50 h at 1250 °C. A significant variation in the microstructural characteristics was achieved by altering substrate temperature and rotation speed. A refined microstructure with smaller intercolumnar gaps and long feather arms lowered the CMAS infiltration by 56%–72%. Garnet phase, which formed as a continuous layer on top of apatite and fluorite, is identified as a beneficial reaction product that improves the CMAS resistance.

Keywords: TBCs; thermal barrier coatings; gadolinium zirconate; GZO; rare earth zirconates; EB-PVD; microstructure; CMAS; volcanic ash; infiltration

Citation: Mikulla, C.; Steinberg, L.; Niemeyer, P.; Schulz, U.; Naraparaju, R. Microstructure Refinement of EB-PVD Gadolinium Zirconate Thermal Barrier Coatings to Improve Their CMAS Resistance. *Coatings* 2023, 13, 905. https://doi.org/10.3390/coatings13050905

Academic Editor: Xiaofeng Zhao

Received: 28 February 2023
Revised: 24 April 2023
Accepted: 5 May 2023
Published: 11 May 2023

Copyright: © 2023 by the authors. Licensee MDPI, Basel, Switzerland. This article is an open access article distributed under the terms and conditions of the Creative Commons Attribution (CC BY) license (https://creativecommons.org/licenses/by/4.0/).

1. Introduction

In modern airplanes and gas turbines, the efficiency of turbine engines is improved by enhancing the operation temperature. To withstand these enormous thermal loads, components within the high temperature section made of nickel-based alloys, such as vanes, combustion liners or turbine blades, are covered with ceramic thermal barrier coatings (TBCs) [1]. In commercial turbines, 7 wt.% Y_2O_3 stabilized ZrO_2 (7YSZ) is widely used, and is the state-of-the-art TBC material. It is typically deposited by atmospheric plasma spraying (APS) or by electron beam physical vapor deposition (EB-PVD). These methods create porous coatings with low thermal conductivity and offer compliance against thermal stresses, which are caused by the high temperature variations in operation.

The degradation of TBCs in aero-engines by calcium-magnesium-aluminum-silicates (CMAS) and other siliceous melts that are found in sand-laden environments or volcanic ashes (VA), is a major challenge in aviation industry. This damage can be erosive or corrosive, depending on whether the particles remain in a solid state or melt [2–5]. The deposits melt in the hotter sections of the turbine (temperatures above 1050 °C), stick to the components' surfaces and infiltrate the porous thermal barrier coatings which lowers the strain tolerance and leads to premature spallation [2] or reduces the lifetime of components. Additionally, the CMAS melt interacts chemically with the TBC material which changes phase composition and properties by means of dissolution and reprecipitation mechanisms [2,6].

Due to the ever-increasing turbine inlet temperatures and the drawbacks of 7YSZ over 1250 °C with respect to phase stability, sintering and CMAS resistance, intense research has been performed on novel TBC materials with low thermal conductivities and high resistance against CMAS. Among many new compositions, rare earth (RE)-

zirconates (RE = gadolinium [7,8], lanthanum [8], yttrium [9,10], etc.) or Al_2O_3 [11,12] have shown high reactivity with the siliceous deposits and proven to be effective in restricting the CMAS attack. These TBCs dissolve into the CMAS melt and form, once the maximum solubility of a specific element is reached, crystalline reaction products, such as apatite, garnet or spinel. With their increased melting points, these phases trap on the one hand the deposit elements in a crystalline product. On the other hand, they seal potential melt flow channels which reach deeper into the TBC material to prevent stiffening and to hinder or to slow down the dissolution of large quantities of the TBC or subjacent layers and materials.

Another strategy to improve the CMAS resistance of TBCs is the optimization of the microstructure. Various studies showed that the CMAS resistance of non-reactive state-of-the-art 7YSZ TBCs could be influenced by different morphologies in coatings produced via APS [13,14], PS-PVD [15] and EB-PVD [16,17]. In the latter, variation of rotation speed and of substrate temperature created different columnar microstructures with different intercolumnar gap width, column width and feather arm lengths. Smaller intercolumnar gaps and longer feather arms significantly slowed down the infiltration of molten CMAS into the 7YSZ TBC. A numerical study [18] simulating the flow of molten CMAS in a non-reactive EB-PVD TBC supported these experimental findings. However, studies about microstructural refinement through EB-PVD process variation using a reactive rare-earth TBC material combined with the examination of its influence on the CMAS resistance have not yet been published.

This approach of microstructural refinement has opened up a new window for the CMAS related research and has been applied on reactive TBCs in this study. EB-PVD process parameters were carefully varied in order to create gadolinium zirconate (GZO) thermal barrier coatings with different columnar microstructures. Their typical microstructural features were analyzed in the as-coated state. Infiltration experiments using CMAS/VA deposits with different chemical compositions at 1250 °C were performed and the infiltration as well as the reaction kinetics were analyzed.

The main aim of this study was to determine the effect of different EB-PVD GZO microstructures on the CMAS infiltration behavior and the reaction kinetics in order to identify factors improving the CMAS resistance of reactive thermal barrier coatings. Besides the microstructure, different deposits were applied to consider the influence of the chemical composition on the reaction kinetics.

2. Materials and Methods

2.1. EB-PVD GZO Thermal Barrier Coatings

Gadolinium zirconate (GZO) thermal barrier coatings were deposited at Germany Aerospace Center (DLR) in Cologne, Germany, using a 150 kW EB-PVD coater (ESPRI, von Ardenne GmbH, Dresden, Germany) with separate loading, pre-heating and coating chambers and single-source evaporation of a GZO ingot of commercial quality (Phoenix coating Resources Inc., Mulberry, FL, USA, now a subsidiary of Saint-Gobain Coating Solutions, Avignon, France) with a diameter of 62.5 mm. As metallic substrates and bondcoats would have been severely oxidized during isothermal infiltration experiments at 1250 °C, the TBCs were instead directly deposited on flat, 1-mm-thick, 100 mm × 40 mm large, densified and sintered Al_2O_3 sheets (Quick-Ohm Küpper, Wuppertal, Germany—99.6% purity) without a bond coat.

All samples were pre-heated for 15 min in a connected, separate chamber prior to deposition. In order to produce different microstructures, the substrate temperature and the rotation speed during the coating process were varied. The main process parameters and layer thicknesses are shown in Table 1.

The coating variants presented in this study were chosen from a larger number of variants and represented the largest changes with respect to the microstructural characteristics and CMAS infiltration kinetics. The applicability as parameters in an industrial coating set up was kept in mind. The temperature steps of 40–60 °C were large enough to withstand the slight temperature fluctuations occurring typically in the course of the coating process in industrial-scale EB-PVD coaters. The rotation speed of 12 rpm is considered a standard parameter widely used in research and industry, whereas 20 rpm implies a perceptible increase that was still bearable from a wear and mechanical load perspective for the coater.

Table 1. Process parameters of the GZO layers deposited by EB-PVD on sintered Al_2O_3 substrates.

Process Parameter	GZO1 (12 rpm, 970 °C)	GZO2 (12 rpm, 1030 °C)	GZO3 (20 rpm, 970 °C)	GZO4 (20 rpm, 930 °C)
Substrate Temperature	970 °C	1030 °C	970 °C	930 °C
Rotation Speed	12 rpm	12 rpm	20 rpm	20 rpm
Pressure	6×10^{-3} mbar			
Beam Power	80–82 kW	62 kW	72–76 kW	72–74 kW
Average Coating Thickness	330 μm	200 μm	190 μm	200 μm

With a surface temperature of 970 °C and a rotation speed of 12 rpm, the coating parameters of GZO1 were close to standard values that were commonly used for various columnar EB-PVD TBCs also produced in other studies at DLR for 7YSZ and GZO [19–21]. This variant was considered as a base case. As a first variation, for GZO2, the surface temperature was increased by 60 to 1030 °C while maintaining the rotation speed at 12 rpm. For the second variant GZO3, the surface temperature was kept similar to that of GZO1 at 970 °C and the rotation speed was increased from 12 to 20 rpm. For GZO4, the rotation speed was increased from 12 to 20 rpm and the beam power was decreased in order to achieve a lower substrate temperature of around 930 °C, as preliminary experiments indicated a better CMAS resistance for lower substrate temperatures.

The deposition speed was in the range between 4.7 and 5.3 μm/min and the thickness of the TBC was between 190 and 200 μm for GZO2-4 and 330 μm for GZO1. The higher TBC thickness of GZO1 was intentionally chosen to have more space for investigations and to better find differences between the versions. All specimens of one variant, i.e., GZO1, GZO2 etc., were coated in the same respective deposition run.

2.2. CMAS Deposits and Infiltration Experiments

CMAS deposits with different chemical compositions were used in this study: One natural volcanic ash sample collected in 2010 from site of the volcano Eyjafjallojökull located in Iceland (63°40′42.10″ N; 19°37′31.75″ W) named IVA, and two artificial CMAS powders, named CMAS1 and CMAS2. Their compositions match those found in aeroengines operating in desert regions. The CMAS1 and CMAS2 powders were synthesized, as described in previous studies [22,23] in an in-house laboratory facility, via co-decomposition of Me-nitrates (Me: Al, Ca, Fe, Mg), SiO_2 and TiO_2 powders (Merck KGaA, Darmstadt, Germany) and a heat treatment for 1 h at 1250 °C. CMAS2, with its higher CaO content, was fabricated by mixing CMAS2 with 20 wt.% anhydrite $CaSO_4$ powder at room temperature.

The chemical compositions, viscosities and melting ranges of the CMAS used are given in Table 2. These three different deposits were already used in previous studies conducted at DLR [12,17,22,23].

For the infiltration experiments, CMAS/VA powder with a concentration of 20 mg/cm^2 was applied on top of the GZO coating (samples sized approximately 15 mm × 15 mm) by mixing the powder with deionized water, putting it in a circular template with a diameter of 6 mm and letting it dry at ambient air. To enable the infiltration, the samples were annealed in an isothermal cyclic furnace at 1250 °C on a pre-heated sample holder with a heating rate of 360 K/min and rapid cooling (quenching) of ~550 K/min to room temperature via ventilated ambient air.

After preliminary experiments of short-term infiltration (10 min at 1250 °C with CMAS2) for all four coatings GZO1-4, the two coating variants with significant differences in their microstructural characteristics (as described in Section 3.1) and infiltration depth after 10 min (see Section 3.2), namely GZO1 and GZO4, were selected for extensive infiltration experiments and analysis with all three deposits and annealing times between 10 min and 50 h in the above-mentioned furnace.

For XRD analysis of the reaction phases, spalled GZO particles from the EB-PVD coating chamber were milled, mixed with CMAS/VA deposits in proportions of 40/60 wt.% (GZO/CMAS) and annealed on a platinum foil for 5 h at 1250 °C in the above-mentioned cyclic furnace. The reacted mixtures were then crushed and milled for subsequent XRD analysis.

Table 2. Chemical composition in mol.% and properties of Iceland volcanic ash (IVA), and two synthetized CMAS powders (CMAS1, CMAS2), as used and described in previous studies [12,17,22,23].

Deposit	Chemical Composition [mol.%]								Melting Range	Viscosity at 1250 °C (exp.)	Main Phases
-	SiO_2	CaO	MgO	Al_2O_3	FeO	TiO_2	Na_2O	K_2O			
Iceland volcanic ash (IVA)	49.7	12.5	6.0	7.4	17.6	4.3	2.0	0.4	1060–1150 °C	250 Pa·s	Amorphous
CMAS1 (synthetic)	41.6	24.7	12.3	11.1	8.7	1.6	-	-	1230–1250 °C	6.9 Pa·s	Pyroxene + Anorthite
CMAS2 (synthetic)	37.3	32.4	11.2	9.9	7.8	1.4	-	-	1235–1240 °C	4.0 Pa·s	Pyroxene + Melillite

2.3. Characterization Methods

The samples were prepared with metallurgical standard preparation methods. The as-coated and infiltrated samples were analyzed by scanning electron microscopy (SEM—DSM ultra 55, Carl Zeiss, Jena, Germany). Energy-dispersive spectroscopy (EDS—Inca, Oxford Instruments, Abingdon, United Kingdom) was used to identify the CMAS reaction products within the coating and to determine the infiltration depth (EDS mapping).

The porosity of the as-coated GZO layers was determined both by image analysis of SEM micrographs using ImageJ software and by gas displacement pycnometry with Helium (AccuPyc II 1340, Micromeritics, Norcross, GA, USA). In the latter, rectangular pieces of GZO on Al_2O_3 substrate with a coating volume of approximately 35 mm^3 were used. Additional pycnometry measurements of uncoated Al_2O_3 substrates allowed the subtraction of the substrate porosity, which increased the accuracy of the measurement of the coating porosity.

XRD analysis of the reacted GZO-CMAS/VA powder mixtures was performed with a D8 Advance diffractometer with Cu-Kα-radiation (Bruker AXS, Karlsruhe, Germany). Powder mixtures of coating material and CMAS/VA deposit were used due to their higher specific contact surface area during the reaction time, forming a higher quantity of reaction products and thus leading to a stronger signal of the relevant peaks.

3. Results

3.1. Microstructure of the as-Coated EB-PVD Gadolinium Zirconate Coating

The SEM micrographs of the unpolished and polished top-view, as well as the cross-sections in different magnifications of the as-coated layers are shown in Figure 1. The polished top views were obtained in the upper few µm of the TBCs by carefully polishing away the column tops.

Table 3 summarizes the microstructural characteristics such as columnar width, intercolumnar gap width and feather arm length, as well as porosity values determined with different methods (image analysis and helium pycnometry). The intercolumnar gaps are divided into two categories (as seen in Figure 2, left) with bigger ones at the intersection of three or more columns and smaller ones along the sides between two adjacent columns.

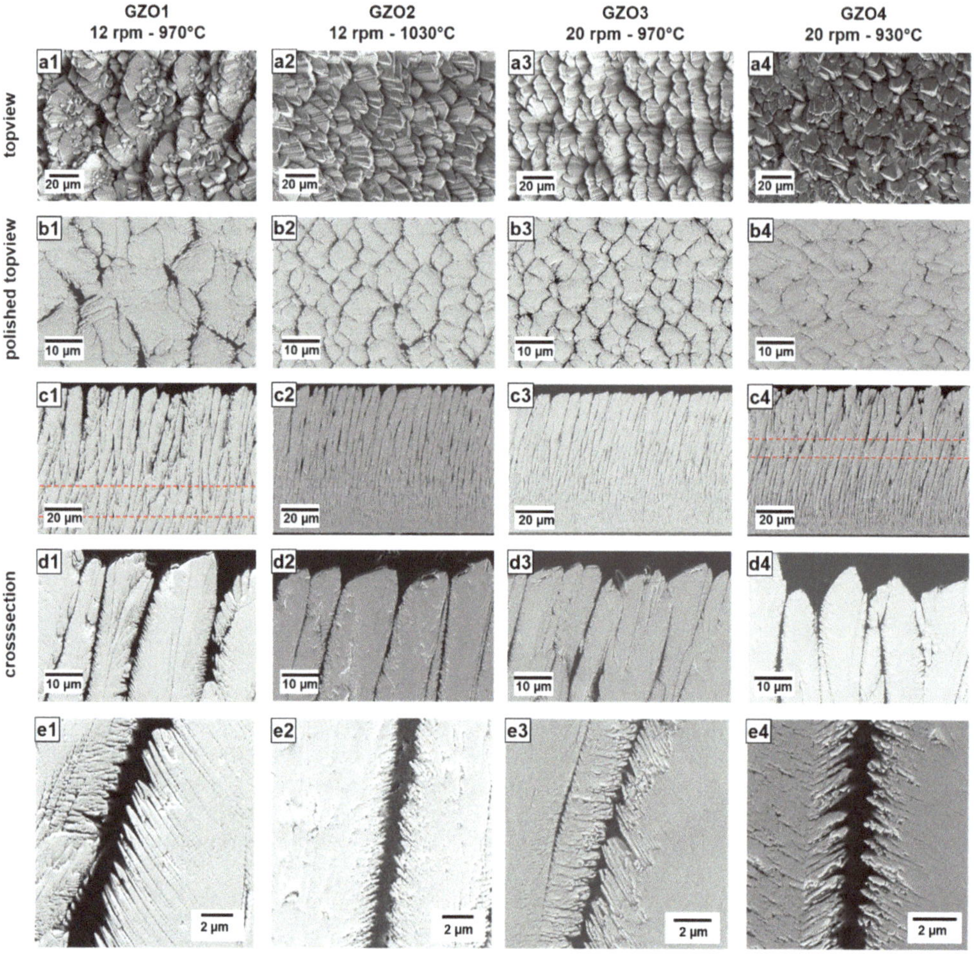

Figure 1. SEM micrographs of the as-coated EB-PVD GZO layers GZO1 to GZO4 as unpolished top-view (**a1–a4**), polished top-view (**b1–b4**), cross-section overview (**c1–c4**) and detail (**d1–d4**,**e1–e4**). Dashed lines: change of dominant columnar growth direction.

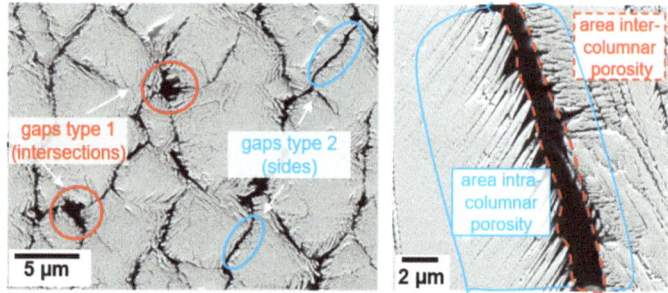

Figure 2. Microstructural characteristics calculated in Table 3 shown in exemplary SEM images. **Left**: Two categories of inter columnar gap width (type 1: intersections between three or more columns, type 2: sides between two adjacent columns) in polished top-views. **Right**: areas considered for the calculation of the geometry factor (g) (ratio of intra- to intercolumnar porosity) in a cross-section.

Table 3. Microstructural characteristics of the GZO layers, determined with cross-sectional and polished top-view SEM images as well as He-pycnometry.

Characteristics	GZO1 (12 rpm, 970 °C)	GZO2 (12 rpm, 1030 °C)	GZO3 (20 rpm, 970 °C)	GZO4 (20 rpm, 930 °C)
Column width	15.6 ± 1.0 µm	7.6 ± 1.5 µm	6.7 ± 1.1 µm	5.3 ± 1.0 µm
Intercolumnar gap width (corners/sides)	3.0 ± 1.4 µm/ 1.5 ± 0.3 µm	0.9 ± 0.26 µm/ 0.3 ± 0.06 µm	1.1 ± 0.3 µm/ 0.2 ± 0.06 µm	0.85 ± 0.16 µm/ 0.2 ± 0.08 µm
Feather arm length	1.6 ± 0.7 µm	1.2 ± 0.3 µm	1.1 ± 0.5 µm	2.2 ± 0.4 µm
Intercolumnar porosity (ImageJ)				
Top-view	8.3 ± 0.6%	6.5 ± 0.9%	8.2 ± 1.0%	3.0 ± 0.4%
Cross-section	7.0 ± 1.6%	6.0 ± 0.9%	8.1 ± 0.5%	5.7 ± 0.8%
Geometry factor (g) (Ratio of intra to intercolumnar porosity/ feather to gap)	0.86 ± 0.09 to 1	0.98 ± 0.410 to 1	1.62 ± 0.36 to 1	1.26 ± 0.22 to 1
Overall porosity (He-pyknometry)	5.5 vol.%	14.6 vol.%	9.6 vol.%	17.3 vol.%

As the EB-PVD coatings experienced a strong variation in porosity across the coating thickness, porosity was measured using different methods at different sections of the coating e.g., the intercolumnar porosity was measured by means of ImageJ in cross sectional and polished top-view SEM images in the upper section of the TBC. In addition to that, helium pycnometry was used to measure the overall porosity of the coating. Another important measurement, namely the geometry factor (g), which is the ratio of intra- to intercolumnar porosity, was estimated by dividing the areas of pores between the feather arm by the area of the intercolumnar gaps (as shown in Figure 2, right). Similar measurements have been used in previous studies [17] to calculate the tortuosity τ.

Note that the difference between the porosity values underlines the difference between the used methodologies and the sections considered for their calculation. While the ImageJ analysis considered only the intercolumnar porosity of the upper part of the coating with its wider gaps, the pycnometry calculation was based on the entire coating, including also the lower TBC section with small columns, narrower gaps and less porosity.

The lowest geometry factor (g) of GZO1 for all four samples indicates that the porosity contributed by the gaps prevailed, compared to the one within the feather arms.

The GZO2 coating had smaller column widths, shorter feather arms and smaller intercolumnar gaps than GZO1. The top-view SEM images (Figure 1(a2,b2)) show that columns grew in a more regular and ordered way compared to GZO1. Little to no re-nucleation of smaller columnar nuclei on the surface of large columns (as it is the case for GZO1) could be observed. Its overall porosity of 14.6 vol.% was found to be significantly higher than that of GZO1, whereas the intercolumnar porosity at the top section was slightly smaller. On average, (g) was calculated as 0.98 to 1, which indicates that the feather arm porosity and intercolumnar porosity were in the same order.

For GZO3, coated at a higher rotation speed of 20 rpm at 970 °C, the microstructure exhibited thinner columns, smaller intercolumnar gap widths and shorter feather arm lengths (Figure 1(a3–e3))—all of which are, within the standard deviation, comparable to those values of GZO2. The unpolished surface was also similar to that of GZO2, whereas the intercolumnar porosity of the top section was larger than for GZO2, with the highest value of all four variants. It experienced a higher feather arm porosity compared to that of intercolumnar porosity (highest (g)-value of 1.62 to 1) and a medium overall porosity of 9.6 vol.%.

The last variant GZO4, deposited with lower substrate temperature (930 °C) and higher rotation speed (20 rpm), exhibited the thinnest columns, the narrowest intercolumnar gaps as well as the longest feather arms of all four microstructures. Its columns, as seen in the top-view SEM micrographs (Figure 1(a4,b4)), were less ordered than GZO2 and GZO3 but did not exhibit the fan-like structure observed in GZO1. Due to its narrower gaps, the intercolumnar porosity of GZO4 was the lowest for all four coating variants. At the same time, it possessed the highest overall porosity and (g) value—both indicating that a large quantity of open, intracolumnar pores prevailed in this microstructure.

A distinct columnar bending (against the rotation axis) was observed in all the microstructures (Figure 1(c1–c4)).

The slope of this curvature was found to increase with decreasing substrate temperature and increasing rotation speed. GZO2 (1030 °C, 12 rpm) has exhibited least bending compared to the other coatings deposited at lower temperature (GZO1, GZO3 and GZO4). For the coatings manufactured at the same substrate temperature (i.e., 970 °C for GZO1 and GZO3), the higher rotation speed (GZO3) caused a stronger bending. The strongest curvature was observed for the high-rotation/low-temperature microstructure GZO4. The reason for the curvature, that was observed in previous investigations and in literature as well [20,24,25], is still under current investigation.

For weakly bended GZO2 and GZO3, the columns were curved throughout the entire height of the coating. Contrarily, for the two other microstructures, GZO1 and GZO4, respectively, the dominant columnar growth direction changed after a specific thickness as nuclei orientated perpendicularly to the surface prevailed. The stronger bent GZO4 exhibits this change already after a thickness of 105–140 µm, whereas in the less curved GZO1, differently orientated nuclei prevailed after 165–200 µm.

XRD analysis of the GZO (respective diffractograms are not presented here) showed that the coatings and powder all consisted of cubic GZO phase (PDF-file 01-080-0471). Deviating from the relative peak intensities for a powder with statistical grain orientation that was provided by the PDF file, the dominant (311) peak indicated a stronger (311) texture for the EB-PVD layers. For higher substrate temperatures (GZO2 versus GZO1 and GZO3 versus GZO4), the peak intensities of (111) and (200/400) decreased with respect to the highest peak (311). With more available energy due to the higher substrate temperature, the surface diffusion might have occurred more extensively and allowed the atoms to settle into more thermodynamically favorable positions, which seems to be, in this case, in the (311) orientation. The similar change in the peak intensities (lower (111) and (200/400) with respect to (311)) could be observed when increasing the rotation speed (GZO3 versus GZO1). This might be explained by the smaller radiation losses that occurred during the shorter rotation cycles at higher speed, resulting in longer times at higher temperature and hence more energy that was available for the above-mentioned enhanced surface diffusion. For the GZO powder, the (111)-peak as strongest reflex correlated to the statistical distribution.

3.2. CMAS Infiltration Experiments

3.2.1. Reaction Products from the GZO-CMAS Interactions Using Powder Mixtures

Reaction products of GZO and CMAS/VA were determined by combining XRD analysis (Figure 3) with EDS spot measurements in the SEM (Figure 4) on 40 wt.% coating −60 wt.% CMAS powder mixtures after 5 h annealing at 1250 °C. The average chemical compositions of the reaction products measured from EDS are presented in Table 4. Among the patterns in Figure 3, dominant peaks from the fluorite phase $(Gd,Zr,Ca)O_{1-x}$ and apatite phase $(Gd,Ca)_4(Gd,Zr)_6(SiO_4)_6O_2$ were found in all powder mixtures. For the GZO + IVA mixture, besides fluorite and apatite, additional peaks corresponding to a zirconolite-type phase $(Ca,Gd,Zr)_2(Fe,Ti,Al,Mg)_2O_7$ were identified. For GZO + CMAS2, apart from fluorite and apatite, other noticeable peaks of Andradite (a species of the garnet group) with the general formula $(Gd,Zr,Ca)_3(Mg,Al,Fe,Ti,Zr)_2(Si,Al,Fe)O_{12}$ were found. In GZO + CMAS1, the weak signals corresponding to the garnet phase were also identified along with spinel $(MgAl_{2-x}Fe_x)O_4$.

Figure 4 depicts a clear difference in the reaction product formation and the sizes of the formed crystalline products for different powder mixtures. Globular fluorite and elongated, prismatic apatite crystals could be seen in all reacted powder mixtures. Larger garnet crystals appeared for both GZO + CMAS1 and GZO + CMAS2 mixtures and their respective quantitative occurrence in the images is in good agreement with the different signal strength of the corresponding XRD peaks in Figure 3, i.e., for CMAS1, the garnet crystals were sporadically present and the garnet diffraction peaks were weak, whereas for CMAS2, the higher peak intensity in the XRD pattern aligns with the higher presence of garnet in the SEM image. Spinel and zirconolite are visible as small grains in GZO + CMAS1 and GZO + IVA, respectively (marked as SP and Z in Figure 4a,c), and the intensities of their XRD peaks are low with respect to the apatite and fluorite peaks.

The EDS spot measurements of the reacted powders (Table 4) show that the chemical composition of the various reaction phases varied slightly with respect to the composition of the CMAS/VA deposit.

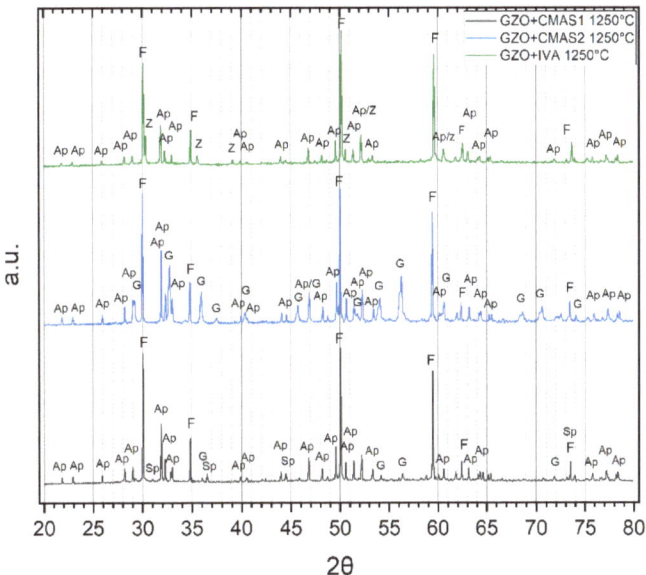

Figure 3. Normalized XRD patterns of the 40/60 wt.% powder mixtures of GZO with CMAS1 (black), CMAS2 (blue) and IVA (green) after 5 h at 1250 °C with peaks attributed to apatite (Ap), fluorite (F), garnet (G) and zirconolite (Z).

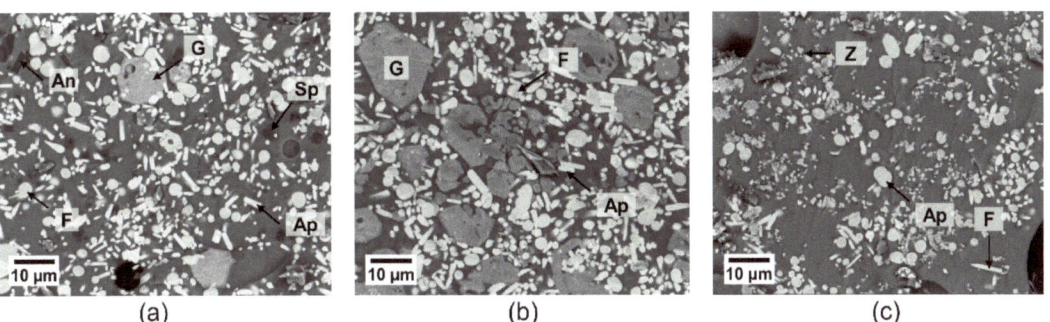

Figure 4. SEM micrographs of the powder mixtures of GZO with (**a**) CMAS1, (**b**) CMAS2, (**c**) IVA after annealing for 5 h at 1250 °C; with An: anorthite, Ap: apatite, F: fluorite, G: garnet, Sp: spinel, Z: zirconolite.

Table 4. Chemical composition of the reaction products in the GZO-CMAS/VA powder mixtures after 5 h at 1250 °C measured via EDS spot measurement. Several spots were measured at different locations and the average was calculated.

Deposit	Phase	MgO	AlO$_{1.5}$	SiO$_2$	CaO	TiO$_2$	FeO	ZrO$_2$	GdO$_{1.5}$	Na$_2$O	K$_2$O
CMAS1	Residue	9.7	17.1	34.5	21.1	1.4	7.1	3.6	5.5	-	-
	Garnet	7.5	11.7	21.1	18.4	1.9	18.1	5.1	16.2	-	-
	Fluorite	0.4	1.8	-	3.4	1.4	3.1	71.4	17.9	-	-
	Apatite	0.5	2.9	35.0	14.3	-	0.5	3.1	43.6	-	-
	Spinel	30.7	51.4	1.2	0.7	-	15.0	1.0	0.1	-	-
CMAS2	Residue	12.6	9.5	37.3	28.7	0.8	3.7	3.5	3.9	-	-
	Garnet	9.2	6.3	26.3	25.0	1.7	13.7	7.8	10.0	-	-
	Fluorit	0.5	1.1	-	6.2	0.9	1.8	74.0	14.7	-	-
	Apatite	-	1.7	34.4	18.4	-	-	6.5	39.0	-	-
IVA	Residue	6.2	14.3	42.7	8.1	1.7	10.2	3.6	6.8	5.4	1.0
	Zirconolite	-	4.4	4.3	2.3	20.1	18.0	28.6	21.5	0.7	-
	Fluorite	-	1.2	0.2	1.0	2.5	5.3	73.3	15.8	-	-
	Apatite	-	2.7	35.8	11.2	0.2	1.5	5.7	42.9	-	-

The fluorite, apatite and garnet phases of the powder mixtures with calcia-rich CMAS2 all showed higher calcia contents in their compositions compared to the GZO + CMAS1 phases with 6.2 mol.% instead of 3.4 mol.% (fluorite), 18.4 mol.% instead of 14.3 mol.% (apatite) and 25.0 mol.% instead of 18.4 mol.% (garnet), respectively. On the other hand, the silica content in the apatite phase is similar, within the error range of EDS measurements, for all the powder mixtures at 34.4–35.8 mol.%. Similarly to the apatites, chemical compositions of the residues orient towards the original compositions of the deposits as well. However, the higher original titania content (4.3 mol.%) of the unreacted IVA deposit compared to CMAS1 and CMAS2 (1.6 and 1.4 mol.%) was reduced in the IVA residue to 1.7 mol.% due to the reprecipitation of titania-rich zirconolite crystals. The zirconia content was found to be similar in all three residues, whereas the gadolinia content was lowest (3.9 mol.%) in the CMAS2 residue. This might be attributed to the fact that garnet as an additional Gadolinia bearing phase formed extensively in CMAS2 compared to the other two powder mixtures.

3.2.2. Effect of Different Microstructures on the Infiltration and Reaction Kinetics

In order to compare the CMAS resistance of the different GZO microstructures, the infiltration depth of the deposits within the coating was determined by tracing the main elements of the CMAS via EDS area mapping. The depth differed significantly between all four coatings. Figure 5 shows exemplarily the cross-sectional SEM images of these variants after 10 min infiltration with CMAS2 and the EDS maps of silicon signals. The molten deposit infiltrated to a depth of 185 ± 10 µm for GZO1, 114 ± 22 µm for GZO2, 92 ± 25 µm for GZO3 and 52 ± 12 µm for GZO4, respectively. The two coatings with the largest difference in microstructure (Table 3) and infiltration depth (Figure 5), namely, GZO1 and GZO4, were selected for further analysis in the course of this study.

Figure 6 shows the infiltration depth of CMAS1, CMAS2 and IVA in GZO1 and GZO4 as a function of annealing time at 1250 °C. The different infiltration kinetics for the two microstructures can be clearly seen: GZO1, with wider intercolumnar gaps, was infiltrated between 185 ± 31 µm deep after 10 min and 211 ± 29 µm after 50 h for CMAS1 and slightly less for CMAS2, namely, 185 ± 10 µm up to 180 ± 25 µm, and the least for IVA with 141 ± 14 µm up to 129 ± 28 µm, respectively. GZO4, possessing a feathery microstructure with smaller intercolumnar gaps and longer feather arms, was infiltrated less: When CMAS1 was being used for the infiltration, the infiltration depth increased from 54 ± 6 µm

(10 min) to 63 ± 10 μm (50 h). For CMAS2, the molten deposit has reached a depth of 52 ± 12 μm (10 min) up to 78 ± 10 μm (50 h). For IVA, the infiltration depth was between 52 ± 11 μm and 56 ± 18 μm. Thus, the reduction of infiltration depth due to the optimized microstructure for GZO4 in comparison to GZO1 is between 56% to 72%, depending on the parameter set of CMAS deposit and annealing time.

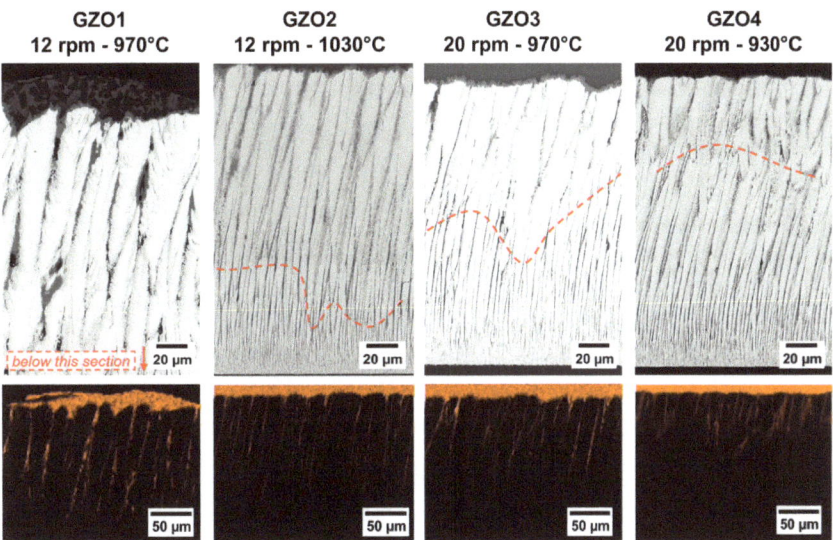

Figure 5. Different GZO microstructures infiltrated with CMAS2 for 10 min at 1250 °C (dashed line: infiltration depth) and EDS mapping of the Si element in the cross section of the same sample.

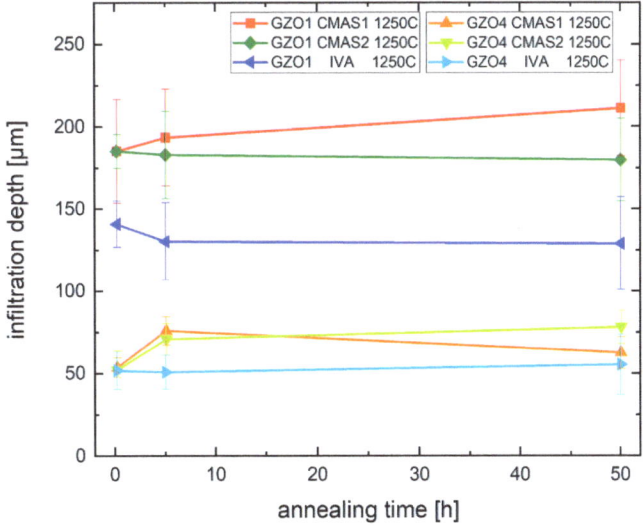

Figure 6. Infiltration depth of the CMAS deposit into the TBC at different annealing times with CMAS1, CMAS2 and IVA at 1250 °C.

Another important observation about the variation in the reaction layer depth versus annealing time with respect to the microstructure is presented in Figure 7. Contrarily to the infiltration depth, the reaction layer thickness did not diverge consistently for all deposits when comparing the two microstructures for the same CMAS deposits and annealing time.

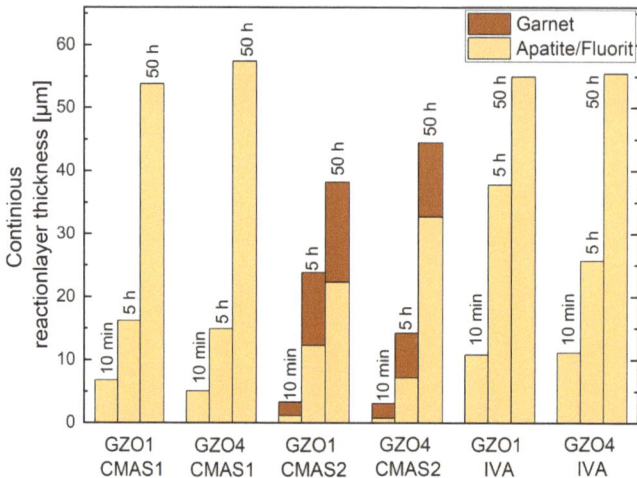

Figure 7. Thickness of the continuous reaction layer on top of the different microstructures formed after interaction with CMAS1, CMAS2 and IVA at 1250 °C.

Another important observation about the variation in the reaction layer depth versus annealing time with respect to the microstructure is presented in Figure 7. Contrarily to the infiltration depth, the reaction layer thickness did not diverge consistently for all deposits when comparing the two microstructures for the same CMAS deposits and annealing time.

Another important observation about the variation in the reaction layer depth versus annealing time with respect to the microstructure is presented in Figure 7. Contrarily to the infiltration depth, the reaction layer thickness did not diverge consistently for all deposits when comparing the two microstructures for the same CMAS deposits and annealing time.

For CMAS1 infiltrated samples, this reaction layer was defined by the apatite and fluorite crystals that entirely transformed the GZO at the specific location. It has an average thickness (GZO1/GZO4) of 6.7 ± 0.7 μm/5.1 ± 1.0 μm, 16.2 ± 3.3 μm/14.9 ± 2.1 μm and 53.8 ± 1.6 μm/57.5 ± 4.0 μm after 10 min, 5 h and 50 h annealing time, respectively. The garnet was not considered as it has formed only sporadically and discontinuously on top of the apatite.

In case of CMAS2, both the apatite–fluorite network and the garnet (continuous layers) were considered. The apatite–fluorite network was thinner than its counterpart formed by CMAS1. The average overall reaction layer thickness consisting of apatite, fluorite and garnet is (GZO1/GZO4) 2.8 ± 0.9 μm/3.0 ± 0.7 μm, 23.8 ± 2.6 μm/14.3 ± 2.6 μm and 38.2 ± 10.0 μm/45.5 ± 4.2 μm after 10 min, 5 h and 50 h annealing time, respectively. Due to the larger intercolumnar gaps for GZO1, a stronger garnet formation could be observed after 5 h compared to GZO4 with its smaller gaps. The thicker garnet layer slowed down the formation of additional apatite after additional, longer annealing times, which can be seen in the comparison of Figure 7 after 50 h, where the apatite layer on GZO1 was thinner than the one on GZO4 with less garnet between the TBC and residue.

With IVA being used as deposit, the apatite–fluorite network has formed, as shown in Figure 7. The average overall reaction layer thickness was (GZO1/GZO4) 10.8 ± 3.7 μm/ 11.2 ± 2.9 μm, 37.8 ± 6.1 μm/25.7 ± 4.5 μm and 55.0 ± 7.4 μm/55.5 ± 6.3 μm after 10 min, 5 h and 50 h annealing time, respectively. The zirconolite layer was not considered in the total reaction layer thickness, as it mainly formed in the CMAS residue on top of the reaction interface.

In GZO1, the combination of differences in microstructure and CMAS composition also influenced the infiltration depth by causing a divergent reaction behavior. Due to the larger intercolumnar gaps, a more extensive garnet formation occurred with CMAS2 compared to CMAS1. This garnet incorporated more CMAS, provided additional sealing of wide gaps and thus restricted the infiltration depth of CMAS2. In contrast for GZO4, the influence of the additional CMAS2 garnet formation was limited, as the smaller intercolumnar gaps of the GZO4 microstructure were sufficiently sealed by the faster forming apatite. As a consequence, the difference in the infiltration depth for GZO4 was smaller.

3.2.3. Effect of Chemical Composition of CMAS and Time on the Reaction Kinetics of EB-PVD GZO at 1250 °C

Both GZO1 and GZO4 reacted quickly (within 10 min) with all CMAS/VA compositions and formed crystalline products within the intercolumnar gaps as well as at the columnar tips, as shown in Figure 8. No profound effect of microstructure was observed on the reaction phases; however, a clear difference in the reaction products was observed with respect to the different CMAS compositions. Hence, only the chemical compositions of reaction products for one microstructure, namely GZO4, that formed with the different deposits, are presented and discussed hereinafter. The average chemical compositions of the reaction products measured with GZO4 are summarized in Table 5.

In the case of CMAS1 (Figure 8a,d), the molten deposit infiltrated the intercolumnar gaps as well as parts of the feather arms and partially dissolved the column tips. A thin, continuous apatite layer $(Gd,Ca)_4(Gd,Zr)_6(SiO_4)_6O_2$ surrounding the GZO columns was formed. At the TBC/CMAS interface, globularly shaped fluorite $(Gd,Zr,Ca)O_{1-x}$ was precipitated. While smaller intercolumnar gaps contained solely apatite, larger gaps could not be sealed completely by the apatite. Instead, CMAS residue and small globular fluorite crystals were visible in these gaps.

With CaO-enriched, less viscous CMAS2 (Figure 8b,e), a similar infiltration scenario could be observed where the intercolumnar gaps and feather arms were infiltrated along with a slight dissolution of the column tips and the formation of apatite at the interface. Additionally, crystalline garnet $((Gd,Zr,Ca)_3(Mg,Al,Fe,Ti,Zr)_2(Si,Al,Fe)O_{12})$ already started to form after 10 min at the TBC column/CMAS residue interface and within the larger intercolumnar gaps. As the dissolution of smaller parts of GZO and apatite within garnet crystals can be seen in Figure 8b,e, it appears that the garnet grew at the expense of both the original coating and the apatite. In medium and larger gaps, additional globular fluorite particles were found.

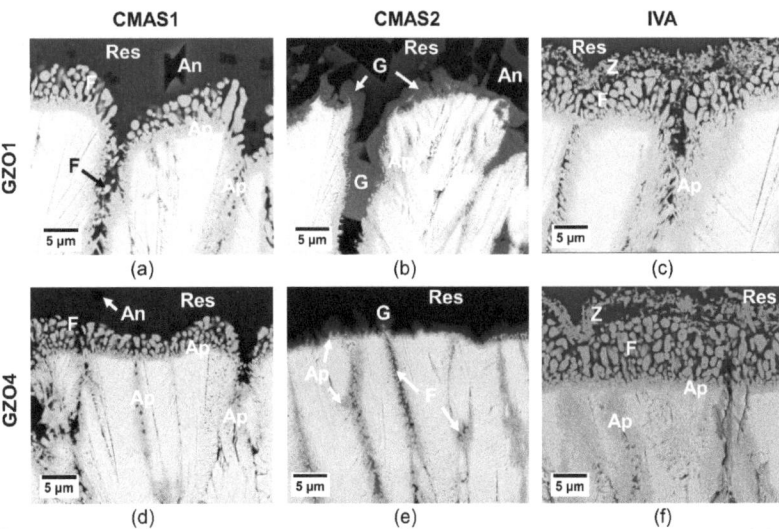

Figure 8. Initial reaction layer of GZO1 (**a**–**c**) and GZO4 (**d**–**f**) after 10 min infiltration at 1250 °C with CMAS1 (**a**,**d**), CMAS2 (**b**,**e**) and IVA (**c**,**f**). Ap: apatite, F: fluorite, res: residue, G: garnet, Z: zirconolite, An: anorthite.

Table 5. Chemical composition of the reaction products of infiltrated GZO4 samples after 10 min at 1250 °C measured via EDS spot measurement. Several spots were measured at different locations and the average was calculated.

			GZO4, 10 min at 1250 °C [mol.%]							
Deposit	Label	Phase	MgO	$AlO_{1.5}$	SiO_2	CaO	TiO_2	FeO	ZrO_2	$GdO_{1.5}$
CMAS1	Res	Residue	12.2	16.4	34.8	19.6	1.4	8.0	3.2	4.5
	Ap	Apatite	0.0	1.7	21.2	11.2	0.8	0.5	30.4	34.3
	F	Fluorite	2.8	3.9	8.3	6.6	1.7	4.0	57.8	14.9
	An	Anorthite	1.2	36.6	40.1	19.2	-	2.0	0.8	0.3
CMAS2	Res	Residue	12.2	7.8	38.6	26.6	1.3	7.5	3.3	2.7
	G	Garnet	9.4	4.5	31.8	28.6	2.3	13.4	5.0	5.0
	Ap	Apatite	0.0	2.0	20.3	11.1	0.5	0.0	32.3	33.8
IVA	Res	Residue	7.8	14.5	44.1	9.3	2.2	11.4	3.8	7.0
	Z	Zirconolite	0.8	2.6	1.4	2.0	19.8	19.2	32.4	21.9
	Ap	Apatite	-	-	17.7	5.4	1.3	2.9	35.0	37.7
	F	Fluorite	0.6	0.7	-	1.1	3.2	5.6	70.6	18.1

Iceland volcanic ash (IVA) (Figure 8c,f) experienced similar infiltration and reaction behavior to that of CMAS1. The column tips dissolved and formed both globular fluorite particles and an apatite layer surrounding the columns. In larger gaps, the apatite grew in an elongated prismatic form accompanied by small, globular fluorite. Additionally, small particles of Fe-Ti-rich zirconolite were continuously precipitated in the residue above the globular fluorite.

After the longer annealing time of 50 h, a clear distinction between different phases was established, as shown in Figure 9. For CMAS1, the column tips were transformed into a two-phase layer of apatite and fluorite. While these were finely dispersed at the interface of the unreacted GZO columns, they were clearly distinguishable as elongated, prismatic apatite and globular fluorite at the CMAS–residue interface. Large garnet crystals of several tens of μm were discretely present on top of the apatite/fluorite reaction layer, covering approximately 20%–25% of the surface. Small spinel crystals (($MgAl_{2-x}Fe_x)O_4$) were found in the residue as well as in the intercolumnar gaps.

A clear difference in the reaction layer sequence was found in CMAS2 where a finely dispersed apatite–fluorite bilayer was formed at the location of the former columns. Contrarily to the CMAS1 sample, the two phases were not separated by residue and thus did not exhibit their characteristic elongated or globular shape. Instead, they formed a finely dispersed, continuous network with a varying image contrast, indicating the compositional differences. The thickness of this bilayer was smaller compared to that of the CMAS1 case. The garnet crystals continued to grow at the column–residue interface forming a continuous dense layer on top of the apatite–fluorite as well as in the larger intercolumnar gaps and pockets. Additionally, smaller, dark particles in the lower reaction zone accumulated those elements that were locally not incorporated in apatite and fluorite, being mainly magnesia, alumina and iron oxide. EDX spot measurements also detected Gd, Zr, Si and Ca. No distinct phase class could be attributed to these particles in this study without further TEM analysis.

Finally, in the case of IVA, GZO has formed mainly a thick apatite and fluorite layer. The two phases formed a marbled and veined network, were larger in size and more clearly distinguishable than for CMAS2. Larger fluorite particles were visible at the very top of the reaction layer, whereas a finer apatite and fluorite network was found in the lower part of the layer. Contrarily to the CMAS1 sample, the two phases were not separated by significant amounts of residue. Additionally, Fe-Ti-rich zirconolite was found on top of this layer and spinel crystals were dispersed in the bilayer network at the lower part of the reaction layer. No residue was left on top of the reaction layer after 50 h, since IVA vigorously reacted with the GZO.

The chemical compositions of the reaction products after 50 h for GZO1 were published in a recent study [26]. The only minor differences to those after 10 min will be reviewed briefly in the discussion Section 4.3.

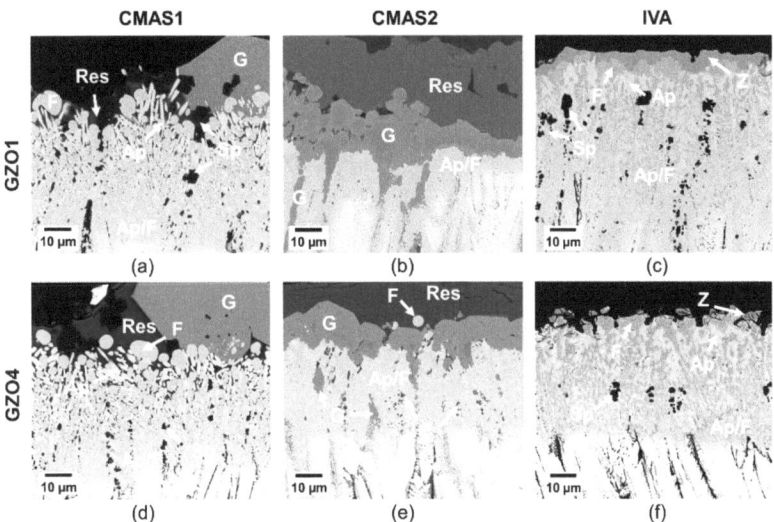

Figure 9. Reaction layer of GZO1 (**a–c**) and GZO4 (**d–f**) after 50 h infiltration at 1250 °C with CMAS1 (**a,d**), CMAS2 (**b,e**) and IVA (**c,f**). Ap: apatite, F: fluorite, Sp: spinel, res: residue, G: garnet, Z: zirconolite, Ap/F: apatite–fluorite mixed phase, x: unknown phase.

4. Discussion

The investigation and improvement of CMAS resistance of thermal barrier coatings has undergone intense research in the past decade [27]. However, most studies focused on the examination and comparison of different compositions of the coating material, especially on those forming reactive phases with the molten deposit. The effect of microstructural optimization was studied mainly on thermal conductivity or cyclic lifetime. However, the potential of microstructural modifications to improve the CMAS resistance was very little explored, especially within the EB-PVD process. Although its positive effect on the infiltration resistance has been shown for non-reactive 7YSZ EB-PVD coatings [16,18], similar examinations have not yet been performed for quick reactive coating materials with rare-earth zirconates such as GZO.

4.1. Effect of the Process Parameters on the EB-PVD Microstructure

The modification of microstructure is an important leverage to change and improve the properties of thermal barrier coatings that significantly influence the performance, efficiency and durability of turbines, such as thermal conductivity [28–30] or lifetime [31–33]. However, the CMAS resistance can also be improved by varying and optimizing the microstructure, as is shown in this study.

Substrate temperature and rotation speed have already proven to be decisive parameters for modifying the microstructures of EB-PVD TBCs [31,34–37]. Jesuraj et al. studied the effect of the substrate temperature on GZO [34] and Lanthanum zirconate [35] TBCs in a non-rotational EB-PVD coating process on nanomechanical properties and could observe an improved coating adhesion, higher hardness and Young's modulus for coatings deposited at 700 °C compared to lower substrate temperatures of 500 °C and 600 °C. However, due to the stationary positioning and the lower substrate temperatures, a significantly different microstructure was achieved compared to the present TBCs.

Table 3 summarizes the microstructural characteristics of the studied coatings. With the variation of process parameters, the microstructural characteristics such as column width and intercolumnar gap width could be significantly altered between 50% and 86% and the porosity changed, depending on the method of measurement, between 62% and 68% (with reference to the respective maximum value).

The substrate temperature, which was varied in this study between 930 and 1030 °C, is linked to the energy available during the coating process. The condensation of a vapor particle on a solid substrate and the morphology of the coating depend mainly on the energy. The activation energy for surface and volume diffusion during condensation is proportional to the homologue substrate temperature. By decreasing the substrate temperature, less energy was available for the surface diffusion of the condensing GZO particles, which hindered the formation of larger columns via diffusion and instead caused a higher imperfection of the columns, a more pronounced feather arm structure and higher porosity.

With a higher rotation speed (GZO1 to GZO3), the porosity and quantity of nanopores was increased, as it has similarly been observed in various studies for EB-PVD 7YSZ coatings [30,38]. At faster rotation speeds, the vapor incident angle (VIA) between the ingot and the substrate changed at a higher frequency, causing a larger variation in the condensation process by a more competitive growth of different growth directions and thus higher porosity. Additionally, the time interval out-of-sight of the ingot, when no vapor reaches the substrate, was shorter for higher rotation speeds. During that interval, less time was available for relaxation and surface diffusion processes of the condensed particles towards free lattice positions which could reduce disturbances in the columns.

By decreasing the substrate temperature and increasing the rotation speed at the same time, the intercolumnar gap width and the column diameter were reduced, while the feather arms were extended and the over-all porosity was increased. The higher tortuosity and geometry factor (g) indicate that the porosity was increased more by the longer feather arms than it was decreased by the smaller intercolumnar gaps. Thus, the microstructure can be referred to as 'feathery'. In this way, the modification of the GZO thermal barrier coatings in this study aligns with similar variations performed in the framework of past studies on 7YSZ [16], where increasing the rotation speed from 12 rpm to 20 rpm and decreasing the substrate temperature from 1000 °C to 840 °C during the coating process reduced the column diameter by approximately 50%, from 15–30 μm to 8–15 μm, and the gap width from 2 μm to 1.2 μm while increasing the feather arm length.

However, the material properties, such as melting point, vapor pressure, condensation enthalpies, theoretical density or emissivity, influence the columnar structure and have to be considered when determining the specific process parameter windows for new compositions.

This indicates why 7YSZ columns grow straightly, whereas the GZO counterparts exhibited a clear curvature against the rotation direction. Moreover, re-nucleation occurred once a thickness of 165–200 μm for GZO1 and 105–140 μm for GZO4 was achieved and different dominant growth direction prevailed, as shown in Figure 1(c1,c4)

The explanation for this bending behavior is still under investigation. However, first observations indicate that it is both coater-specific and material-specific. While in some studies [20,24,25] the EB-PVD GZO TBCs were also bent, other studies [39–41] report straight GZO columns without a comparable behavior at similar deposition parameters (substrate temperature: 800–1030 °C, rotation speed: 7–25 rpm). At the same time, other TBC layers such as 7YSZ [16], Al$_2$O$_3$ [11], DySZ [8] or 65YZ [9] produced by the same EB-PVD coater did not exhibit a comparable curvature.

Further variations of the EB-PVD process parameters for GZO conducted by the authors (not shown here) indicate that the bending of the columns was more pronounced when lowering the substrate temperature, whereas straight columns can be created with substrate temperatures above 1030 °C. At an increased substrate temperature, more thermal energy is provided that could facilitate the diffusion of the newly condensing particles on the surface into energetically favorable positions, meaning a growth perpendicular to the surface might be naturally favorable. At lower temperatures, the samples are thermally less activated and the diffusion, that reduces the curvature, might be less pronounced for GZO.

The large discrepancy of porosity values with respect to the different measurement procedure indicates the inhomogeneous character of the porosity in EB-PVD coatings (intercolumnar versus intracolumnar, upper section versus entire coating) and the attention that needs to be taken when choosing the method for further interpretation of the derived porosity values. He-pycnometry, which covers the entity of the open pores at a large scale and takes the three-dimensional character of the pores into account, gives a good idea about the overall porosity of the coating. However, as can be seen in Figure 1(c1–c4), the porosity varies with respect to the coating height. The limited incline of the infiltration depth shows that with reactive TBCs, such as GZO, only the porosity of the upper coating is decisive for the infiltration behavior. Hence, He-pycnometry is less suitable for this aspect than an ImageJ picture analysis of the upper TBC section.

4.2. Effect of the Microstructure on the Infiltration and Reaction Kinetics

'Feathery' 7YSZ microstructures with higher porosities compared to standard microstructures with less porosity have been proven to slow down the CMAS infiltration kinetics due to narrower intercolumnar gaps and longer feather arms that reduce the capillary forces. These studies [16–18] have led to a conclusion that the pore network is more influential than the actual total porosity on the CMAS infiltration resistance. By careful processing, this pore network can be tailored. However, the state-of-the-art material 7YSZ exhibits a reduced reactivity with CMAS, which is needed to seal the porosity with reaction products.

The microstructural effect was clearly observed in yttria-rich zirconia coatings within two studies by the same research group where different microstructures of the same TBC material were created due to different ingot systems used in the EB-PVD coating process. Exposing those two microstructures (dual ingots 7YSZ and yttria [10] and single ingot 65YZ [26]), to molten CMAS resulted in different infiltration behavior and reaction kinetics, with infiltration depths being 9%–12% higher after only 40% of the annealing time (20 h versus 50 h at 1250 °C).

GZO is known for its CMAS resistance and it was shown that microstructure sometimes can greatly influence its protective nature [42]. In a recent comparative study [26] with EB-PVD GZO and yttria rich zirconia (65YZ), the microstructure of the GZO coating (same as GZO1 in this study) was proven to be a defining factor for the poor CMAS resistance (by a factor of 2.6–3.2 less protective compared to the 65YZ). Based on that study, microstructural refinement of GZO was executed in this work. In this study, it is now clearly demonstrated that the modification of the pore network of EB-PVD GZO enhanced the CMAS infiltration resistance.

The presence of smaller intercolumnar gaps and longer feather arms (GZO4) is beneficial to reduce the infiltration depth. The narrower gaps were sealed faster and more effectively by the primary reaction product apatite slowing down the flow of the molten residue into the coating even after 10 min, as shown in Figure 8d–f. Larger gaps (GZO1), however, could not always be sealed immediately by the apatite, as it can be seen in Figure 8a–c. Instead, CMAS residue could be found in pockets under those larger gaps after 10 min at 1250 °C, consequently allowing deeper infiltration. The reprecipitation of a reactive phase is reached at local oversaturation of the melt with zirconia or gadolina, which is reached faster at a high solute–solvent ratio. As in large intercolumnar gaps, the more solvent was available for the same amount of solute, the more time was needed to reach the point of local oversaturation compared to smaller gaps.

The slight increase in the infiltration depth between 10 min and 50 h in both microstructures indicates that the major part of the sealing process was achieved during the initial few minutes. A similar non-linear infiltration behavior with respect to time was also observed [43] for APS 7YSZ TBCs infiltrated with volcanic ashes. This corroborates the importance of the intercolumnar gap width as a defining microstructural feature with respect to the CMAS resistance.

In contrast, the reaction layer thickness was partially influenced by the microstructures. For CMAS1, it was approximately the same for GZO1 and GZO4, as can be seen in Figure 7. Once the intercolumnar gaps were sealed by the crystalline reaction product, the diffusion of the elements needed for the formation of additional reaction products was performed through the residue–reaction-product interface via the reaction phases (presumably the grain boundaries between the crystalline reaction products). The intercolumnar gaps and feather arms no longer play a relevant role for the transport of the necessary elements for the reaction layer growth at the interface.

In contrast, with the formation of more than one reaction phase, the microstructure could indirectly influence the reaction layer formation, when one phase (garnet in this case) grew stronger due to microstructural effects and could affect the formation of the other phase after additional annealing time. In the case of CMAS2, a stronger garnet formation in the initial stages (after 5 h) due to larger intercolumnar gaps could slow down the formation of apatite in the further reaction process (as seen after 50 h in Figure 7).

The results derived from this study indicate that a microstructure advantageous for extended CMAS resistance consists of a coating with small column diameters (creating a larger density of intercolumnar gaps), narrow intercolumnar gaps (able to be sealed quickly), long feather arms (reducing the capillary forces), and increased porosity, which consists mainly of intracolumnar porosity (meaning a ratio of intra to intercolumnar porosity larger than one). Those desirable tendencies are marked with the yellow arrows in Figure 10, which graphically summarizes the main finding of the current work for all investigated versions GZO1 to GZO4. From this graph it is obvious that GZO4 is the most CMAS-resistant microstructure of the current version, and as measured, GZO1 is the worst. GZO2 and GZO3 are situated in-between but were not investigated in the current work and are a topic of ongoing research.

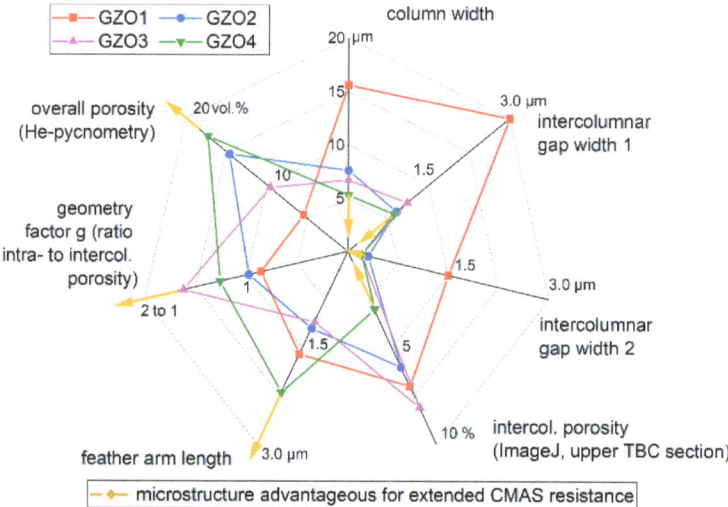

Figure 10. Comparison of the microstructural characteristics of GZO1-4 and the changes in those characteristics advantageous for extended CMAS resistance indicated by the directions of the yellow arrows.

4.3. Influence of the Chemical Composition of the CMAS on the Reaction Kinetics

The chemical composition of the CMAS or the volcanic ash and the properties derived from its composition, such as the viscosity, play a major role in the types of reaction products that form, their kinetics and hence on the overall CMAS infiltration depth.

While apatite and fluorite formed within all the studied CMAS compositions, zirconolite was exclusively present in the case of IVA. Spinel formed only in the case of CMAS1 and IVA, while garnet was present for CMAS1, CMAS2. As can be seen from Table 4, the major difference in between Ca-rich and Ca-lean CMAS compositions was the presence of garnet as discrete crystals or continuous layer and the absence of spinel for the Ca-rich deposits. For the Si-rich volcanic ash (IVA), zirconolite was present as an outermost reaction product.

The viscosity is one of the factors that governs the infiltration kinetics. The viscosities of the deposits at 1250 °C (see Table 2) were experimentally determined in the previous studies [12,17]. In this study, their influence on the infiltration kinetics within GZO can be seen more effectively in the case of high-viscous IVA compared to the less viscous CMAS1 and CMAS2. Similar behavior was observed for non-reactive TBC material such as 7YSZ [16,17], where low-viscosity deposits infiltrated faster into the TBC. However, once a reactive component (in the case of GZO) is added, this effect becomes only evident with larger differences in the viscosity: e.g., viscosity difference between CMAS and IVA is rather higher than between CMAS1 and CMAS2, as can be seen in Figure 6. Hence, the viscosity effect is negligible between CMAS1 and CMAS2.

The reaction of IVA with GZO shows a vigorous initial apatite reaction layer compared to its CMAS counterparts. The reason can be explained on the bases of the higher viscosity of IVA: i.e., the higher viscosity allows a longer time for the reaction and hence more reaction layer thickness.

The identification/classification of the reaction phases was achieved by combining EDS and XRD analysis in this study with reference to the results of TEM analysis of the similar phases published in various studies [7,26,41]. As the occurring reaction phases can exist in a certain range of compositions [44,45], those results can be transferred to studies with slightly varying CMAS compositions.

Additionally, the changes in the individual phases will be discussed more in detail in the following section:

Apatite and Fluorite: The formation of apatite ($Ca_2Gd_8(SiO_4)_6O_2$) as an initial reaction product due to its high reaction kinetics was observed for various rare-earth zirconates with a large variety of CMAS and VA compositions and has been described intensively in the literature [7,39,41,44–46]. In this study, it formed with all three deposits, both at an initial state (10 min) and at long annealing times (up to 50 h).

The ratio of the three main components (calcia, silica and gadolinia) varied with respect to the used deposit. The apatite formed with IVA contains 50% less calcia than the apatite formed with CMAS1 and CMAS2. On the contrary, despite a higher silica in IVA, the apatites formed with all three deposits have approximately the same silica content.

Fluorite $(Gd,Zr,Ca)O_{1-x}$ formed next to the apatite for all deposits and annealing times. As the apatite consumes more gadolina than zirconia and more silica than calcia, the melt was enriched in zirconia and calcia, which consequently resulted in the formation of fluorite close to the apatite. While for CMAS1 and IVA the spherical-shaped fluorites formed both in the intercolumnar gaps and at the coating–residue interface at a size of several µm (even after 10 min), for CMAS2 they exist sporadically in the intercolumnar gaps with a sub-µm size. The chemical composition of the fluorite phase is similar for all deposits within a range of a few mol.%, mainly consisting of 70–75 mol.% zirconia, 15–20 mol.% gadolinia, 1–6 mol.% calcia and 1–5 mol.% iron oxide. Only the CMAS1 10 min infiltration experiments with GZO4 contained higher traces of magnesia, alumina, silica and titania.

A comparison of the initial apatite and fluorite (after 10 min) with the reaction products after 50 h annealing [26] shows a slight difference in the chemical composition with a higher zirconia content for initial apatite and low percentages of magnesia, alumina, silica and titania for initial fluorite. This might be linked to the different local availability of deposit elements in the initial stage of the annealing process, the transitory, non-equilibria nature of the short-term reaction products and the higher error for EDS spot measurements of smaller particles.

The morphologies of the apatite–fluorite bilayers after longer annealing times have varied with respect to CMAS type, i.e., they were found to be finely dispersed for CMAS2, marbled and veined for IVA and in a distinct elongated-prismatic or round shape for CMAS1 (see Figure 9). This appearance mainly depends on the additional reaction phases specific to the respective CMAS chemistry e.g., garnet and zirconolite. In the case of CMAS2 and IVA, garnet and zirconolite formed extensively and integrated several elements that were barely or not dissolved in apatite and fluorite, eventually reducing the quantity of residue between apatite and fluorite. In contrast, with CMAS1, the garnet formed sporadically, i.e., more residue remains between the apatite and fluorite. For CMAS1 and IVA, the two phases grew in size and became more distinguishable with longer annealing times (Figure 9). While the finely dispersed bilayer at the interface of the unreacted coating just formed recently, the phases at the former column tips formed initially and could grow more extensively in the course of the longer annealing time. The findings described above were nearly independent of the GZO microstructure.

Spinel and Zirconolite: Due to the formation of apatite and fluorite, the CMAS/VA melts were primarily depleted in silica and calcia and thus enriched in magnesia, alumina, iron oxide and titania. This composition of the melt then promoted the formation of spinel ($(MgAl_{2-x}Fe_x)O_4$), garnet ($(Gd,Zr,Ca)_3(Mg,Al,Fe,Ti,Zr)_2(Si,Al,Fe)_3O_{12}$) or zirconolite ($(Ca,Gd,Zr)_2(Fe,Ti,Al,Mg)_2O_7$), depending on the chemical composition of the applied CMAS/VA deposit.

While it was not yet visible after the short annealing experiments, spinel formed as a secondary reaction product after longer annealing times in the intercolumnar gaps for CMAS1 and IVA and close to the surface for CMAS1. In CMAS2, no spinel was detected. Instead, these elements of the deposit were incorporated into garnet and the smaller, dark particles in the lower reaction zone. For IVA, instead of garnet or spinel, zirconolite formed on the surface due to the higher titania content in the melt that accumulated at 19–20 mol.% in the reaction phase. This behavior was also observed in previous studies [10]. As between 16 and 20 mol.% iron oxide was bonded in the zirconolite, the melt was depleted in this element, which hindered the formation of spinel and garnet. The same formation of those reaction products was observed by Mechnich and Braue [7], when infiltrating EB-PVD GZO with artificial volcanic ash having a bulk composition close to that of IVA [47]. Although the zirconolite precipitated within the CMAS residue over the entire length of the coating on top of the apatite, it did not offer any protection against infiltration. The larger (for 10 min and 5 h) or similar (for 50 h) apatite thickness of IVA compared to CMAS1 (as seen in Figure 7) indicates that the zirconolite did not exhibit a sealing or protective behavior that positively influenced the reaction kinetics of subjacent layers, as did the garnet in case of CMAS2.

Garnet: Garnet has a complex crystal structure with the general formula $A_3B_2C_3O_{12}$ and can accommodate a multitude of different cations in the three different positions A, B and C. The dodecahedral site A (coordination number, CN = 8) can be occupied by Ca, Zr or Gd, the octahedral B position (CN = 6) by Al, Mg, Fe, Ti and Zr and the tetrahedral site C (CN = 4) by Si, Al and Fe. With the elements used in this study the garnet formula can be generalized as $(Gd,Zr,Ca)_3(Mg,Al,Fe,Ti,Zr)_2(Si,Al,Fe)O_{12}$.

Iron has proved to be a key element in the formation of the garnet phase for GZO. Various studies using FeO-free CMAS, such as the widely used $33CaO-9MgO-13AlO_{1.5}-45SiO_2$ [6], did not detect any garnet formation during the interaction of GZO with CMAS [39,41,45,48]. When Fe-containing $33CaO-13AlO_{1.5}-10FeO_x-44SiO_2$ was used in the infiltration experiments [39], dendritic garnets were found within the residual melt and thought to be formed during cooling. The CMAS used in this study contained 7.8 and 8.7 mol.% FeO, respectively. Early and quick formation of garnet crystals, even after 10 min at the TBC/CMAS interface, and the increasing garnet quantity and reaction layer thickness with respect to time, prove that it formed as a reaction product. In the case of IVA, despite its high FeO content, garnet formation did not take place, as the majority of FeO was preferentially incorporated into zirconolite due to the higher TiO_2 content in IVA. As a result, no sufficient quantities of free FeO remained in the residue to form additional garnet.

Besides iron oxide, calcia is another key component promoting the formation of the garnet phase in the case of GZO. With the larger calcia content of 27.0 mol.% in CMAS2 compared to 24.7 mol.% in CMAS1, garnet formed a continuous layer instead of sporadic garnet crystals. The fact that the garnet containing more Zr^{4+} and Ca^{2+} formed more extensively than one with more Gd^{3+} under identical annealing conditions, indicates that the dodecahedral site A preferentially accommodates those first two cations.

The extensive formation of a continuous garnet phase as a part of the reaction layer was beneficial to the CMAS resistance. Comparing the chemical composition of the various reaction phases shows that garnet contained less $GdO_{1.5}$ and ZrO_2 and bound more CMAS constituents than apatite; i.e., garnet consumes less coating material GZO and more CMAS deposit. While apatite contained approximately 50 mol.% of the oxides $GdO_{1.5}$ and ZrO_2, the garnet phase only contained around 10–20 mol.% $GdO_{1.5}$ and ZrO_2 (Tables 4 and 5 and [26]). In this way, a higher quantity of CMAS was bonded in crystalline products with high melting point and increased chemical stability which left less glassy residue for further infiltration.

Furthermore, as can be seen in Figure 7, in the presence of a continuous garnet layer (formed with CMAS2), the reaction layer thickness was reduced significantly compared to the reaction layer without continuous garnet (formed with CMAS1), between 24% and 85% considering only the apatite–fluorite layer, and up to 52% when the entire continuous layer including the garnet is taken into account.

Additionally, a continuous garnet layer structurally stabilizes the subjacent apatite and fluorite network (see Figure 9b,e). In the case of CMAS2, the apatite and fluorite were existent as a marbled, veined biphase network with no visible residue in between, which was however the case for the CMAS1 samples with only discontinuous garnet. This behavior was exhibited for all time intervals once the garnet layer forms as a continuous manner (5 h: not shown here, 50 h: Figure 9).

No distinct variation of the chemical composition of reaction products with respect to the microstructures was observed within the scope of this study.

5. Conclusions

Gadolinium zirconate thermal barrier coatings with various microstructures were deposited by the EB-PVD technique. After being exposed to three CMAS/VA deposits for time intervals between 10 min and 50 h, different infiltration and reaction kinetics could be observed, as well as various reaction products. The intercolumnar gaps could be effectively sealed by quick formation of the reaction products apatite and garnet. Depending on the chemical composition of the CMAS, phases such as zirconolite, spinel and garnet, were reprecipitated in different proportions.

A high rotation speed of 20 rpm combined with a low substrate temperature of 930 °C created a microstructure that reduced the infiltration depth by 56%–72% in comparison to the less favorable GZO1 microstructure. By this, the infiltration depth could be reduced from around 180–210 μm for GZO1 to only 52–78 μm for the optimized GZO4. The following conclusions can be drawn from this study:

- EB-PVD GZO microstructure can be significantly modified by altering the substrate temperature and rotation speed during the coating process. Its microstructural characteristics (such as column diameters, intercolumnar gap width or feather arm length) varied within the range between 50% and 86% and its porosity between 62% and 68% among all the produced microstructures (with reference to the respective maximum value).

- Microstructures that are advantageous for extended CMAS resistance consist of thin columns, narrow intercolumnar gaps, long feather arms, high porosity and a high geometry factor (g), i.e. ratio of intra- to intercolumnar porosity.
- The infiltration kinetics is strongly dependent on the EB-PVD microstructure. For suitable EB-PVD microstructures, the majority of the gap sealing process is achieved within the initial few minutes. Subsequently, the infiltration is stopped or continues only non-linearly.
- The reaction characteristics are partially influenced by the microstructure. Growth of the reaction layer is rather determined by the phase stability, the interchange of ions between reaction phases and the CMAS residue and the morphology of the reaction products. In the case of multi-reaction layers, the microstructure can significantly change the reaction dynamics and phase formations.
- The melt composition plays a major role for the reaction product and kinetics. Different reaction products form as a function of the chemical composition of the CMAS and the residual melt. While apatite and fluorite are formed for all deposits used in the study, the emergence of garnet, spinel and zirconolite is chemistry-dependent.
- The garnet phase is beneficial to the CMAS resistance of TBCs as it incorporates more deposit constituents in a crystalline product and consumes less coating material than the apatite phase. Iron and calcium are key elements in forming garnet and promote its development as a continuous sealing layer. Garnet formation occurs as early as after 10 min at 1250 °C.

Author Contributions: Conceptualization, C.M., L.S. and R.N.; methodology, C.M. and R.N.; validation, C.M.; formal analysis, C.M. and P.N.; investigation, C.M. and P.N.; data curation, C.M.; writing—original draft preparation, C.M.; writing—review and editing, R.N., U.S. and L.S.; visualization, C.M.; supervision U.S. and R.N.; project administration R.N. and C.M.; funding acquisition: R.N., L.S, U.S. and C.M. All authors have read and agreed to the published version of the manuscript.

Funding: The work was performed in the framework of the Research Project DFG SCHU 1372/5 and LE 1373/34 funded by the Deutsche Forschungsgemeinschaft (DFG, Germany Research Foundation). The authors acknowledge the financial support.

Institutional Review Board Statement: Not applicable.

Informed Consent Statement: Not applicable.

Data Availability Statement: The data present in this study are available from the corresponding author upon reasonable request. The data are not publicly available as at this time as the data also forms part of an ongoing study.

Acknowledgments: The authors would like to express their gratitude to J. Brien, A. Handwerk and D. Peters from DLR Cologne for manufacturing the EB-PVD GZO layers, as well as for technical support and advice. Additionally, the authors would like to thank P. Mechnich for technical assistance and scientific discussions.

Conflicts of Interest: The authors declare no conflict of interest. The funders had no role in the design, execution, interpretation, or writing of the study.

References

1. Padture, N.P.; Gell, M.; Jordan, E.H. Thermal Barrier Coatings for Gas-Turbine Engine Applications. *Science* **2002**, *296*, 280. [CrossRef]
2. Levi, C.G.; Hutchinson, J.W.; Vidal-Setif, M.H.; Johnson, C.A. Environmental degradation of thermal-barrier coatings by molten deposits. *MRS Bull.* **2012**, *37*, 932–941. [CrossRef]
3. Naraparaju, R.; Lau, H.; Lange, M.; Fischer, C.; Kramer, D.; Schulz, U.; Weber, K. Integrated testing approach using a customized micro turbine for a volcanic ash and CMAS related degradation study of thermal barrier coatings. *Surf. Coat. Technol.* **2018**, *337*, 198–208. [CrossRef]
4. Wellman, R.G.; Nicholls, J.R. Erosion, corrosion and erosion–corrosion of EB PVD thermal barrier coatings. *Tribol. Int.* **2008**, *41*, 657–662. [CrossRef]

5. Steinberg, L.; Naraparaju, R.; Heckert, M.; Mikulla, C.; Schulz, U.; Leyens, C. Erosion behavior of EB-PVD 7YSZ coatings under corrosion/erosion regime: Effect of TBC microstructure and the CMAS chemistry. *J. Eur. Ceram. Soc.* **2018**, *38*, 5101–5112. [CrossRef]
6. Krämer, S.; Yang, J.; Levi, C.G.; Johnson, C.A. Thermochemical Interaction of Thermal Barrier Coatings with Molten CaO–MgO–Al_2O_3–SiO_2 (CMAS) Deposits. *J. Am. Ceram. Soc.* **2006**, *89*, 3167–3175. [CrossRef]
7. Mechnich, P.; Braue, W. Volcanic Ash-Induced Decomposition of EB-PVD $Gd_2Zr_2O_7$ Thermal Barrier Coatings to Gd-Oxyapatite, Zircon, and Gd, Fe-Zirconolite. *J. Am. Ceram. Soc.* **2013**, *96*, 1958–1965. [CrossRef]
8. Schulz, U.; Braue, W. Degradation of $La_2Zr_2O_7$ and other novel EB-PVD thermal barrier coatings by CMAS (CaO-MgO-Al_2O_3-SiO_2) and volcanic ash deposits. *Surf. Coat. Technol.* **2013**, *235*, 165–173. [CrossRef]
9. Gomez Chavez, J.J.; Naraparaju, R.; Mechnich, P.; Kelm, K.; Schulz, U.; Ramana, C.V. Effects of yttria content on the CMAS infiltration resistance of yttria stabilized thermal barrier coatings system. *J. Mater. Sci. Technol.* **2020**, *43*, 74–83. [CrossRef]
10. Naraparaju, R.; Chavez, J.J.G.; Schulz, U.; Ramana, C.V. Interaction and infiltration behavior of Eyjafjallajokull, Sakurajima volcanic ashes and a synthetic CMAS containing FeO with/in EB-PVD ZrO_2-65 wt% Y_2O_3 coating at high temperature. *Acta Mater.* **2017**, *136*, 164–180. [CrossRef]
11. Naraparaju, R.; Pubbysetty, R.P.; Mechnich, P.; Schulz, U. EB-PVD alumina (Al_2O_3) as a top coat on 7YSZ TBCs against CMAS/VA infiltration: Deposition and reaction mechanisms. *J. Eur. Ceram. Soc.* **2018**, *38*, 3333–3346. [CrossRef]
12. Mikulla, C.; Naraparaju, R.; Schulz, U.; Toma, F.-L.; Barbosa, M.; Steinberg, L.; Leyens, C. Investigation of CMAS Resistance of Sacrificial Suspension Sprayed Alumina Topcoats on EB-PVD 7YSZ Layers. *J. Therm. Spray Technol.* **2020**, *29*, 90–104. [CrossRef]
13. Shan, X.; Luo, L.; Chen, W.; Zou, Z.; Guo, F.; He, L.; Zhang, A.; Zhao, X.; Xiao, P. Pore filling behavior of YSZ under CMAS attack: Implications for designing corrosion-resistant thermal barrier coatings. *J. Am. Ceram. Soc.* **2018**, *101*, 5756–5770. [CrossRef]
14. Gildersleeve, E.; Viswanathan, V.; Sampath, S. Molten silicate interactions with plasma sprayed thermal barrier coatings: Role of materials and microstructure. *J. Eur. Ceram. Soc.* **2019**, *39*, 2122–2131. [CrossRef]
15. Rezanka, S.; Mack, D.E.; Mauer, G.; Sebold, D.; Guillon, O.; Vaßen, R. Investigation of the resistance of open-column-structured PS-PVD TBCs to erosive and high-temperature corrosive attack. *Surf. Coat. Technol.* **2017**, *324*, 222–235. [CrossRef]
16. Naraparaju, R.; Huttermann, M.; Schulz, U.; Mechnich, P. Tailoring the EB-PVD columnar microstructure to mitigate the infiltration of CMAS in 7YSZ thermal barrier coatings. *J. Eur. Ceram. Soc.* **2017**, *37*, 261–270. [CrossRef]
17. Naraparaju, R.; Chavez, J.J.G.; Niemeyer, P.; Hess, K.U.; Song, W.J.; Dingwell, D.B.; Lokachari, S.; Ramana, C.V.; Schulz, U. Estimation of CMAS infiltration depth in EB-PVD TBCs: A new constraint model supported with experimental approach. *J. Eur. Ceram. Soc.* **2019**, *39*, 2936–2945. [CrossRef]
18. Kabir, M.R.; Sirigiri, A.K.; Naraparaju, R.; Schulz, U. Flow Kinetics of Molten Silicates through Thermal Barrier Coating: A Numerical Study. *Coatings* **2019**, *9*, 332. [CrossRef]
19. Munawar, A.U.; Schulz, U.; Cerri, G.; Lau, H. Microstructure and cyclic lifetime of Gd and Dy-containing EB-PVD TBCs deposited as single and double-layer on various bond coats (vol 245C, pg 92, 2014). *Surf. Coat. Technol.* **2015**, *279*, 53. [CrossRef]
20. Schulz, U.; Nowotnik, A.; Kunkel, S.; Reiter, G. Effect of processing and interface on the durability of single and bilayer 7YSZ/gadolinium zirconate EB-PVD thermal barrier coatings. *Surf. Coat. Technol.* **2020**, *381*, 125107. [CrossRef]
21. Schulz, U.; Terry, S.G.; Levi, C.G. Microstructure and texture of EB-PVD TBCs grown under different rotation modes. *Mater. Sci. Eng. A* **2003**, *360*, 319–329. [CrossRef]
22. Naraparaju, R.; Schulz, U.; Mechnich, P.; Döbber, P.; Seidel, F. Degradation study of 7wt.% yttria stabilised zirconia (7YSZ) thermal barrier coatings on aero-engine combustion chamber parts due to infiltration by different CaO–MgO–Al_2O_3–SiO_2 variants. *Surf. Coat. Technol.* **2014**, *260*, 73–81. [CrossRef]
23. Naraparaju, R.; Mechnich, P.; Schulz, U.; Rodriguez, G.C.M. The Accelerating Effect of $CaSO_4$ Within CMAS (CaO-MgO-Al_2O_3-SiO_2) and Its Effect on the Infiltration Behavior in EB-PVD 7YSZ. *J. Am. Ceram. Soc.* **2016**, *99*, 1398–1403. [CrossRef]
24. Schmitt, M.P.; Rai, A.K.; Bhattacharya, R.; Zhu, D.; Wolfe, D.E. Multilayer thermal barrier coating (TBC) architectures utilizing rare earth doped YSZ and rare earth pyrochlores. *Surf. Coat. Technol.* **2014**, *251*, 56–63. [CrossRef]
25. Munawar, A.U.; Schulz, U.; Cerri, G. Microstructural Evolution of GdZ and DySZ Based EB-PVD TBC Systems After Thermal Cycling at High Temperature. *J. Eng. Gas Turbines Power* **2013**, *135*, 6. [CrossRef]
26. Gomez Chavez, J.J.; Naraparaju, R.; Mikulla, C.; Mechnich, P.; Kelm, K.; Ramana, C.V.; Schulz, U. Comparative Study of EB-PVD Gadolinium-Zirconate and Yttria-rich Zirconia Coatings Performance against Fe-containing Calcium-Magnesium-Aluminosilicate (CMAS) Infiltration. *Corros. Sci.* **2021**, *190*, 109660. [CrossRef]
27. Nieto, A.; Agrawal, R.; Bravo, L.; Hofmeister-Mock, C.; Pepi, M.; Ghoshal, A. Calcia–magnesia–alumina–silicate (CMAS) attack mechanisms and roadmap towards Sandphobic thermal and environmental barrier coatings. *Int. Mater. Rev.* **2021**, *66*, 451–492. [CrossRef]
28. Renteria, A.F.; Saruhan, B.; Schulz, U.; Raetzer-Scheibe, H.J.; Haug, J.; Wiedenmann, A. Effect of morphology on thermal conductivity of EB-PVD PYSZ TBCs. *Surf. Coat. Technol.* **2006**, *201*, 2611–2620. [CrossRef]
29. Wolfe, D.E.; Singh, J.; Miller, R.A.; Eldridge, J.I.; Zhu, D.-M. Tailored microstructure of EB-PVD 8YSZ thermal barrier coatings with low thermal conductivity and high thermal reflectivity for turbine applications. *Surf. Coat. Technol.* **2005**, *190*, 132–149. [CrossRef]
30. Jang, B.K.; Matsubara, H. Influence of rotation speed on microstructure and thermal conductivity of nano-porous zirconia layers fabricated by EB-PVD. *Scr. Mater.* **2005**, *52*, 553–558. [CrossRef]

31. Schulz, U.; Fritscher, K.; Leyens, C.; Peters, M.; Kaysser, W.A. Thermocyclic Behavior of Differently Stabilized and structured EB-PVD thermal barrier coatings. *Mater. Werkst.* **1997**, *28*, 370–376. [CrossRef]
32. Matsumoto, M.; Wada, K.; Yamaguchi, N.; Kato, T.; Matsubara, H. Effects of substrate rotation speed during deposition on the thermal cycle life of thermal barrier coatings fabricated by electron beam physical vapor deposition. *Surf. Coat. Technol.* **2008**, *202*, 3507–3512. [CrossRef]
33. Ganvir, A.; Joshi, S.; Markocsan, N.; Vassen, R. Tailoring columnar microstructure of axial suspension plasma sprayed TBCs for superior thermal shock performance. *Mater. Des.* **2018**, *144*, 192–208. [CrossRef]
34. Jesuraj, S.A.; Kuppusami, P.; Dharini, T.; Panda, P.; Devapal, D. Effect of substrate temperature on microstructure and nanomechanical properties of $Gd_2Zr_2O_7$ coatings prepared by EB-PVD technique. *Ceram. Int.* **2018**, *44*, 18164–18172. [CrossRef]
35. Jesuraj, S.A.; Kuppusami, P.; Ajith Kumar, S.; Panda, P.; Udaiyappan, S. Investigation on the effect of deposition temperature on structural and nanomechanical properties of electron beam evaporated lanthanum zirconate coatings. *Mater. Chem. Phys.* **2019**, *236*, 121789. [CrossRef]
36. Park, C.; Choi, S.; Chae, J.; Kim, S.; Kim, H.; Oh, Y.-S. Effect of Substrate Rotation on the Phase Evolution and Microstructure of 8YSZ Coatings Fabricated by EB-PVD. *J. Korean Ceram. Soc.* **2016**, *53*, 81–86. [CrossRef]
37. Renteria, A.F. A Small-Angle Scattering Analysis of the Influence of Manufacture and Thermal Induced Morphological Changes on the Thermal Conductivity of EBPVD PYSZ Thermal Barrier Coatings. Ph.D. Dissertation, RWTH Aachen, Aachen, Germany, 2007.
38. Yamaguchi, N.; Wada, K.; Kimura, K.; Matsubara, H. Microstructure modification of yttria-stabilized zirconia layers prepared by EB-PVD. *J. Ceram. Soc. Jpn.* **2003**, *111*, 883–889. [CrossRef]
39. Jackson, R.W.; Zaleski, E.M.; Hazel, B.T.; Begley, M.R.; Levi, C.G. Response of molten silicate infiltrated $Gd_2Zr_2O_7$ thermal barrier coatings to temperature gradients. *Acta Mater.* **2017**, *132*, 538–549. [CrossRef]
40. Karaoglanli, A.C.; Doleker, K.M.; Ozgurluk, Y. Interface failure behavior of yttria stabilized zirconia (YSZ), $La_2Zr_2O_7$, $Gd_2Zr_2O_7$, YSZ/$La_2Zr_2O_7$ and YSZ/$Gd_2Zr_2O_7$ thermal barrier coatings (TBCs) in thermal cyclic exposure. *Mater. Charact.* **2020**, *159*, 110072. [CrossRef]
41. Kramer, S.; Yang, J.; Levi, C.G. Infiltration-inhibiting reaction of gadolinium zirconate thermal barrier coatings with CMAS melts. *J. Am. Ceram. Soc.* **2008**, *91*, 576–583. [CrossRef]
42. Lavigne, O.; Rio, C.; Vidal-Setif, M.H.; Jaquet, V.; Joulia, A.; Tillard, L. Influence of morphology on thermal properties and CMAS resistance of gadolinium zirconate EB-PVD coatings. In Proceedings of the HELSMAC Symposium, Cambridge, UK, 7–8 April 2016.
43. Rivera-Gil, M.A.; Gomez-Chavez, J.; Ramana, C.V.; Naraparaju, R.; Schulz, U.; Munoz Saldana, J. High Temperature Interaction of Volcanic Ashes with 7YSZ TBC's produced by APS: Infiltration behavior and phase stability. *Surf. Coat. Technol.* **2019**, *378*, 124915. [CrossRef]
44. Poerschke, D.L.; Levi, C.G. Phase equilibria in the calcia-gadolinia-silica system. *J. Alloys Compd.* **2017**, *695*, 1397–1404. [CrossRef]
45. Perrudin, F.; Rio, C.; Vidal-Setif, M.H.; Petitjean, C.; Panteix, P.J.; Vilasi, M. Gadolinium oxide solubility in molten silicate: Dissolution mechanism and stability of $Ca_2Gd_8(SiO_4)_6O_2$ and $Ca_3Gd_2(Si_3O_9)_2$ silicate phases. *J. Eur. Ceram. Soc.* **2017**, *37*, 2657–2665. [CrossRef]
46. Guijosa-Garcia, C.Y.; Rivera-Gil, M.A.; Ramana, C.V.; Naraparaju, R.; Schulz, U.; Muñoz-Saldaña, J. Reaction Products from High Temperature Treatments of $(La_xGd_{1-x})_2Zr_2O_7$ System and Volcanic Ash Powder Mixtures. *JOM* **2022**, *74*, 2791–2808. [CrossRef]
47. Mechnich, P.; Braue, W.; Schulz, U. High-Temperature Corrosion of EB-PVD Yttria Partially Stabilized Zirconia Thermal Barrier Coatings with an Artificial Volcanic Ash Overlay. *J. Am. Ceram. Soc.* **2011**, *94*, 925–931. [CrossRef]
48. Poerschke, D.L.; Levi, C.G. Effects of cation substitution and temperature on the interaction between thermal barrier oxides and molten CMAS. *J. Eur. Ceram. Soc.* **2015**, *35*, 681–691. [CrossRef]

Disclaimer/Publisher's Note: The statements, opinions and data contained in all publications are solely those of the individual author(s) and contributor(s) and not of MDPI and/or the editor(s). MDPI and/or the editor(s) disclaim responsibility for any injury to people or property resulting from any ideas, methods, instructions or products referred to in the content.

Article

Automatic Recognition of Microstructures of Air-Plasma-Sprayed Thermal Barrier Coatings Using a Deep Convolutional Neural Network

Xiao Shan [1], Tianmeng Huang [1], Lirong Luo [2], Jie Lu [1], Huangyue Cai [1,*], Junwei Zhao [3], Gang Sheng [3] and Xiaofeng Zhao [1,*]

[1] Shanghai Key Laboratory of Advanced High-Temperature Materials and Precision Forming, School of Materials Science and Engineering, Shanghai Jiao Tong University, Shanghai 200240, China
[2] Engineering Research Center of Nano-Geo Materials of Ministry Education, Faculty of Materials Science and Chemistry, China University of Geosciences, Wuhan 430074, China
[3] Chengdu Chengfa Taida Aviation Technology Co., Ltd., Chengdu 610000, China
* Correspondence: caihy27@sjtu.edu.cn (H.C.); xiaofengzhao@sjtu.edu.cn (X.Z.)

Abstract: Either to obtain desirable microstructures by adjusting processing parameters or to predict the properties of a thermal barrier coating (TBC) according to its microstructure, fast and reliable quantitation of the microstructure is imperative. In this research, a machine-learning-based approach—a deep convolution neural network (DCNN)—was established to accurately quantify the microstructure of air-plasma-sprayed (APS) TBCs based on 2D images. Four scanning electron microscopy (SEM) images (view field: 150 μm × 150 μm, image size: 3072 pixel × 3072 pixel) were taken and labeled to train the DCNN. After training, the DCNN could recognize correctly 98.5% of the pixels in the SEM images of typical APS TBCs. This study demonstrated that a small dataset of SEM images could be enough to train a DCNN, making it a powerful and feasible method for quantitively characterizing the microstructure osf APS TBCs.

Keywords: thermal barrier coatings (TBC); plasma sprayed; microstructural characterization; machine learning; convolution neural network

1. Introduction

Air-plasma-sprayed (APS) thermal barrier coatings (TBCs) are widely used to protect the metals in hot sections of gas-turbine engines [1]. They have a unique and complex microstructure: impingement of molten spray particles onto substrates results in "splats", and the successive build-up of the "splats" results in a layered microstructure [2]. Several types of microstructure defect exist in APS TBCs: cracks, pores, and unmelted particles [3,4]. Cracks can be further classified into two categories: intersplat cracks caused by imperfect contacts between splats, and intrasplat cracks caused by relaxation of the quenching stress [5]. Pores, also called globular pores in some papers, are formed by incomplete contact or partially molten particles [6].

Both for aircraft and for industrial gas-turbine engines, APS TBCs require low thermal conductivity, high strain tolerance, high fracture toughness, and high sintering resistance [7,8]. All these properties are influenced largely by the microstructure and have a big impact on the lifetime of an APS TBC [9]. Therefore, investigation of the microstructure is of great significance in the area of TBC research. Either to obtain desirable microstructures by adjusting processing parameters or to predict the properties of an APS TBC according to its microstructure, quantitatively characterizing the microstructures of APS TBCs is imperative [10], and a fast and reliable characterization method will be beneficial to the whole APS TBC research community.

Image analysis with scanning electron microscopy (SEM, Mira, TESCAN, Czech Republic) images plays a crucial role in the quantitative characterization of the microstructures of APS TBCs, due to its feasibility and reliability [3,10,11]. Figure 1 illustrates the procedures of conventional image analysis [10,12]. First, thresholding of a grayscale SEM image (Figure 1A) is performed, which yields a binary image (Figure 1B) wherein microstructural defects (e.g., equiaxed pores and cracks) are separated from the coating material. Thresholding is effective because the gray values of microstructural features are usually lower than those of the coating material. Second, morphological filtering of the opening operation is performed on the binary image, which can separate cracks (Figure 1C) from equiaxed pores (Figure 1D). That opening operation can perform such segmentation by relying on the fact that cracks are usually thinner than equiaxed pores.

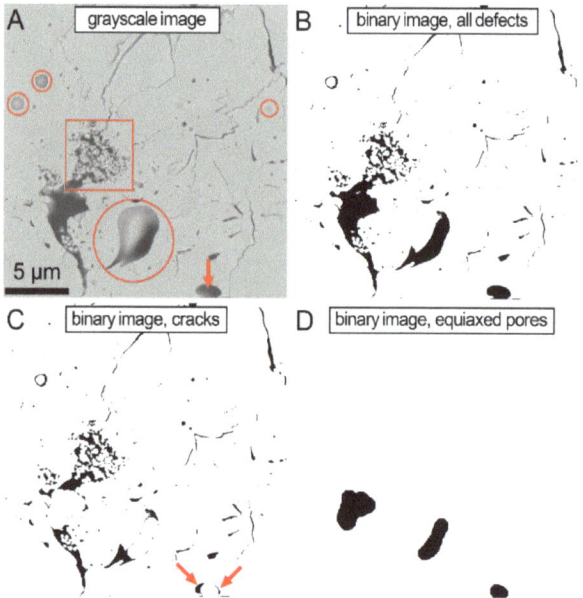

Figure 1. Images illustrating the procedures of conventional images analysis for SEM images of APS TBCs. (**A**) gray-scale BSE image, (**B**) binary image with all pores, (**C**) binary image with cracks and (**D**) binary image with equiaxed pores. The circles in (**A**) indicate pores whose gray values are close to the coating material; the squares indicate unmelted regions. The arrow in (**A**) indicates an equiaxed pore, some of which tend to be classified as cracks using an opening operation. The arrow in (**C**) indicate an equiaxed pore's protrusions that were wrongly classified as cracks during opening operation.

Although the above-mentioned image analysis method has been widely used, both the thresholding and the opening operations have inherent issues. For thresholding, some defects—for example, the equiaxed pores indicated by the circles in Figure 1A—can hardly be recognized if they have gray values close to the coating material. In opening operations, the protrusions of some equiaxed pores are often classified as cracks (indicated by the arrows in Figure 1C). In addition to these issues, the conventional image analysis method for APS TBCs is based on explicit programming. Therefore, specific code must be developed if a certain type of microstructure—for example, unmelted regions—needs to be recognized, which is possibly difficult to realize.

Recently, machine learning—an approach to realizing artificial intelligence—has become increasingly important in materials science and engineering [13,14]. It has been applied in research areas such as property prediction, discovery and design of materials, characterization of materials, knowledge extraction via text mining, and molecular simulation [13]. Even in the area of TBC research, the last few years have witnessed its applications [15–21]. Previous studies have demonstrated that some machine learning models (e.g., random forest and convolution neural network) can be used to analyze optical microscopy images [22,23]. Compared with conventional image analysis, machine learning does not need explicit code for recognizing a certain microstructure. Feeding SEM and labeled images into a proper machine learning model may realize automatic recognition. So far, however, whether machine learning can realize pixel-wise automatic microstructure recognition of SEM images of APS TBCs is still unclear. In this research, a deep convolution neural network (DCNN) model was built. Four SEM images were taken and labeled to train the model. After training, the model's performance was evaluated, and the predicted images were examined in detail, through which possible reasons for wrong recognitions were analyzed.

2. Research Methodology

The core components of the model and the training process are illustrated in Figure 2. First, input (SEM images) and true targets (labeled SEM images) are needed, because supervised learning was adopted in this research. The input is fed into the model. Next, the model makes a prediction, which is then compared with the true target. Taking the prediction and the true target as inputs, the loss function computes a loss score, which quantifies the difference between them. Based on the score, the optimizer adjusts the model's parameters in the correct direction, aiming at lowering the score in the next training loop. The above depicts one training loop. After all the training images have been used, one epoch finishes. After many epochs, the prediction might be very close to the true target.

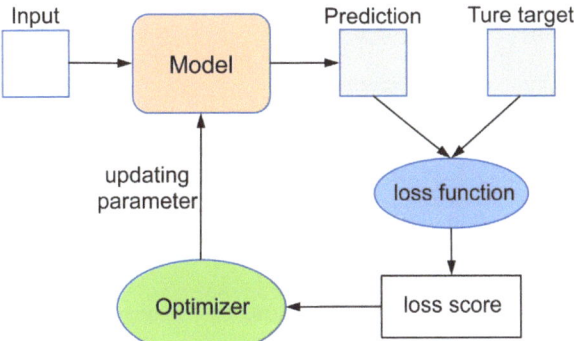

Figure 2. Schematic showing the core components of the machine learning model and the training process.

2.1. Dataset Preparation

2.1.1. Acquiring SEM Images

Four backscatter electron (BSE) SEM images were taken on the cross-section of a yttria-stabilized zirconia (YSZ) APS TBC. As can be seen in Figure 3A, this APS TBC contains not only typical APS TBC microstructures—cracks and equiaxed pores—but also unmelted regions. The view field of the BSE image is 150 µm × 150 µm, and the image size is 3072 pixels × 3072 pixels. The numerical resolution reaches 0.046 µm/pixel, which is enough for quantitatively characterizing the microstructure of plasma-sprayed TBCs [12].

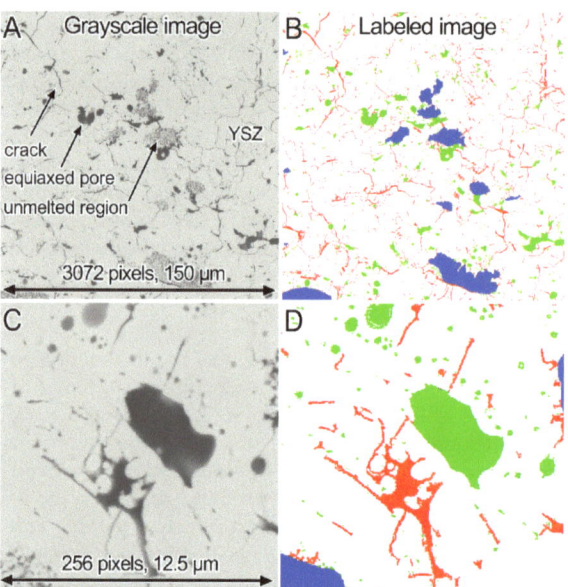

Figure 3. (**A**) Initial grayscale SEM image, (**B**) manually labeled image, (**C**) cropped grayscale image used for training, and (**D**) manually labeled image of the image in (**C**). Cracks, equiaxed pores, unmelted regions, and YSZ are indicated by red, green, blue, and white, respectively.

2.1.2. Labeling SEM Images

The task of the model is pixel-wise semantic segmentation—to classify each pixel in the images—so pixel-wise labeling is warranted. The four SEM images contain $3072 \times 3072 \times 4$ = 37,748,736 pixels. If all the pixels were labeled by hand, it would be a labor-heavy and time-consuming process. In addition to manual labeling, therefore, automatic labeling was also adopted. Figure 4 shows the labeling process used in this research. First, the unmelted regions were manually labeled using Photoshop software (version Photoshop CC2018). Second, non-YSZ regions (also referred to as defects in this research, including cracks, equiaxed pores, and unmelted regions) were automatically extracted from the remaining region using thresholding. However, some non-YSZ regions could not be extracted; these problematic regions were restored manually. Afterwards, isolated equiaxed pores were automatically identified based on their shapes. Next, the remaining equiaxed pores (e.g., those connected to other types of microstructures) were manually selected using Photoshop software. Finally, the labeled images were obtained. A truly labeled SEM image is shown in Figure 3B. After labeling, the area percentages of different classes of microstructures were calculated, as shown in Figure 5. One point worth stressing here is that the total area percentage of the three classes of defects is not necessarily equal to the porosity of an APS TBC, because the unmelted regions also contain YSZ (see the region indicated by the square in Figure 1A). However, the two are equal when no unmelted regions are involved.

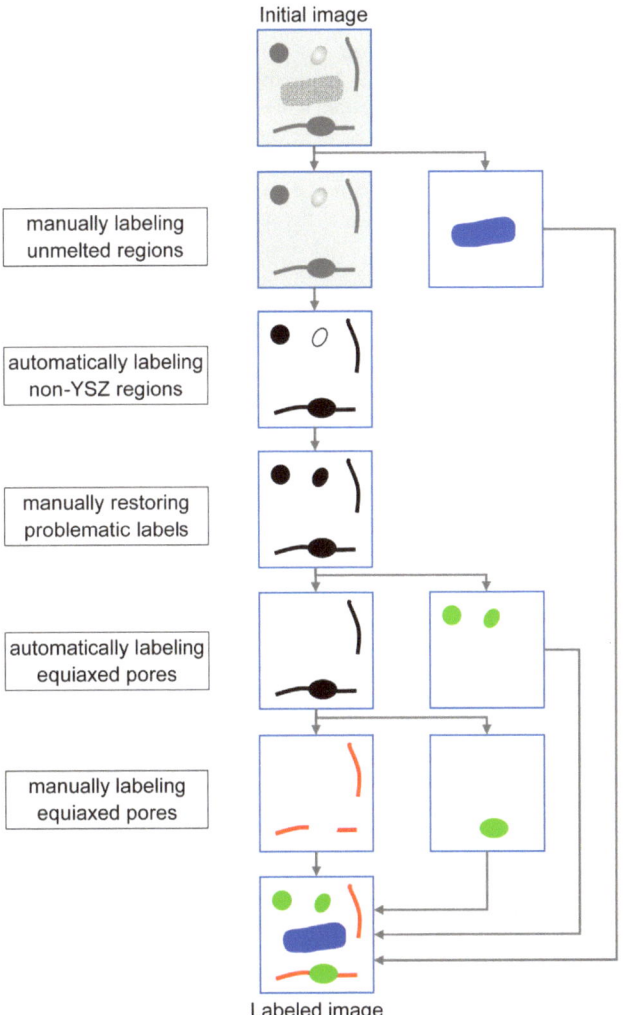

Figure 4. Schematic showing the labeling process.

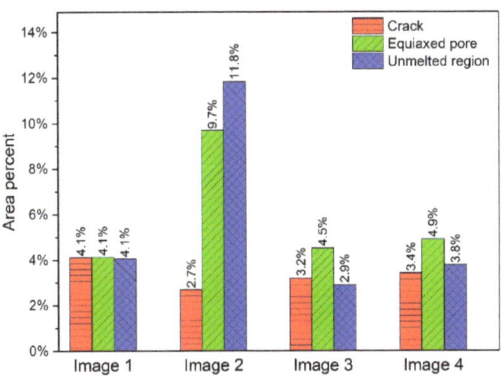

Figure 5. Area percentages of different classes of microstructures in the four SEM images.

2.1.3. Dataset Splitting

The image dataset was split into a training set, a validation set, and a test set. The training set was used to train the model. During the training process, the performance of the model on training data will always improve, but this does not mean that the model will perform well on data it has never seen. Therefore, a validation set was reserved; its purpose is to monitor how the model performs on new data. After training, the final performance of the model was evaluated using the test set. For dataset splitting, each of the initial 3072-pixel-wide images was divided into 144 256-pixel-wide images, so an image dataset containing 576 small images was obtained. The dataset was then split into three subsets according to Table 1.

Table 1. The numbers of small images in the training, validation, and testing sets.

Dataset	Image Number
Training	462
Validation	57
Test	57

2.2. Model Building

2.2.1. Workspace Setup

The DCNN model was built using Python in a Jupyter notebook. To enable fast experimentation with deep neural networks, Keras—a widely used deep learning framework—was used.

2.2.2. Model Architecture

Figure 6 shows the overall architecture of the DCNN model; its detailed specifications are presented in Table 2. The model has an encoder–decoder structure with skip connections. It is composed of 23 layers (Table 2), including convolution layers, pooling layers, up-sampling layers, and dropout layers. The convolution layers are to extract features of the images. They are composed of a certain number of kernels, whose parameters (commonly known as weights) are random values before training and need to be adjusted by model training. Pooling layers perform down-sampling operations, aiming at reducing the sizes of feature-maps to process and induce spatial-filter hierarchies. Contrary to pooling layers, up-sampling layers increase the sizes of feature-maps. Dropout layers were introduced to the model because they can reduce overfitting.

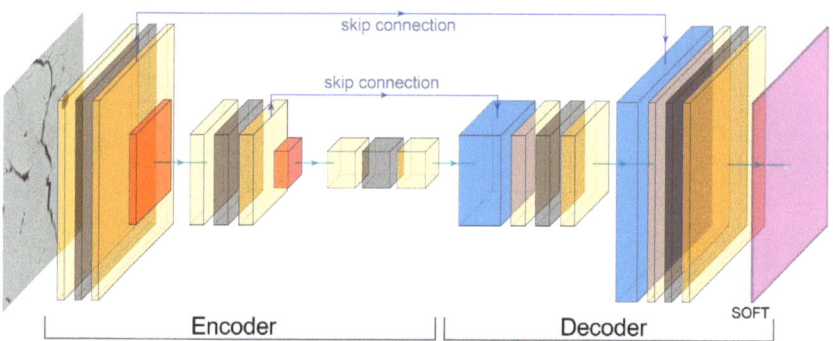

Figure 6. 3D schematic showing the overall architecture of the DCNN.

Table 2. The detailed specifications of the DCNN.

No.	Layer	Kernel	Parameter	Input Size	Output Size
1	Input	-	0	(N, 256, 256, 3)	(N, 256, 256, 3)
2	Convolution	(3, 3) × 32	896	(N, 256, 256, 3)	(N, 256, 256, 32)
3	Dropout	-	0	(N, 256, 256, 32)	(N, 256, 256, 32)
4	Convolution	(3, 3) × 32	9248	(N, 256, 256, 32)	(N, 256, 256, 32)
5	Pooling	-	0	(N, 256, 256, 32)	(N, 128, 128, 32)
6	Convolution	(3, 3) × 64	18,496	(N, 128, 128, 32)	(N, 128, 128, 64)
7	Dropout	-	0	(N, 128, 128, 64)	(N, 128, 128, 64)
8	Convolution	(3, 3) × 64	36,928	(N, 128, 128, 64)	(N, 128, 128, 64)
9	Pooling	-	0	(N, 128, 128, 64)	(N, 64, 64, 64)
10	Convolution	(3, 3) × 128	73,856	(N, 64, 64, 64)	(N, 64, 64, 128)
11	Dropout	-	0	(N, 64, 64, 128)	(N, 64, 64, 128)
12	Convolution	(3, 3) × 128	147,584	(N, 64, 64, 128)	(N, 64, 64, 128)
13	Up sampling	-	0	(N, 64, 64, 128)	(N, 128, 128, 128)
14	Concatenate	-	0	(N, 128, 128, 128) (N, 128, 128, 64)	(N, 128, 128, 192)
15	Convolution	(3, 3) × 64	110,656	(N, 128, 128, 192)	(N, 128, 128, 64)
16	Dropout	-	0	(N, 128, 128, 64)	(N, 128, 128, 64)
17	Convolution	(3, 3) × 64	36,928	(N, 128, 128, 64)	(N, 128, 128, 64)
18	Up sampling	-	0	(N, 128, 128, 64)	(N, 256, 256, 64)
19	Concatenate	-	0	(N, 256, 256, 64) (N, 256, 256, 32)	(N, 256, 256, 96)
20	Convolution	(3, 3) × 32	27,680	(N, 256, 256, 96)	(N, 256, 256, 32)
21	Dropout	-	0	(N, 256, 256, 32)	(N, 256, 256, 32)
22	Convolution	(3, 3) × 32	9248	(N, 256, 256, 32)	(N, 256, 256, 32)
23	Convolution	(1, 1) × 4	132	(N, 256, 256, 32)	(N, 256, 256, 4)

Total trainable parameters: 471,652

2.2.3. Loss and Metrics

There are four classes of microstructures in the SEM images, and most of the pixels are YSZ, so the task of the model is an imbalanced multiclass segmentation problem. Therefore, the generalized Dice loss [24], which is effective for this kind of segmentation, was used in this research.

Metrics are functions that are used to intuitively judge a model's performance. Three of the most commonly used metrics for semantic segmentation tasks were adopted in this research: (pixel) accuracy, mean (pixel) accuracy, mean intersection over union (mean IoU). Accuracy is the percentage of pixels in an image that are classified correctly. Accuracy can be computed using the following equation:

$$\text{accuracy} = \frac{\sum_{i=1}^{k} N_{ii}}{\sum_{i=1}^{k} \sum_{j=1}^{k} N_{ij}} \quad (1)$$

where k is the number of classes in a certain image, N_{ij} is the number of pixels of class i predicted to be class j. Classes 1–4 represent cracks, equiaxed pores, unmelted regions, and YSZ, respectively.

Accuracy is one of the easiest to understand conceptually. However, it is not an ideal metric when class imbalance—one class dominating the image—occurs. For example, suppose that the area of the non-YSZ region (including cracks, equiaxed pores, and unmelted regions) were 5%; the accuracy could reach up to 95% if all the pixels were classified as YSZ. To compensate for this issue, mean accuracy was introduced. It is the average of the accuracies of all the classes. The accuracy for class i is the number of pixels of a class that are predicted correctly, divided by the number of all the pixels of the class in the labeled image, whose formula is as follows:

$$\text{accuracy}_i = \frac{N_{ii}}{\sum_{j=1}^{k} N_{ij}} \quad (2)$$

Based on this equation, the mean accuracy can be calculated using the following formula:

$$\text{mean accuracy} = \frac{1}{k}\sum_{i=0}^{k} \text{accuracy}_i \quad (3)$$

where k is the number of classes in a certain image. Mean interaction-over-union (IoU) is also a popular metric when class imbalance occurs. It is the average of the IoUs of all the classes. IoU is stricter than accuracy. For class i, it is the number of pixels of a class that are predicted correctly divided by the sum of the number of all the pixels of the class in the labeled image and the number of pixels that are wrongly classified as class i. The following formula shows the way calculating it:

$$\text{IoU}_i = \frac{N_{ii}}{\sum_{j=1}^{k} N_{ij} + \left(\sum_{j=1}^{k} N_{ji} - N_{ii}\right)} \quad (4)$$

Based on this equation, the mean IoU can be calculated using the following formula:

$$\text{mean IoU} = \frac{1}{k}\sum_{i=1}^{k} \text{IoU}_i \quad (5)$$

2.3. Model Training

The model was trained using four graphic processing units (GPU, NVIDIA Tesla V100 32 GB) for 100 epochs, and the loss scores were recorded after each epoch.

3. Results

3.1. Evolution of the Model during the Training Process

Figure 7 shows the evolutions of the losses (the generalized Dice loss) during training. Overall, the losses changed significantly during the first 15 epochs, and then changed slowly in the following epochs. The training loss kept decreasing till the end. The validation loss stopped improving from about the 83ed epoch, indicating that the model began to overfit and would not obtain better results on never-before-seen data with more adjustment. Therefore, 100 epochs are enough for training the model. The model with the lowest loss represented the best model and was used for the following analysis.

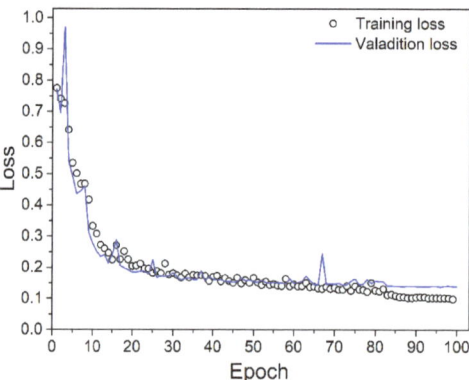

Figure 7. Training and validation losses as a function of epoch.

3.2. Model's Performance on All the Test Images

Figure 8 presents about half of the images predicted by the trained model, together with the initial grayscale images and the manually labeled images. These images give an overall visual impression of the model's performance. As can be seen in this figure, in most cases, the predicted images are very similar to the labeled images.

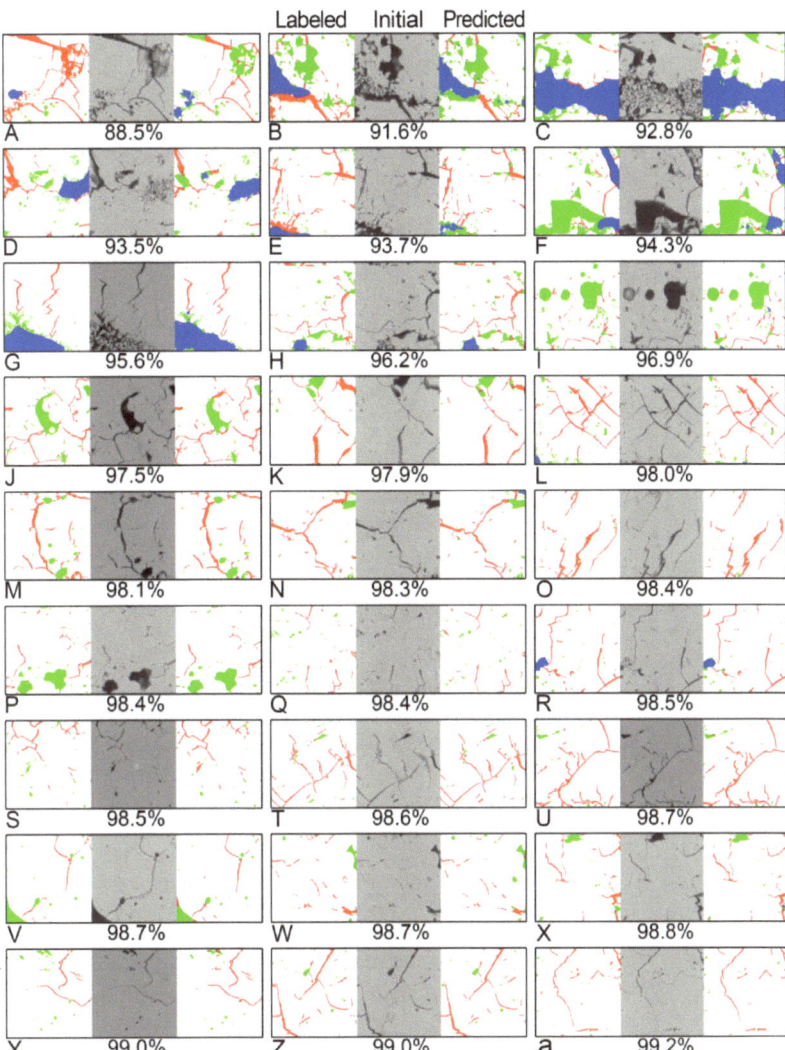

Figure 8. Images predicted by the model, together with the initial grayscale images, and the manually labeled images. These images were selected this way: all the predicted images were ranked according to accuracy, and then the even-numbered images were selected. The accuracy for each image is presented below each grayscale image. (**A–Z,a**) Images predicted by the model, together with the initial grayscale images, and the manually labeled images. These images were selected this way: all the predicted images were ranked according to accuracy, and then the even-numbered images were selected. The accuracy for each image is presented below each grayscale image.

Figure 9A shows the metrics of the trained model calculated based on the 57 test images. The accuracy reached up to 97.0%, meaning that only 3.0% of pixels in the grayscale images were wrongly predicted. The mean accuracy and mean IOU dropped to 80.1% and 71.4%, respectively. The reason for this is that the accuracies of cracks, equiaxed pores, and unmelted regions were not high enough, as can be seen in Figure 9B.

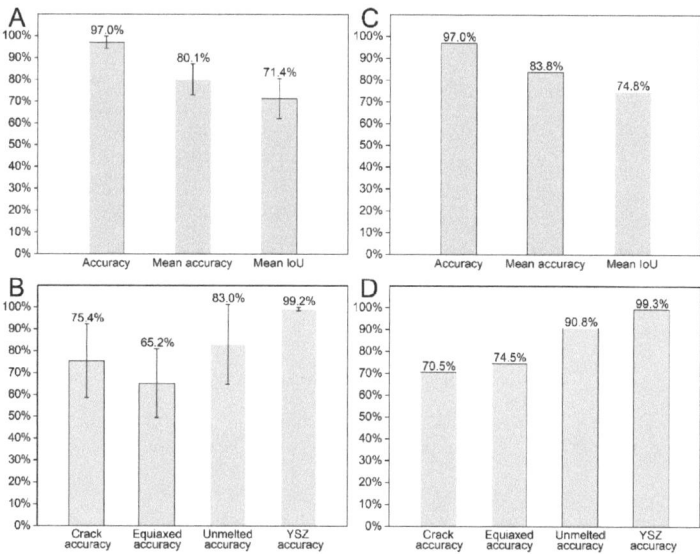

Figure 9. (**A**,**B**) Metrics of the trained model evaluated using 57 test images: (**A**) Overall metrics and (**B**) accuracy for each class of microstructure; the error bars represent standard deviations. (**C**,**D**) Metrics of the trained model evaluated using one hypothetical image containing all the test images: (**A**) overall metrics and (**B**) accuracy for each class of microstructure.

The data in Figure 9A,B were obtained based on small images (256 pixels × 256 pixels, 12.5 µm × 12.5 µm), but sometimes evaluations on small images can be misleading. To illustrate, assume that there is only one test image which contains all the 57 test images. Then, the model's performance was evaluated using this large hypothetical image (1933 pixel × 1933 pixel, 94 µm × 94 µm). The results are shown in Figure 9C,D. As can be seen in Figure 9C, the accuracy did not change, but both the mean accuracy and mean IoU increased slightly. In Figure 9D, note that the unmelted accuracy increases from 83.0% (Figure 9B) to 90.8%. This indicates that for the 57 test images, the images resulting in high unmelted accuracy tended to contain large areas of unmelted regions, which in turn means that a larger fraction of unmelted region pixels can be recognized if they are in larger unmelted regions. The same phenomenon occurred for equiaxed pores. The comparison between Figure 9B,D indicates that the size of the test image plays an important role in the model evaluation process, and the metrics evaluated using the test images whose sizes are closer to a usual size may be more representative.

In addition to pixel-level microstructure recognition, sometimes a trained model may also be used to calculate the area percentages of different microstructures. Figure 10 shows the true and predicted area percentages of different microstructures in the test images. It is shown that for some test images, the predicted area percentages of cracks (Figure 10A) and equiaxed pores (Figure 10B) deviate far from the true area percentages. In contrast, the predicted area percentage of the defect for each image is almost equal to the true value.

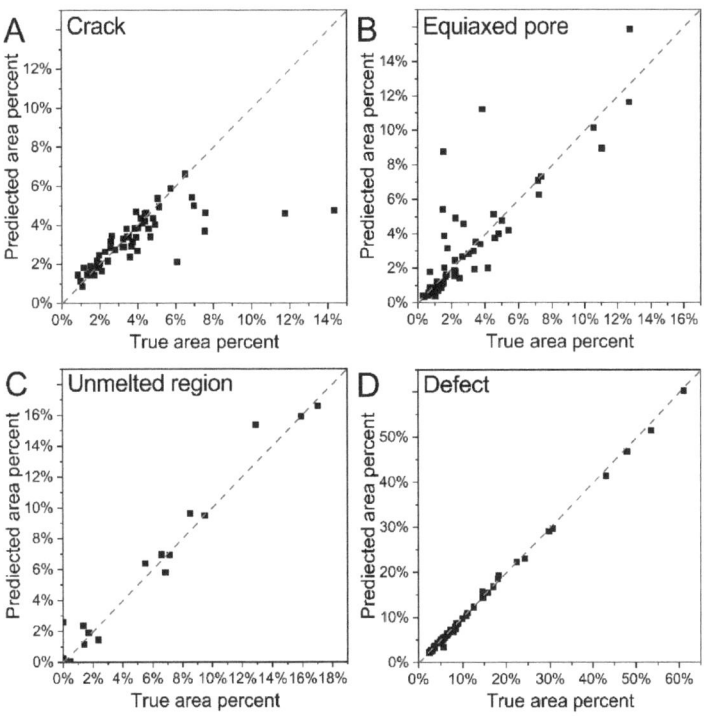

Figure 10. True and predicted area percentages of different microstructures in the test images: (**A**) crack, (**B**) equiaxed pore, (**C**) unmelted region, and (**D**) defect.

The large difference between true and predicted values for crack and equiaxed pore may have been caused by the small image size (256 pixels × 256 pixels, 12.5 μm × 12.5 μm), which is unlikely to be used in practice.

The model was also evaluated using the aforementioned hypothetical image (1933 pixel × 1933 pixel, 94 μm × 94 μm), and the results are shown in Figure 11. For this large image, the trained model exhibited good ability of estimating the area percentages of different microstructures.

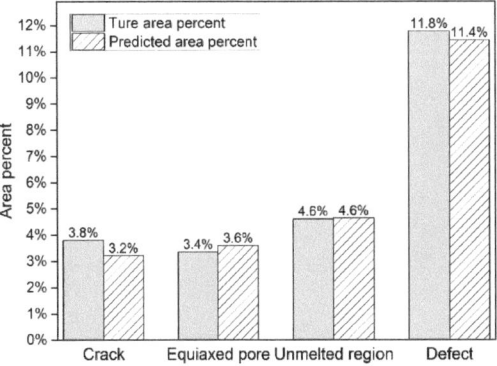

Figure 11. True and predicted area percentages for each class of microstructure in a hypothetical image containing all 57 test images.

3.3. Model's Performance on Typical APS TBC Images

It can be deduced from Figure 5 that some test images contain larger unmelted regions than usual, so the model's performance in Section 3.2 is indeed not representative enough of typical APS TBCs. To obtain more representative data, test images having area percentages of unmelted regions larger than 1% were removed, by which 39 typical APS TBC images remained and were used to evaluate the model. The results are shown in Figure 12. The accuracy reached up to 98.5%, and the crack accuracy increased to 82.6% (75.4% in Figure 9B). Note that the unmelted accuracy decreased to 0.0%, meaning that no unmelted region was successfully recognized, presumably because the features of unmelted regions become vague such areas are too small.

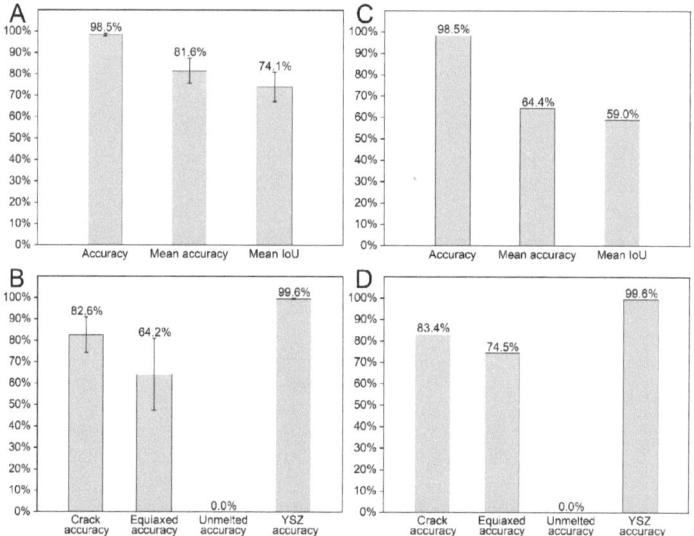

Figure 12. (**A**,**B**) Metrics of the trained model evaluated using 39 more typical APS TBC images selected from all the 57 test images: (**A**) Overall metrics and (**B**) accuracy for each class of microstructure; the error bars represent standard deviations. (**C**,**D**) Metrics of the trained model evaluated using one hypothetical image containing all 39 more typical APS TBC images: (**A**) overall metrics and (**B**) accuracy for each class of microstructure.

Figure 12C,D present the metrics of the model evaluated using one hypothetical image (1599 pixels × 1599 pixels, 78 µm × 78 µm) containing all 39 test images. The accuracy is the same as that in Figure 12A. However, the mean accuracy and mean IoU are significantly lower; the reason is that no unmelted region pixels were successfully recognized (Figure 12D: unmelted accuracy = 0.0%). Despite the decreases in the mean accuracy and the mean IoU, accuracies of typical microstructures (cracks, equiaxed pores, and YSZ) did not decline; in particular, the equiaxed accuracy increased from 64.2% to 74.5%, reaching the same value as that in Figure 9D.

Figure 13 shows the true and predicted area percentages of different microstructures in the 39 typical APS TBC images. Compared with Figure 10, it seems that the model exhibits better performance when evaluating the area percentages of different microstructures. Figure 14 shows the data for the large hypothetical image containing all 39 test images. For this large image, the trained model also exhibited a good ability to estimate the area percentages of different microstructures.

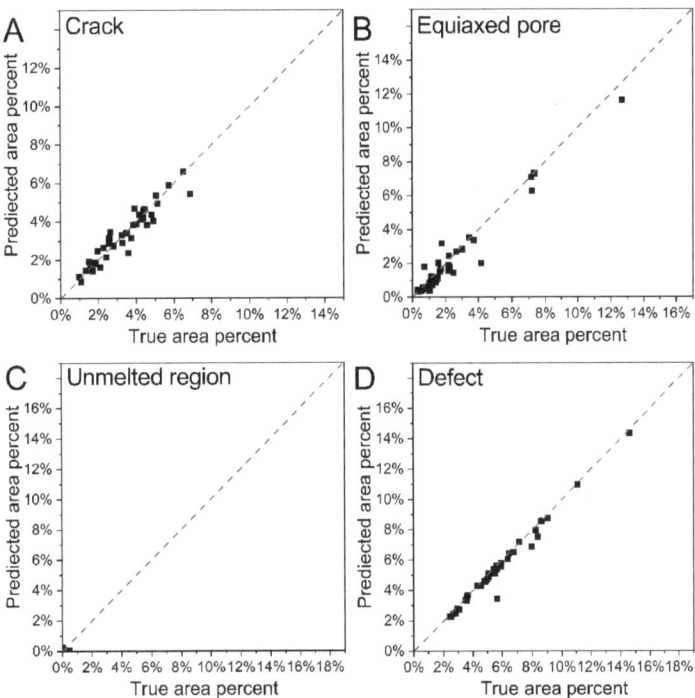

Figure 13. True and predicted area percentages of different objects in the 39 more typical APS TBC images selected from all the 57 test images: (**A**) crack, (**B**) equiaxed pore, (**C**) unmelted region, and (**D**) defect.

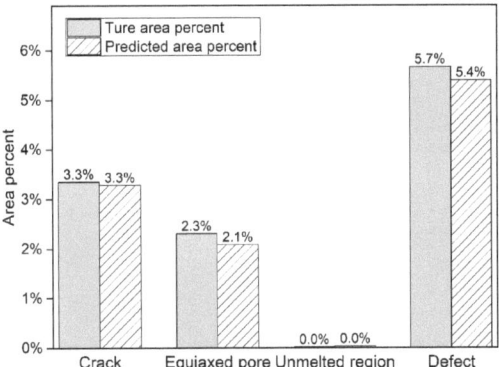

Figure 14. True and predicted area percentages for each class of microstructure in a hypothetical image containing all the 39 typical APS TBC images.

4. Discussion

Although the overall pixel accuracy of the model can reach up to 97.0%, the accuracies for some classes are still not high enough. Possible reasons are discussed in this section. Figure 15 shows the predicted amount of each class of microstructure in a hypothetical image containing all 57 test images. For YSZ, the model has very high accuracy, such that only a tiny fraction (0.3% + 0.2% + 0.2% = 0.7%) of YSZ was predicted as other classes. Non-negligible fractions of the non-YSZ classes can be wrongly predicted as YSZ, accounting for ~30%, ~50%, and ~60% of the wrong predictions of the cracks, equiaxed pores, and unmelted regions, respectively. The most likely reason is that the all the three non-YSZ classes are immediately next to YSZ, but sometimes the boundaries between YSZ and non-YSZ regions are difficult to precisely locate at the pixel level.

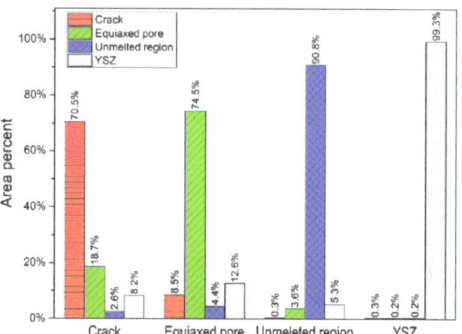

Figure 15. Predicted amount each class of microstructure in a hypothetical image containing all 57 test images.

In addition to YSZ, cracks tended to be wrongly predicted as equiaxed pores, accounting for as many as ~60% of the wrong predictions. Conversely, equiaxed pores also tended to be wrongly predicted as cracks, accounting for ~30% of the wrong predictions. A detailed examination of the predicted image revealed that there are two common situations in which misclassifications between cracks and equiaxed pores can occur. The first is that some cracks next to the unmelted regions tend to be predicted as equiaxed pores, as can be seen in Figure 16. The second occurred when cracks were linked to equiaxed pores. As can be seen in Figure 17, the arrow points to an equiaxed pore, but the model classified it as a crack.

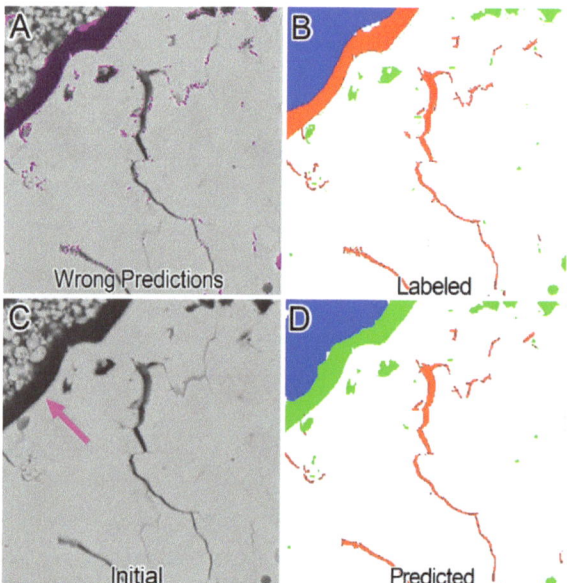

Figure 16. Test image for which cracks were wrongly predicted as equiaxed pores. (**A**) The grayscale image with wrong predictions are marked by purple, (**B**) the labeled image, (**C**) the initial grayscale image The arrow indicates a crack next to an unmelted region that was wrongly predicted as an equiaxed pore, and (**D**) the predicted image.

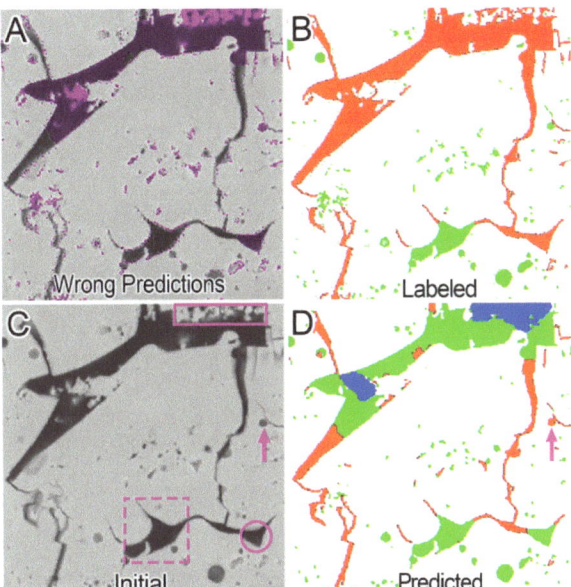

Figure 17. The lowest-accuracy (87.0%) test image (**A**), in which wrong predictions are marked by purple, together with the labeled image (**B**), the initial grayscale image (**C**) (violet solid rectangle indicates a region where wrong predictions occurred due to wrong label, violet dashed square indicates a region where an equiaxed pore and cracks are connected, circle indicates a region where both the wrong prediction and the label seem to be reasonable), and the predicted image (**D**) (arrow indicates an equiaxed pore that was wrongly predicted as a crack).

In addition to the routine wrong predictions, there were also other types of wrong predictions that are worth noting. First, although some predictions are wrong, a comparison of the predicted image with the labeled image reveals that the model seems to be more correct. For example, the region in the rectangle in Figure 17 is unmelted, and the model predicted it as unmelted (Figure 17D), but the region was labeled as cracked and YSZ (Figure 17B). Such "wrong" predictions are actually caused by labeling errors. Therefore, it is expected the model's performance will be further improved by minimizing labeling error.

Second, sometimes both the wrong prediction and the label seem to be reasonable. For example, the defect in the circle in Figure 17 was labeled as a crack, but the model predicted it as an equiaxed pore. Connections between equiaxed pores and cracks are very common in APS TBCs. In this research, manual labeling was used to separate them. For example, the coarse part of the defect in the dashed square in Figure 17C was labeled as equiaxed pores. Therefore, predicting the defect in the circle as an equiaxed pore is also reasonable.

The above analysis shows that the misclassification between cracks and equiaxed pores is an important factor causing wrong predictions. One reason is that so far there are no widely accepted rigorous definitions of cracks and equiaxed pores, so sometimes it is hard to determine what a pixel belongs to, even for a human expert, during the image annotation process. In such situations, misclassification between crack and equiaxed pores can hardly be avoided.

In addition to image-related factors, model configuration may also play a significant role in influencing the model's performance. The study is only a preliminary attempt to assess the possibility of utilizing a machine-learning-based model to quantitively characterize the microstructure of APS TBCs, so only one specific model was adopted. To obtain a high-performance model, future work will investigate the influence of model configuration (e.g., model architecture and loss function) on the model's performance. Additionally, once a high-performance model has been obtained, a comparative study of the model and conventional image analysis method will then be conducted.

5. Summary

This research aimed to answer a key question: whether machine learning can realize pixel-wise automatic microstructure recognition of SEM images of APS TBCs. To answer this question, a machine-learning-based model—a deep convolution neural network (DCNN)—was established. Four SEM images of APS TBCs (containing not only cracks and equiaxed pores, but also unmelted regions) were taken and labeled for training the model. The results showed that the (pixel) accuracy of the trained model reached up to 97.0%—that is, 97.0% of the pixels in the SEM images can be recognized correctly; for SEM images of typical APS TBCs (containing less unmelted regions), the accuracy increased to 98.5%. Despite the high overall accuracy, the accuracies for some classes of microstructure were not high enough. Possible reasons were analyzed. It was found that wrong predictions were mainly due to the misclassification between YSZ and non-YSZ, and between cracks and equiaxed pores. The major finding of this research is that a small dataset of SEM images could be enough to train a DCNN, which may provide a powerful and feasible method for quantitively characterizing the microstructures of APS TBCs.

Author Contributions: Conceptualization, X.S.; methodology, X.S., L.L., J.L. and H.C.; data curation, X.S.; funding acquisition, X.S.; project administration, X.S.; writing-original draft, X.S.; Formal analysis, T.H.; investigation, T.H.; validation, L.L.; resources, L.L., J.L., H.C., J.Z. and G.S.; visualization, H.C.; software, X.Z.; supervision, X.Z., writing—reviewing and editing, X.Z. All authors have read and agreed to the published version of the manuscript.

Funding: The authors would acknowledge the financial support from the National Natural Science Foundation of China (No. 52102071 and 52102072).

Institutional Review Board Statement: Not applicable.

Informed Consent Statement: Not applicable.

Data Availability Statement: The full training datasets and all validation and test cases are available from the corresponding author upon reasonable request. The data are not publicly available due to the reason that at this time as the data also forms part of an ongoing study.

Acknowledgments: We would like to thank Guo Wei from Northwestern Polytechnic University for stimulating discussions.

Conflicts of Interest: The authors declare no conflict of interest.

References

1. Bakan, E.; Vaßen, R. Ceramic top coats of plasma-sprayed thermal barrier coatings: Materials, processes, and properties. *J. Therm. Spray Technol.* **2017**, *26*, 992–1010. [CrossRef]
2. McPherson, R. A review of microstructure and properties of plasma sprayed ceramic coatings. *Surf. Coat. Technol.* **1989**, *39–40*, 173–181. [CrossRef]
3. Heberlein, J.; Fauchais, P.; Boulos, M. *Thermal Spray Fundamentals: From Powder to Part*; Springer: New York, NY, USA, 2014.
4. Dwivedi, G.; Viswanathan, V.; Sampath, S.; Shyam, A.; Lara-Curzio, E. Fracture toughness of plasma-sprayed thermal barrier ceramics: Influence of processing, microstructure, and thermal aging. *J. Am. Ceram. Soc.* **2014**, *97*, 2736–2744. [CrossRef]
5. Kulkarni, A.; Wang, Z.; Nakamura, T.; Sampath, S.; Goland, A.; Herman, H.; Allen, J.; Ilavsky, J.; Long, G.; Frahm, J.; et al. Comprehensive microstructural characterization and predictive property modeling of plasma-sprayed zirconia coatings. *Acta Mater.* **2003**, *51*, 2457–2475. [CrossRef]
6. Wang, Z.; Kulkarni, A.; Deshpande, S.; Nakamura, T.; Herman, H. Effects of pores and interfaces on effective properties of plasma sprayed zirconia coatings. *Acta Mater.* **2003**, *51*, 5319–5334. [CrossRef]
7. Cao, X.; Vassen, R.; Stoever, D. Ceramic materials for thermal barrier coatings. *J. Eur. Ceram. Soc.* **2004**, *24*, 1–10. [CrossRef]
8. Frommherz, M.; Scholz, A.; Oechsner, M.; Bakan, E.; Vassen, R. Gadolinium zirconate/YSZ thermal barrier coatings: Mixed-mode interfacial fracture toughness and sintering behavior. *Surf. Coat. Technol.* **2016**, *286*, 119–128. [CrossRef]
9. Clarke, D.; Oechsner, M.; Padture, N. Thermal-barrier coatings for more efficient gas-turbine engines. *MRS Bull.* **2012**, *37*, 891–898. [CrossRef]

10. Deshpande, S.; Kulkarni, A.; Sampath, S.; Herman, H. Application of image analysis for characterization of porosity in thermal spray coatings and correlation with small angle neutron scattering. *Surf. Coat. Technol.* **2004**, *187*, 6–16. [CrossRef]
11. Drexler, J.; Ortiz, A.; Padture, N. Composition effects of thermal barrier coating ceramics on their interaction with molten Ca-Mg-Al-sillicate (CMAS) glass. *Acta Mater.* **2012**, *60*, 5437–5447. [CrossRef]
12. Lavigne, O.; Renollet, Y.; Poulain, M.; Rio, C.; Moretto, R.; Brannvall, P.; Wigren, J. Microstructural characterisation of plasma sprayed thermal barrier coatings by quantitative image analysis. In *Quantitative Microscopy of High Temperature Materials*; IOM Communication: London, UK, 2001; pp. 131–146.
13. Morgan, D.; Jacobs, R. Opportunities and challenges for machine learning in materials science. *Annu. Rev. Mater. Res.* **2020**, *50*, 71–103. [CrossRef]
14. Butler, K.; Davies, D.; Cartwright, H.; Isayev, O.; Walsh, A. Machine learning for molecular and materials science. *Nature* **2018**, *559*, 547–555. [CrossRef] [PubMed]
15. Chen, W.; Lu, Y.; Li, J.; Zimmerman, B. Automatic classification of microstructures in thermal barrier coating images. In Proceedings of the 2017 IEEE International Symposium on Multimedia, Taichung, Taiwan, 11–13 December 2017; pp. 99–106.
16. Lu, Y.; Chen, W.; Wang, X.; Alisworth, Z.; Tsui, M.; Al-Ghaib, H.; Zimmerman, B. Deep learning-based models for porosity measurement in thermal barrier coating images. *Int. J. Multimed. Data* **2020**, *11*, 20–35. [CrossRef]
17. Yunus, M.; Alsoufi, M. Prediction of mechanical properties of slasma Sprayed thermal barrier coatings (TBCs) with genetic programming (GP). *Int. J. Eng. Trends Technol.* **2017**, *47*, 139–145. [CrossRef]
18. Ye, D.; Wang, W.; Zhou, H.; Fang, H.; Huang, J.; Li, Y.; Gong, H.; Li, Z. Characterization of thermal barrier coatings microstructural features using terahertz spectroscopy. *Surf. Coat. Technol.* **2020**, *394*, 125836. [CrossRef]
19. Ye, D.; Wang, W.; Xu, Z.; Yin, C.; Zhou, H.; Li, Y. Prediction of thermal barrier coatings microstructural features based on support vector machine optimized by cuckoo search algorithm. *Coatings* **2020**, *10*, 704. [CrossRef]
20. Ma, Z.; Zhang, W.; Luo, Z.; Sun, X.; Li, Z.; Lin, L. Ultrasonic characterization of thermal barrier coatings porosity through BP neural network optimizing Gaussian process regression algorithm. *Ultrasonics* **2020**, *100*, 105981. [CrossRef] [PubMed]
21. Padture, N.; Gell, M.; Jordan, E. Thermal barrier coatings for gas-turbine engine applications. *Science* **2002**, *296*, 280–284. [CrossRef] [PubMed]
22. Bulgarevich, D.S.; Tsukamoto, S.; Kasuya, T.; Demura, M.; Watanabe, M. Pattern recognition with machine learning on optical microscopy images of typical metallurgical microstructures. *Sci. Rep.* **2018**, *8*, 2078. [CrossRef] [PubMed]
23. Chowdhury, A.; Kautz, E.; Yener, B.; Lewis, D. Image driven machine learning methods for microstructure recognition. *Comp. Mater. Sci.* **2016**, *123*, 176–187. [CrossRef]
24. Sudre, C.; Li, W.; Vercauteren, T.; Ourselin, S.; Cardoso, M.J. Generalised dice overlap as a deep learning loss function for highly unbalanced segmentations. In *Deep Learning in Medical Image Analysis and Multimodal Learning for Clinical Decision Support*; Cardoso, M., Arbel, T., Carneiro, G., Syeda-Mahmood, T., Tavares, J., Moradi, M., Bradley, A., Greenspan, H., Papa, J., Madabhushi, A., et al., Eds.; Springer: Cham, Switzerland, 2017; pp. 240–248.

Disclaimer/Publisher's Note: The statements, opinions and data contained in all publications are solely those of the individual author(s) and contributor(s) and not of MDPI and/or the editor(s). MDPI and/or the editor(s) disclaim responsibility for any injury to people or property resulting from any ideas, methods, instructions or products referred to in the content.

Article

Fabrication of Nanostructures Consisting of Composite Nanoparticles by Open-Air PLD

Anna Og Dikovska [1,*], Daniela Karashanova [2], Genoveva Atanasova [3], Georgi Avdeev [4], Petar Atanasov [1] and Nikolay N. Nedyalkov [1]

1. Institute of Electronics, Bulgarian Academy of Sciences, 72 Tsarigradsko Chaussee, 1784 Sofia, Bulgaria; paatanas@ie.bas.bg (P.A.); nned@ie.bas.bg (N.N.N.)
2. Institute of Optical Materials and Technologies "Acad. J. Malinowski", Bulgarian Academy of Sciences, Acad. G. Bonchev Str., bl. 109, 1113 Sofia, Bulgaria; dkarashanova@yahoo.com
3. Institute of General and Inorganic Chemistry, Bulgarian Academy of Sciences, Acad. G. Bonchev Str., bl. 11, 1113 Sofia, Bulgaria; genoveva@svr.igic.bas.bg
4. Rostislaw Kaischew Institute of Physical Chemistry, Bulgarian Academy of Sciences, Acad. G. Bonchev Str., bl. 11, 1113 Sofia, Bulgaria; g_avdeev@ipc.bas.bg
* Correspondence: dikovska@ie.bas.bg

Abstract: We present a two-step physical method for the fabrication of composite nanoparticle-based nanostructures. The proposed method is based on the pulsed laser deposition (PLD) technique performed sequentially in vacuum and in air. As a first step, thin-alloyed films of iron with noble metal were deposited by PLD in vacuum. The films were prepared by ablation of a mosaic target formed by equal iron and gold sectors. As a second step, the as-prepared alloyed films were ablated in air at atmospheric pressure as the laser beam scanned their surface. Two sets of experiments were performed in the second step, namely, by applying nanosecond (*ns*) and picosecond (*ps*) laser pulses for ablation. The structure, microstructure, morphology, and optical properties of the samples obtained were studied with respect to the laser ablation regime applied. The implementation of the ablation process in open air resulted in the formation of nanoparticle and/or nanoparticle aggregates in the plasma plume regardless of the ablation regime applied. These nanoparticles and/or nanoaggregates deposited on the substrate formed a complex porous structure. It was found that ablating FeAu films in air by *ns* pulses resulted in the fabrication of alloyed nanoparticles, while ablation by *ps* laser pulses results in separation of the metals in the alloy and further oxidation of Fe. In the latter case, the as-deposited structures also contain core–shell type nanoparticles, with the shell consisting of Fe-oxide phase. The obtained structures, regardless of the ablation regime applied, demonstrate a red-shifted plasmon resonance with respect to the plasmon resonance of pure Au nanoparticles.

Keywords: *ns*-PLD; *ps*-PLD; open air; nanocomposites; alloys

1. Introduction

Iron-oxide-containing magnetic nanoparticles and nanostructures have found a variety of practical applications, including high-density data storage media, electronic elements and sensors, and controlled drug delivery and cancer diagnostics/treatment systems [1–6]. Magnetite as well as maghemite, because of their unique physical properties in nanosized form, are among the most investigated oxides, with numerous current and potential applications. These oxides exhibit unique magnetic properties, biocompatibility, and biodegradability, which in turn make them suitable for biomedical application [2,3,7]. Such uses usually require a specific size distribution and a desired shape of the nanoparticles or ensembles of nanoparticles, since these define their magnetic properties and, respectively, their efficiency. This is why such nanosized objects should be fabricated through precise size- and shape-controlling synthesis, on the one hand, and, if possible, without the use

of additional toxic chemicals, on the other. In general, conventional approaches such as chemical, template-assisted, and lithographic have been extensively investigated and applied for fabrication of a wide variety of magnetic nanostructures such as iron oxides, pure metals, metal alloys, and core–shell structures [2–4,8–11]. Therefore, the challenge and the efforts of scientists and engineers are focused on devising ways of environmentally friendly fabrication of magnetic nanoparticles or nanostructures, utilizing simple and flexible methods and conventional low-cost equipment.

Pulsed laser deposition is a well-established physical method for fabrication of thin films and structures based on laser ablation of a target material [12]. When the process is performed in the air at atmospheric pressure (PLD in open air or atmospheric PLD), the ablated material directly forms nanoparticles in the plasma plume [13,14]. The method has been successfully applied for production of oriented nanowires of magnetic materials [15,16]. Further, the proposed method is easily modified for fabrication of more complicated systems such as composite nanostructures made of magnetic and nonmagnetic materials [17,18]. This technology was developed and applied for ultrashort (*ps* and *fs*) and later for short (*ns*) laser pulses [15,19,20]. The main difference between short and ultrashort ablation in open air is that, in the case of *ns*-laser pulses, the nanoparticles are formed via condensation of ablated material [15,16], while using ultrashort laser pulses leads to a direct ejection of nanoparticles thanks to the specific mechanisms of material removal [19–21]. The PLD technology in open air is generally applied for ablation of bulk target material. However, the same technology could also be applied if the bulk target material is replaced by a thin film deposited from the desired material in advance. The laser ablation of thin films is an area of rapidly growing interest. Numerous researchers have focused their efforts on clean micropatterning of thin films or selective ablation of metal films, even in the case of complicated multilayer structures [22–25]. Such application of laser ablation on thin films requires the use of ultrashort laser pulses for precise processing of the materials. Also, this technology has been successfully applied for fabrication of nanoparticles and/or nanostructures in liquid [16–28]. In general, optical absorption, thermal, and structural properties of the film, and adhesion to the substrate influence and determine the ablation process. The mechanism of synthesis as well as the advantages and possible applications of a variety of complex materials and structures produced in liquid are summarized by Amendola et al. [26]. Similarly, laser ablation from a thin-film target performed in air could be used for production of nanoparticles and nanoaggregates on a substrate. In the latter case, the attention will be focused on the material removed by the ablation of the thin film. The advantage of using a thin-film target versus a bulk target stands out for the materials for which preparation as bulk material is a hard and/or expensive task, for example, the fabrication of bulk alloyed targets from hardly miscible, as well as expensive, metals such as iron and noble metals. Fe has a body-centered cubic cell (bcc), while noble metals have a face-centered cubic cell (fcc). Such metal alloyed thin films with a desirable thickness could be easily and less expensively fabricated by well-developed conventional technologies for thin film synthesis.

Ultrashort pulse lasers, especially *fs* lasers, are yet to find wide industrial applications because of their high price, expensive maintenance, and specific requirements to the working environment. Picosecond laser systems with a pulse duration of up to tens of picoseconds, at the same time, are of considerable scientific and commercial interest. The *ps*-laser pulses preserve the primary features of *fs*-laser pulses regarding their interaction with the material. The picosecond laser systems are simpler and more stable than *fs*-laser systems, leading to their lower cost, which is comparable to that of popular and widely used nanosecond systems. Such lasers could be easily applied in industries. Using *ps* pulses in PLD will allow for even wider commercial applications due to maintaining the process characteristics of ultrashort laser–matter interaction.

In this work, we aimed to fabricate nanostructures consisting of composite nanoparticles of iron-containing alloy, namely FeAu alloy, by implementing a physical method such as PLD in open air. We studied the material removed from a thin-film target when laser ablation was performed by nanosecond and by picosecond laser pulses. We were thus able to emphasize the differences in the structure, microstructure, and morphology of the structures produced by *ns* and *ps* ablation. It was found that nanostructures composed of alloyed nanoparticles were produced by *ns* ablation of thin FeAu film, while ablation by *ps* laser pulses resulted, rather, in separation of the alloy into Au and Fe metals, and further oxidation of iron to oxide phase. In the latter case, the fabricated nanoparticles had a clearly expressed bimodal size distribution, as larger-sized particles exhibited a core–shell type structure with an Fe-oxide phase as a shell. The obtained structures, regardless of the ablation regime applied, demonstrated a plasmon resonance related to the presence of Au or Au-containing nanoparticles. Further, since the ablation from a thin-film target is a specific case of laser–matter interaction where the geometric parameters of the film could significantly influence the processes involved, we performed a numerical simulation of laser–matter interaction with *ns* and *ps* laser pulses. A theoretical study of the temporal temperature distribution on the film surface and film substrate was also carried out to clarify the potential for application of the presented technology.

2. Materials and Methods

2.1. Experimental

Nanostructures consisting of composite nanoparticles were produced by a two-step physical deposition method. It was based on pulsed laser deposition (PLD) performed sequentially in vacuum and in air. A schematic view of the experimental setup is shown in Figure 1. As a first step, thin films were deposited by classical on-axis PLD in a vacuum.

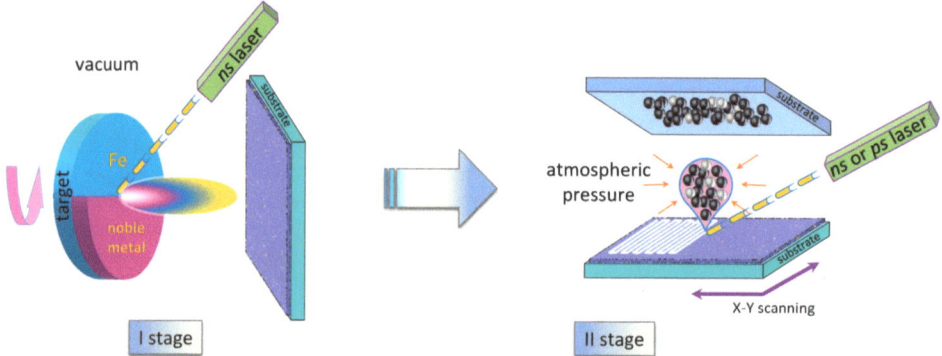

Figure 1. Schematic view of the experimental setup.

The depositions were implemented using a nanosecond Nd: YAG laser emitting at its third harmonic wavelength, 355 nm, with 15 ns pulse duration and 10 Hz repetition rate. The laser fluence applied on the target was 4 J/cm^2. A pure iron target or a mosaic target formed by iron and gold metal equal sectors was used for ablation. The ablated material was deposited on a glass substrate. The experiments were conducted at a target–substrate distance of 30 mm at room substrate temperature and a base pressure of 10^{-4} Torr. All depositions were performed for 15 min. Under these conditions, the maximum measured film thickness was approximately 300 nm. As a second step, the as-prepared thin films were ablated in air at atmospheric pressure (in open air) as the laser beam scanned their surface. The speed of the X–Y scan table was chosen so as to prevent the laser spots from overlapping. Two sets of experiments were performed in the second step based on ablation with nanosecond (*ns*) and picosecond (*ps*) laser pulses, respectively. The *ns* ablation was performed by a Nd: YAG (LS-2147, Lotis TII, Minsk, Belarus) laser system delivering 15 ns laser pulses. The *ps* ablation was carried out by a picosecond Nd: YAG laser (PS-A1-1064, CNL laser, Changchun, China) with a pulse duration of 10 ps. In both cases, the third harmonic frequency of Nd: YAG lasers at wavelength of 355 nm was used with a 10 Hz repetition rate. The laser fluence applied on the thin films for *ns* and *ps* ablation was 1.5 J/cm² and 0.8 J/cm², respectively. The ablated material was deposited on a silicon or quartz substrate at room substrate temperature. The distance between the target and substrate was kept at 5 mm. Further, using the experimental configuration shown in Figure 1, additional experiments were performed on *ns* and *ps* ablation of a bulk Fe in order to compare samples deposited from bulk and thin-film targets.

2.2. Theoretical

In order to estimate the heating evolution and the spatial distribution of the temperature during laser processing of the thin film, as used in the experimental part, a numerical model was applied. For the case of ablation with nanosecond pulse, a one-dimensional heat diffusion equation was applied:

$$C\frac{\partial T}{\partial t} = k\frac{\partial^2 T}{\partial z^2} + S, \qquad (1)$$

The source term is $S(z,t) = I(t)(1 - R)\alpha \, exp(-\alpha z)$, where R is the reflectivity of the material and I is the laser intensity. $C = c\rho$, where c and ρ are the specific heat capacity and material density, respectively, k is the thermal conductivity, and α is the absorption coefficient. The laser intensity is considered to be Gaussian in time; z is the coordinate in the direction parallel to the film surface normal.

In the case of *ps* ablation, a two-temperature model was applied. It takes into account that during the time of the laser pulse duration, the electron system in the film absorbs the laser energy and its temperature increases rapidly. At the same time, the lattice temperature remains close to the initial one since the time for energy transfer between the electrons and the lattice is of the same order as the pulse duration. The model solves the head diffusion equations for both systems:

$$\begin{aligned} C_e\frac{\partial T_e}{\partial t} &= k_e\frac{\partial^2 T_e}{\partial z^2} - \gamma(T_e - T_i) + S \\ C_i\frac{\partial T_i}{\partial t} &= k_i\frac{\partial^2 T_i}{\partial z^2} + \gamma(T_e - T_i) \end{aligned} \qquad (2)$$

where the notations are the same as in Equation (1); γ is the electron–phonon coupling parameter, T_e and T_i are the temperatures of the election and lattice systems, respectively; C and k are the heat capacity and the thermal conductivity for both systems. Equation (1) and System (2) were solved using a classical finite difference scheme [29], as the simulated system was divided into slices in z direction with a thickness of 1 nm. The parameters for Fe used in the model were taken from Refs. [30–32]. The reflection of the film was estimated to be 60% using optical transmission measurement.

2.3. Sample Characterization

The crystalline structure and phase composition of the samples were analyzed by an Empyrean diffractometer (PANalytical, Malvern, UK) through a goniometric X-ray diffraction scan using CuKα radiation. Bright field transmission electron microscopy (BF TEM) and high-resolution (HR) TEM images, and selected area electron diffraction (SAED) patterns were taken by a JEOL JEM 2100 microscope (Akishima-Shi, JEOL Ltd., Tokyo, Japan) to reveal the samples' microstructure and their crystallinity. Samples for TEM analyses were prepared using a direct deposition on carbon-coated TEM Cu grids for a shorter deposition time (3 min) to prevent the ablated materials overlapping. Scanning electron microscopy (SEM) equipped with an energy-dispersive X-ray (EDX) spectrometer was conducted using a LYRA I XMU system (Tescan, Brno, Czech Republic) to study the samples' morphology. The physicochemical state of the sample surface was determined by X-ray photoelectron spectroscopy (XPS) using an AXIS Supra electron spectrometer (Kratos Analytical Ltd., Manchester, UK). The optical properties of the samples were investigated using light from a standard white-light source (DH-2000 Ocean Insight, Dunedin, FL, USA) transmitted through the sample and recorded by a UV–Vis spectrometer (Ocean Optics HR 4000, Ocean Insight, Dunedin, FL, USA).

3. Results

3.1. Experimental

In view of clarifying the processes taking place and for further comparison, experiments were initially performed on laser ablation of a bulk Fe target using the experimental configuration shown in Figure 1. The results obtained from *ns* and *ps* laser ablation from bulk Fe, thin Fe, and FeAu alloyed film are discussed below.

3.1.1. Laser Ablation of Bulk Fe

TEM images of the samples deposited by laser ablation of bulk Fe are shown in Figure 2. The microstructure and the phase composition of the samples produced by *ns* ablation of iron are presented in Figure 2a. Separated and aggregated nanoparticles were produced by direct deposition on the substrate using the experimental configuration shown in Figure 1 at II stage. As seen, the nanoparticles are well-defined and spherically shaped with size distribution in the range of 3–12 nm and a mean Ferret's diameter of 7 nm. This result is in accordance with previous reports on *ns* laser ablation in open air [18]. The nanoparticles are crystalline, as seen from the SAED image in Figure 2a, with interplanar distances predominantly identified as appropriate of the magnetite phase of the iron oxide (Fe_3O_4, cubic, a = 8.4000 Å, 98-003-6314). Figure 2b reports the microstructure of the sample deposited by *ps* ablation of bulk Fe. In this case, the ablated material is represented mostly by agglomerated nanoparticles with sizes between 1 and 7 nm and a mean Ferret's diameter of 3 nm. The shape of the nanoparticles is close to spherical. The SAED pattern presented in Figure 2b shows that the sample has a polycrystalline structure, and the pattern was assigned to the maghemite phase of iron oxide (γ-Fe_2O_3, tetragonal, a = 8.33200 Å c = 25.11300 Å, 96-152-8613).

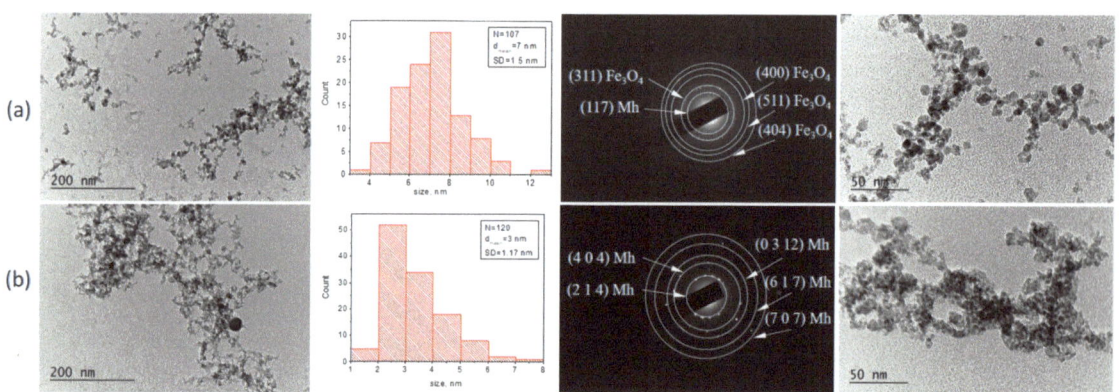

Figure 2. TEM images, size distributions, and SAED patterns of the samples produced by (**a**) *ns* and (**b**) *ps* ablation of Fe bulk. In the respective SAED pattern, the maghemite phase of iron oxide is marked as "Mh".

3.1.2. Laser Ablation of Pure Fe Film

TEM images of the samples deposited by laser ablation of thin Fe films are presented in Figure 3. The morphology of the sample deposited by *ns* ablation is shown in Figure 3a. As seen, *ns* ablation of thin films in air at atmospheric pressure leads to direct formation of nanoparticles. Similar results were already reported and discussed for *ns* ablation of a bulk target [15,16,18]. The morphology reveals that the nanoparticles formed have a polygonal shape, with most of them having a size in the range of 2–20 nm, as seen from the size distribution. The mean diameter of the nanoparticles was estimated to be 10 nm with a standard deviation (SD) of 4.2 (Figure 3a). The nanoparticles are crystalline, as demonstrated from the SAED pattern presented in Figure 3a. The SAED pattern indexation reveals the formation of iron oxide phase magnetite. It should be mentioned that no core–shell structure was observed in the fabricated nanoparticles, regardless of their size. TEM images of the sample deposited using *ps* ablation are shown in Figure 3b. As with the *ns* pulses, laser ablation of thin films with *ps* laser pulses in open air caused formation of nanoparticles. In the case of *ps* ablation, the microstructure of the sample consisted of spherical as well as polygon-shaped nanoparticles. There were clearly distinguishable particles with sizes from approximately 20 nm to 100 nm, as well as smaller ones with sizes in the range of 2–18 nm (size distribution presented in Figure 3b). It should be pointed out that the number of particles with size below 5 nm was considerably higher than in the case of *ns* ablation of thin Fe film (Figure 3a). The nanoparticles are crystalline with a mean diameter of 9 nm and SD of 12.5 (size distribution in Figure 3b). The SAED pattern demonstrates that the interplanar distances can be assigned to the Fe-oxide phase maghemite.

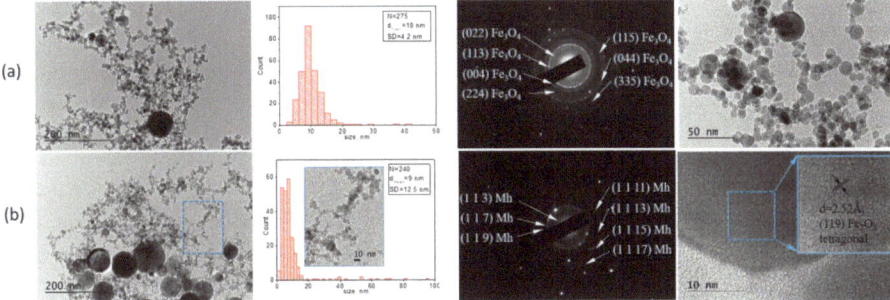

Figure 3. TEM images, size distributions, and SAED patterns of the samples produced by (**a**) *ns* and (**b**) *ps* ablation of thin Fe film. In the respective SAED pattern, the maghemite phase of iron oxide is marked as "Mh".

The surface morphology of the samples prepared by laser ablation of thin Fe film in air is presented in Figure 4. As seen, the depositions performed by *ns* laser pulses have a morphology consisting of distinct features formed by nanoparticles of different sizes (Figure 4a). Such morphology is typical for *ns* laser deposition of bulk materials in air at atmospheric pressure [15–18]. In the case of *ps* laser ablation of a thin film (Figure 4b), a structure consisting of aggregated nanosized particles is also observed. This result is in accordance with our previous reports on *ps* ablation of bulk materials in open air [21]. In addition, separate droplet/particle formations with sizes ranging from 140 to 350 nm are seen across the entire structure.

Figure 4. SEM images of the samples produced by (**a**) *ns* and (**b**) *ps* ablation of thin Fe film.

3.1.3. Laser Ablation of Alloyed FeAu Metal Film

Figure 5 presents the TEM images of the samples produced by ablation of alloyed FeAu film. In the case of *ns* ablation of alloyed film, particles with a nearly spherical shape were formed. The material ablated by *ns* laser pulses has a crystalline structure (Figure 5a).

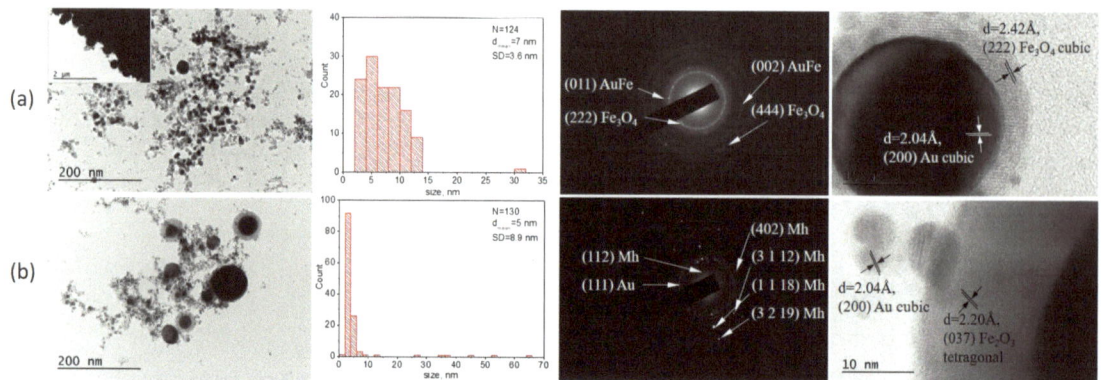

Figure 5. TEM images, size distributions, and SAED patterns of the samples produced by (**a**) *ns* and (**b**) *ps* ablation of alloyed FeAu film. In the respective SAED pattern, the maghemite phase of iron oxide is marked as "Mh".

The indexing of the corresponding SAED pattern identified FeAu alloy ($Au_{0.5}Fe_{0.5}$, cubic, a = 3.8850 Å, 98-010-7985) and Fe_3O_4 phases. The size distribution of the nanoparticles is from 2 to 14 nm with a mean Ferret's diameter of 7 nm. Larger in-size nanoparticles were also observed, as their diameter ranges from 100 to 500 nm (inset in Figure 5a). Here, it should be noted that some of the larger particles exhibited a core–shell structure. A high-resolution TEM image of such a particle reveals the lattice fringe spacing of 2.42 Å, which matches most closely to the (222) planes of the Fe_3O_4 phase. The microstructure of the material ablated by *ps* laser pulses is reported in Figure 5b. As seen, it consists of spherically shaped nanoparticles with different sizes. The morphology of samples produced by ablation of an Fe film and an alloyed FeAu film is similar (Figures 4b and 5b). Again, small-sized nanoparticles ranging from 1 to 10 nm with mean diameter of 5 nm, as well as larger-sized particles, can be clearly distinguished (Figure 5b). It should be noted that all larger-sized particles have a core–shell structure with shell thickness ranging from 2 to 16 nm. High-resolution TEM images of such particles reveal the lattice fringe spacing of 2.20 Å, which matches most closely to the (037) plane of the γ-Fe_2O_3 phase. In such a manner, the shell of the particles was identified as an iron-oxide phase. The core of the particles was difficult to analyze but the data implied either pure metal or metal oxide composition. Further, the separated nanoparticles in high-resolution TEM images were identified as a cubic phase of pure Au (Figure 5b). The SAED image of the ablated material corresponds well to the results obtained by high-resolution TEM images. The sample has a polycrystalline structure, as revealed from the SAED pattern (Figure 5b), for which indexing reveals the presence of pure Au and γ-Fe_2O_3 phases.

XRD patterns of the samples produced by *ns* and *ps* ablation from alloyed FeAu film are shown in Figure 6. The composition of the material deposited by *ns* ablation was identified as a combination of Au-enriched FeAu alloys with a precise phase composition of $Fe_{0.03}Au_{0.97}$ and $Fe_{0.22}Au_{0.88}$ estimated by Vegard's law (Figure 6a). The estimated crystallite size of the two phases was 55 nm and 23 nm, respectively. In the case of *ps* ablation of alloyed FeAu film, the diffraction pattern was identified as a pure Au phase with crystallite size of 17 nm (Figure 6b).

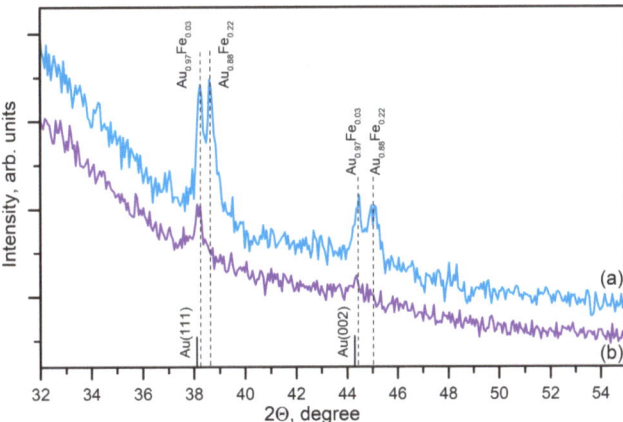

Figure 6. XRD of the samples produced by (a) *ns* and (b) *ps* ablation of alloyed FeAu film.

Figure 7 presents SEM images of the samples deposited by *ns* and *ps* ablation from alloyed FeAu film. The morphology of the sample deposited by *ns* ablation represents an assembly of spheres with a diameter in the range of 150–800 nm (Figure 7a). It is worth noting that such structures composed of spheres/particles of sizes of hundreds of nanometers have not been previously observed. The typical morphology of the samples produced by *ns* ablation of a bulk target in open air is a porous structure composed of nanoparticles [15]. Moreover, the morphology of the sample deposited by *ns* ablation of alloyed FeAu film differs from the morphology of the sample deposited by *ns* ablation of pure Fe film (Figures 4a and 7a). In the case of *ps* ablation from alloyed FeAu film, the sample morphology represents a structure consisting of well-differentiated nanoparticles (Figure 7b). Here, the morphology looks similar to that previously reported for sample deposition in open air [21]. However, separated spheres with a size of 160–370 nm are also observed (Figure 7b).

(a)   (b)

Figure 7. SEM image of the samples produced by (**a**) *ns* and (**b**) *ps* ablation of thin FeAu film.

Figure 8 reports optical properties of the structure produced by *ns* and *ps* ablation of alloyed FeAu film. Both structures demonstrate a pronounced peak in the optical absorbance spectra with a maximum of around 600 nm. We associate such features in the optical absorbance spectra of the samples with the presence of Au and/or Au-containing alloy nanoparticles and their plasmon excitation in the structures [33]. In addition, the surface plasmon resonance of the sample deposited by *ns* ablation (Figure 8a) is broader than that of the samples deposited by *ps* ablation (Figure 8b).

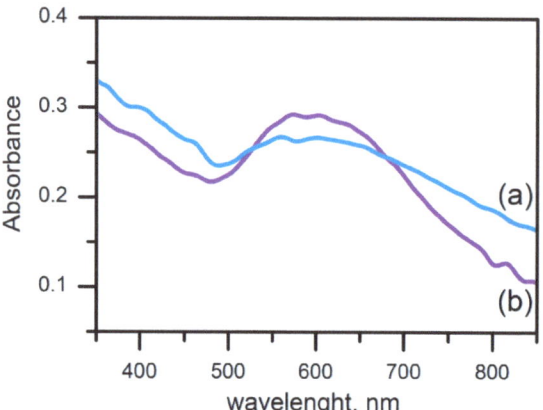

Figure 8. Optical properties of the samples produced by (a) *ns* and (b) *ps* ablation of thin FeAu film.

3.2. Theoretical

Figure 9 shows the estimated distribution of the temperature on the surface of the Fe film and its substrate when a 15 *ns* laser pulse with fluence of 1.5 J/cm^2 is applied. The evolution of the temperature on the Fe film surface with time is presented in Figure 9a. The maximum value ~4000 K is reached at about 30 ns; then, the temperature exponentially decreases to room temperature after ~ 500 ns. The substrate starts heating 20 ns after the laser pulse and the maximum temperature reached is one order of magnitude lower than that of the Fe film's surface (Figure 9b). At the repetition rate of 10 Hz, both the Fe film's and the substrate's surfaces cool down to room temperature before the next pulse *ns* arrives. The temperature distribution in the simulated system at the moment when the film surface temperature is maximal is shown in Figure 9c. The yellow line marks the substrate surface. As seen, when the film surface temperature is maximal, the substrate temperature is lower than the melting temperature of glass substrate (1675 °C according to the manufacturer's data).

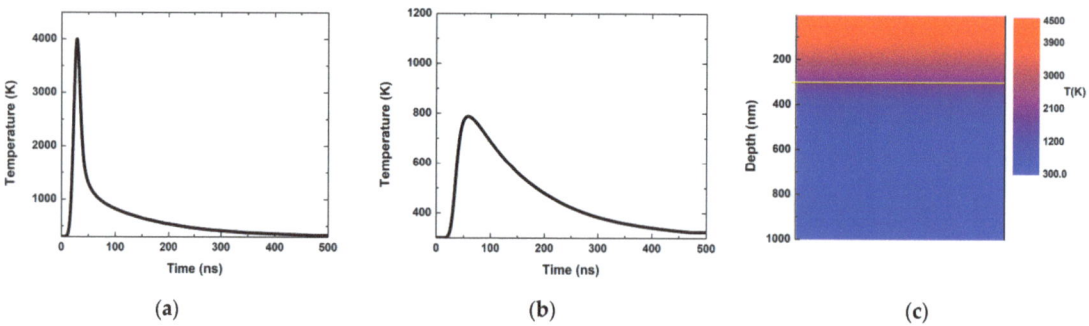

Figure 9. Estimated temperature distribution on the surface of the Fe film and substrate when a 15 ns laser pulse with fluence of 1.5 J/cm^2 is applied. (**a**) Time evolution of the Fe film surface temperature. (**b**) Time evolution of the glass substrate surface temperature. (**c**) Temperature distribution in the simulated system at the moment when the film surface temperature is maximal. The yellow line marks the substrate surface.

Figure 10 shows the estimated distribution of the temperature on the surface of the Fe film and its substrate when a 10 ps laser pulse is applied with fluence of 0.8 J/cm^2. The time evolution of the Fe-film surface temperature is presented in Figure 10a. Instantly, the surface temperature rises to its maximum value of about 4500 K and exponentially decreases with time, reaching room temperature after 50 ns. The temperature of the Fe film's substrate follows a similar behavior. A few ns suffice for the temperature of the substrate to reach its maximum value of about 1000 K (Figure 10b). More than 50 ns are needed for the substrate to cool to approximately half of the maximum value reached. Obviously, both the Fe film's and the substrate's surfaces cool down to room temperature before the next *ps* pulse arrives. The temperature distribution in the simulated system when the film surface temperature is maximum is shown in Figure 10c. The yellow line again delineates the substrate surface. It can be seen that the substrate temperature is lower than the melting point of the glass substrate when the film surface temperature is at its maximum.

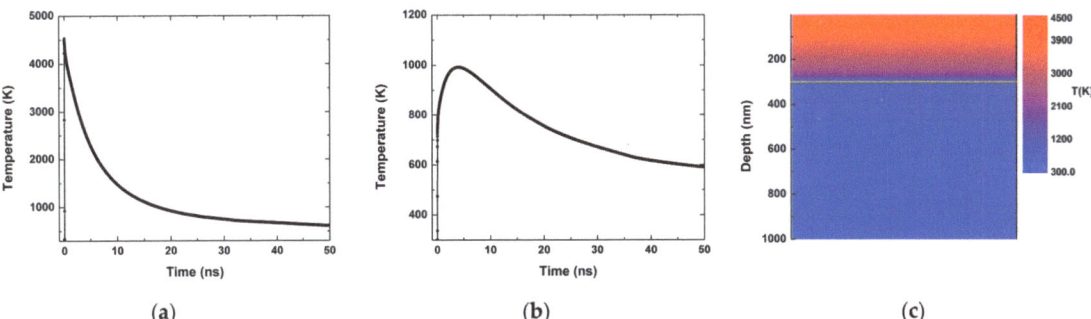

Figure 10. Estimated temperature distribution on the surface of the Fe film and substrate when a 10 ps laser pulse with fluence of 0.8 J/cm^2 is applied. (**a**) Time evolution of the Fe film surface temperature. (**b**) Time evolution of the glass substrate surface temperature. (**c**) Temperature distribution in the simulated system at the moment when the temperature of the film surface is maximal. The yellow line represents the substrate surface.

4. Discussion

Regardless of the ablation regime (*ns* or *ps*), the material ablated from a bulk target in open air forms nanoparticles of various sizes, as previously reported [15–18,20,21]. The mechanism responsible for nanoparticles formation is condensation of the ablated species in the high-pressure (atmospheric) environment—the plasma plume formed due to the interaction of the laser pulse with the target material is compressed, which, in turn, favors the aggregation of the ablated species. In the case of *ns* ablation, the process taking place in open air results in nanoparticles/nanoparticle aggregates formation in the plasma plume due to condensation of the ablated material (atoms, ions, and small clusters) [15–17]. In the case of *ps* ablation, the ultrashort laser pulses directly produce nanoparticles through fragmentation and/or phase explosion, leading to material removal [19–21]. The use of *ns* laser pulses for formation of nanoparticle or structures composed of nanoparticles is a well-developed and widely applied technology [15–18]. A detailed look at the microstructure of nanoparticles deposited by *ns* and *ps* ablation reveals some differences, as seen in Figure 2. Separated nanoparticles are rarely observed by direct ablation with *ps* laser pulses. The ablation by *ps* laser pulses leads to the formation of nanoparticles with slightly smaller diameters than those obtained by *ns* laser pulses (Figure 2). This confirms our previously reported results [20]. Further, we compared the results obtained by the experimental configuration shown in Figure 1 at II stage with those obtained by classical on-axis PLD configuration. We thus found that, independently of the laser pulse duration, the experimental setup for ablation used in this paper does not affect the deposited samples' crystalline structure and phase composition.

Ablation from thin film is a specific case of laser–matter interaction where the geometric parameters of the film could significantly influence the processes involved. The film thickness determines the amount of absorbed energy, while the substrate properties impact the temperature evolution and the ablation mechanism. The processing conditions should be carefully chosen to prevent ablation from the substrate that would change the composition of the ablated material. The temperature evolutions (shown in Figures 9 and 10) indicate that at the used laser fluences the temperature in the film exceeds the evaporation one. Furthermore, at the moment when the film surface temperature is maximal, the substrate temperature is lower than the melting temperature of glass substrate. These conditions define that the film would be ejected without evaporation of material from the substrate. The time lag between reaching the maximal temperatures in the film surface and the substrate also shows that the film decomposition would have occurred before significant heating of the substrate surface. This allows us to conclude that material from the substrate is not expected to be ejected.

Thin metal films were deposited by classical *ns*-PLD (Figure 1, I stage) as a first step of the fabrication of nanostructures consisting of composite nanoparticles. It should be mentioned that such films could be prepared by other deposition methods. After that, these films were used as targets for laser ablation. It should be mentioned that the experimental setup for ablation (Figure 1, II stage) used in this study considerably decreases the droplet formation typical for classical PLD technology.

The targets used for ablation were either pure Fe or Au-enriched FeAu films (Figures S1 and S3). Since Fe easily and fully oxidizes in air, an iron-oxide phase is formed on the surface of the Fe thin-film target. The chemical state on the film surface indicates that the binding of oxygen to iron results in the formation of Fe_2O_3 (Figure S2) [34]. When the laser beam starts scanning the thin-film surface, the initial material for ablation will be a combination of Fe and Fe-oxide. Furthermore, the experiments were performed in air. In this light, it is not surprising that the structure of the deposited material independent of the ablation regime applied is some kind of iron oxide (Figure 3). The phase composition of the material ablated from bulk and thin-film targets is identical regardless of the ablation regime applied (Figures 2a,b and 3a,b). This experimental result confirms the theoretical predictions discussed above. Considering the size of the nanoparticles formed during ablation, it seems that the nanoparticles ablated from the bulk target have a smaller size than those ablated from the thin-film target. Here, it should be pointed out that the ablation threshold for the bulk target is substantially higher than the ablation threshold for the same material in the thin-film form, which makes the direct comparison inappropriate. In order to keep the same laser energy applied on the target (bulk and thin-film target), the size of the laser spot on the bulk target was decreased to obtain the necessary laser fluence for ablation. However, the size of the laser spot on the target to a certain degree predetermines the size of the nanoparticles produced in air. The use of a smaller laser spot size leads to a smaller quantity of ablated material which subsequently condenses in smaller-sized nanoparticles. Bearing the above in mind, one should not be surprised that the mean diameter of the nanoparticles produced by the ablation of a bulk target in this work is smaller than the one produced by the ablation of a thin-film target (Figures 2a,b and 3a,b).

The phase composition of the nanostructure produced by *ns* ablation of thin Fe film was identified as the magnetite phase of iron oxide (Figure 3a). This result corroborates those previous reports [18]. When the Fe film was ablated by *ps* laser pulses, the structure of the deposited sample was identified predominantly as a maghemite phase of iron oxide (Figure 3b). The maghemite phase is a polymorph of hematite but has a similar ferrimagnetic behavior to magnetite [35]. Comparing the structures produced by ablation of Fe film with *ns* and *ps* laser pulses, a difference in the phase composition was observed. In this sense, the ablation regime applied influences the phase composition of the deposited samples. Further, a difference was also observed in the morphology of the structures produced by *ns* ablation of Fe film (Figure 4a,b).

In the case of alloyed FeAu film, by changing the ratio of the Au and Fe slice forming a mosaic target initially, we can alter the ratio of Fe and Au in the alloyed FeAu film. Here, it should be mentioned that the ablation threshold for Fe and Au is different, which means that the ablation efficiency for the two metals will differ. Hence, although the mosaic target was formed by equal parts Au and Fe, the quantity of Au in the deposited film was higher due to the lower ablation threshold of Au and subsequently higher ablation efficiency of this metal (Figure S4, Table S1). However, the surface of the alloyed film was also covered by the iron oxide phase (Figure S5) [34,36–38].

One of the main advantages of *ns* pulsed laser deposition is the stoichiometric transfer of material from the target to the substrate. However, since the process was performed in open air, some deviations in sample phase composition could be observed [20]. In the case of *ns* ablation of Au-enriched FeAu alloy film, two different alloyed phases of Fe and Au could be clearly distinguished (Figure 6a). We assume that the energetically favorable phases according to the Fe–Au diagram were formed in the plasma plume. In such a manner, composite alloyed nanoparticles were formed by *ns* ablation of FeAu film, as a variation in their phase composition could be observed (Figure 6a). The formation of composite nanoparticles was also confirmed by the position of the surface plasmon resonance observed from such nanoparticles (Figure 8a). The theoretical prediction shows that pure Au nanoparticles with similar dimensions demonstrate more than one surface plasmon resonance and their position is at the wavelength of about 520 nm as well as shifted to longer wavelengths [33]. Moreover, our results obtained from samples deposited by *ns* ablation of pure Au film (see Figure S8) well confirm the above. While iron nanoparticles do not show a characteristic absorption in the visible range of the spectrum (Figure S8), the presence of Fe can influence the plasmon resonance of Au. Such a shift of the plasmon resonance of Au-containing Fe alloyed nanoparticles has been previously reported [39,40]. Amendola et al. suggested that the redshift is most probably related to the heterogeneous composition of the particles or particle aggregations [40]. In our case, we assumed that the redshift in the surface plasmon resonance of the *ns*-deposited samples is due to the presence of particles with different alloy composition as well as to their size and size distribution (Figures 6a, 7a and S8).

In the case of *ps* ablation of the alloyed FeAu film, a decomposition was achieved of the alloy followed by formation of Au and Fe-oxide nanoparticles in open air (Figure 5b). The presence of Au nanoparticles was also confirmed by XRD analysis (Figure 6b). Further, the interaction of *ps* laser pulses with the material in open air led to the formation of nanoparticles with bimodal distribution (Figures 3b and 5b). The small-sized nanoparticles were identified as pure Au and Fe-oxide phases (Figure 5b). Furthermore, the larger particles have a core–shell structure, as the shell was found to have different Fe-oxide phases (Figure 5b). The optical properties of this structure demonstrate the presence of plasmon resonance shifted to the longer wavelengths (Figure 8b). The redshift in the resonance position of Au nanoparticles is probably because the Au nanoparticles exist in the Fe-oxide nanoparticle environment and their interaction shifts the resonance at longer wavelengths (see Figure S8).

It should be noted that based on the results obtained by the XRD analyses of the structures deposited by *ns* and *ps* ablation of alloyed FeAu film, it seems that the iron and/or iron compounds are missing (Figure 6a,b). However, the presence of Fe in the deposited structures was confirmed by the EDX analyses (Figures S6 and S7, Tables S2 and S3). This means that a significant portion of iron oxide nanoparticles are amorphous (Figure 6a). Further, the XRD method does not recognize nanoparticles below 5 nm in size as crystalline, so they are registered as an amorphous halo in the XRD pattern (Figure 6b).

It could be summarized that structures consisting of composite alloyed nanoparticles could be produced by classical *ns* ablation of alloyed FeAu film in open air, and that varying the Fe: Au ratio in the film changes the composition of alloyed particles produced. The use of *ps* ablation of the same alloyed film in open air generally decomposes the existing alloy. This ultrashort laser–matter interaction allows the fabrication of structures comprising composite core–shell nanoparticles with some kind of Fe-oxide phase forming the shell. To the best of our knowledge, composite core–shell particles produced by physical methods such as laser ablation in air have not been previously reported. Further, different alloyed films consisting of iron with other noble metals (FePt and FePd) could be used for the fabrication of nanoparticle-composed structures, as the ablation regime applied acts in a similar way to the case of FeAu.

5. Conclusions

Nanostructures consisting of composite nanoparticles were fabricated by ns and ps ablation in open air from thin Fe and Au-enriched FeAu alloy films. It was found that ablation by ps laser pulses leads to formation of nanoparticles with slightly smaller diameters than those obtained by ns laser pulses. Different phase composition was observed when an Fe film was ablated with ns and ps laser pulses, indicating that the ablation regime applied influences the phase composition of the deposited samples. Structures consisting of composite alloyed nanoparticles could be produced by classical ns ablation of alloyed FeAu film in open air and varying the Fe: Au ratio in the film changes the composition of alloyed particles produced. The use of ps ablation of the same alloyed film in open air generally decomposes the existing alloy to Au and Fe metals and causes oxidation of iron to oxide phase. In the latter case, the fabricated nanoparticles have a clearly expressed bimodal size distribution as larger-sized particles demonstrate a core–shell type structure with an Fe-oxide phase as shell. To the best of our knowledge, composite core–shell particles produced by physical methods such as laser ablation in air have not been previously reported. The obtained structures, regardless of the ablation regime applied, demonstrate a plasmon resonance attributed to the presence of Au or Au-containing nanoparticles. A theoretical study of the temperature distribution on the film surface and the film's substrate with time was carried out to clarify the potential for application of the presented technology. The calculations show that at certain experimental conditions, material from the substrate is not expected to be ejected.

Supplementary Materials: The following supporting information can be downloaded at: https://www.mdpi.com/article/10.3390/coatings14050527/s1, Figure S1: XRD pattern of the iron film; Figure S2: XPS spectra of the iron film; Figure S3: XRD pattern of the alloyed FeAu film; Figure S4: EDX analysis of the alloyed FeAu film; Table S1: The estimated quantity of Fe and Au elements in the alloyed FeAu film; Figure S5: XPS spectra of the alloyed FeAu film; Figure S6: EDX analysis of the structure deposited by ns ablation of the alloyed FeAu film; Table S2: The estimated Fe and Au quantity in the structure deposited by ns ablation; Figure S7: EDX analysis of the structure deposited by ps ablation of the alloyed FeAu film; Table S3: The estimated Fe and Au quantity in the structure deposited by ps ablation; **Figure S8**. Optical properties of the samples deposited from pure Fe and Au films by (a) ns and (b) ps ablation.

Author Contributions: Conceptualization, A.O.D. and P.A.; methodology, A.O.D. and N.N.N.; formal analysis, D.K., G.A. (Genoveva Atanasova), and G.A. (Georgi Avdeev); investigation, A.O.D.; data curation, D.K., G.A. (Genoveva Atanasova) and G.A. (Georgi Avdeev); writing—original draft preparation, A.O.D.; writing—review and editing, N.N.N.; visualization, D.K. and G.A. (Genoveva Atanasova); project administration, A.O.D. All authors have read and agreed to the published version of the manuscript.

Funding: The authors wish to thank the financial support of project KP-06-N37/20 "Formation and physical properties of composite nanostructures of metal oxides and noble metals" under the "Competition for Financial Support of Basic Research Projects—2019" Program of the Bulgarian National Science Fund.

Institutional Review Board Statement: Not applicable.

Informed Consent Statement: Not applicable.

Data Availability Statement: Data are contained within the article and Supplementary Materials.

Acknowledgments: We acknowledge the use of the research equipment supported by National Roadmap for Scientific Infrastructure, financially coordinated by the Ministry of Education and Science of the Republic Bulgaria under project D01-351 (ELI-ERIC BG). Research equipment of Distributed Research Infrastructure INFRAMAT, part of Bulgarian National Roadmap for Research Infrastructures, supported by Bulgarian Ministry of Education and Science was used in this investigation.

Conflicts of Interest: The authors declare no conflicts of interest.

References

1. Jungwirth, T.; Marti, M.; Wadley, P.; Wunderlich, J. Antiferromagnetic spintronics. *Nat. Nanotechnol.* **2016**, *11*, 231–241. [CrossRef] [PubMed]
2. Ali, A.; Zafar, H.; Zia, M.; ul Haq, I.; Phull, A.R.; Ali, J.S.; Hussain, A. Synthesis, characterization, applications, and challenges of iron oxide nanoparticles. *Nanotechnol. Sci. Appl.* **2016**, *9*, 49–67. [CrossRef] [PubMed]
3. Meng, Y.Q.; Shi, Y.N.; Zhu, Y.P.; Liu, Y.Q.; Gu, L.W.; Liu, D.D.; Ma, A.; Xia, F.; Guo, Q.Y.; Xu, C.C.; et al. Recent trends in preparation and biomedical applications of iron oxide nanoparticles. *J. Nanobiotechnol.* **2024**, *22*, 24. [CrossRef]
4. Ali, A.; Shah, T.; Ullah, R.; Zhou, P.; Guo, M.; Ovais, M.; Tan, Z.; Rui, Y. Review on Recent Progress in Magnetic Nanoparticles: Synthesis, Characterization, and Diverse Applications. *Front. Chem.* **2021**, *9*, 629054. [CrossRef]
5. Revathy, R.; Sajini, T.; Augustine, C.; Joseph, N. Iron-based magnetic nanomaterials: Sustainable approaches of synthesis and applications. *Results Eng.* **2023**, *18*, 101114. [CrossRef]
6. Chen, L.; Xie, J.; Wang, Z.; Zhao, Y.; Gou, J.; Wu, J. Amorphous Pt-decorated α-Fe_2O_3 sensor with superior triethylamine sensing performance prepared by a one-step impregnation method. *J. Alloys Compd.* **2024**, *976*, 173330. [CrossRef]
7. Mikaeili Ghezeljeh, S.; Salehzadeh, A.; Ataei-e Jaliseh, S. Iron oxide nanoparticles coated with Glucose and conjugated with Safranal (Fe_3O_4@Glu-Safranal NPs) inducing apoptosis in liver cancer cell line (HepG2). *BMC Chem.* **2024**, *18*, 33. [CrossRef] [PubMed]
8. Ling, D.; Hyeon, T. Chemical Design of Biocompatible Iron Oxide Nanoparticles for Medical Applications. *Small* **2012**, *9*, 1450–1466. [CrossRef]
9. Nana, A.B.A.; Marimuthu, T.; Kondiah, P.P.D.; Choonara, Y.E.; Du Toit, L.C.; Pillay, V. Multifunctional magnetic nanowires: Design, fabrication, and future prospects as cancer therapeutics. *Cancers* **2019**, *11*, 1956. [CrossRef]
10. Andrade, R.G.D.; Veloso, S.R.S.; Castanheira, E.M.S. Shape Anisotropic Iron Oxide-Based Magnetic Nanoparticles: Synthesis and Biomedical Applications. *Int. J. Mol. Sci.* **2020**, *21*, 2455. [CrossRef]
11. Mondal, P.; Anweshan, A.; Purkait, M.K. Green synthesis and environmental application of iron-based nanomaterials and nanocomposite: A review. *Chemosphere* **2020**, *259*, 127509. [CrossRef] [PubMed]
12. Eason, R. *Pulsed Laser Deposition of Thin Films: Applications-Led Growth of Functional Materials*; John Wiley & Sons, Inc.: New York, NY, USA, 2007; ISBN 9780470052129.
13. Boutinguiza, M.; Comesaña, R.; Lusquiños, F.; Riveiro, A.; del Val, J.; Pou, J. Production of silver nanoparticles by laser ablation in open air. *Appl. Surf. Sci.* **2015**, *336*, 108–111. [CrossRef]
14. Białous, A.; Gazda, M.; Grochowska, K.; Atanasov, P.; Dikovska, A.; Nedyalkov, N.; Reszczyńska, J.; Zaleska-Medowska, A.; Śliwiński, G. Nanoporous TiO_2 electrode grown by laser ablation of titanium in air at atmospheric pressure and room temperature. *Thin Solid Films* **2016**, *601*, 41–44. [CrossRef]
15. Nikov, R.G.; Dikovska, A.O.; Atanasova, G.B.; Avdeev, G.V.; Nedyalkov, N.N. Magnetic field-assisted formation of oriented nanowires produced by PLD in open air. *Appl. Surf. Sci.* **2018**, *458*, 273–280. [CrossRef]
16. Nikov, R.G.; Dikovska, A.O.; Avdeev, G.V.; Amoruso, S.; Ausanio, G.; Nedyalkov, N.N. PLD fabrication of oriented nanowires in magnetic field. *Appl. Surf. Sci.* **2019**, *471*, 368–374. [CrossRef]
17. Nikov, R.G.; Dikovska, A.O.; Avdeev, G.V.; Atanasova, G.B.; Nedyalkov, N.N. Composite magnetic and non-magnetic oxide nanostructures fabricated by a laser-based technique. *Appl. Surf. Sci.* **2021**, *549*, 49204. [CrossRef]
18. Nikov, R.G.; Dikovska, A.O.; Avdeev, G.V.; Atanasova, G.B.; Karashanova, D.B.; Amoruso, S.; Ausanio, G.; Nedyalkov, N.N. Single-step fabrication of oriented composite nanowires by pulsed laser deposition in magnetic field. *Mater. Today Commun.* **2021**, *26*, 101717. [CrossRef]
19. Nedyalkov, N.; Nakajima, Y.; Terakawa, M. Magnetic nanoparticle composed nanowires fabricated by ultrashort laser ablation in air. *Appl. Phys. Lett.* **2016**, *108*, 043107. [CrossRef]
20. Dikovska, A.; Atanasova, G.; Nikov, R.; Avdeev, G.; Cherkezova-Zheleva, Z.; Paneva, D.; Nedyalkov, N. Formation of Oriented Nanowires from Mixed Metal Oxides. *Materials* **2023**, *16*, 6446. [CrossRef]
21. Dikovska, A.; Atanasova, G.; Dilova, T.; Baeva, A.; Avdeev, G.; Atanasov, P.; Nedyalkov, N. Picosecond Pulsed Laser Deposition of Metals and Metal Oxides. *Materials* **2023**, *16*, 6364. [CrossRef]
22. Kim, B.; Nam, H.K.; Watanabe, S.; Park, S.; Kim, Y.; Kim, Y.-J.; Fushinobu, K.; Kim, S.-W. Selective Laser Ablation of Metal Thin Films Using Ultrashort Pulses. *Int. J. Precis. Eng. Manuf.-Green. Technol.* **2021**, *8*, 771–782. [CrossRef]
23. Hallum, G.E.; Kurschner, D.; Redka, D.; Niethammer, D.; Schulz, W.; Huber, H.P. Time-resolved ultrafast laser ablation dynamics of thin film indium tin oxide. *Opt. Express* **2021**, *29*, 30062. [CrossRef] [PubMed]

24. Kim, B.; Nam, H.-K.; Kim, Y.-J.; Kim, S.-W. Lift-Off Ablation of Metal Thin Films for Micropatterning Using Ultrashort Laser Pulses. *Metals* **2021**, *11*, 1586. [CrossRef]
25. Domke, M.; Nobile, L.; Rapp, S.; Eiselen, S.; Sotrop, J.; Huber, H.P.; Schmidt, M. Understanding thin film laser ablation: The role of the effective penetration depth and the film thickness. *Phys. Procedia* **2014**, *56*, 1007–1014. [CrossRef]
26. Amendola, V.; Amans, D.; Ishikawa, Y.; Koshizaki, N.; Scirè, S.; Compagnini, G.; Reichenberger, S.; Barcikowski, S. Room-Temperature Laser Synthesis in Liquid of Oxide, Metal-Oxide Core-Shells, and Doped Oxide Nanoparticles. *Chem. Eur. J.* **2020**, *26*, 9206. [CrossRef] [PubMed]
27. Amendola, V. Laser-Assisted Synthesis of Non-Equilibrium Nanoalloys. *Chem. Phys. Chem.* **2021**, *22*, 622–624. [CrossRef] [PubMed]
28. Coviello, V.; Forrer, D.; Amendola, V. Recent Developments in Plasmonic Alloy Nanoparticles: Synthesis, Modelling, Properties and Applications. *Chem. Phys. Chem.* **2022**, *23*, e202200136. [CrossRef]
29. Press, W.H.; Teukolsky, S.A.; Vetterling, W.T.; Flannery, B.P. *Numerical Recipes in Fortran*, 1st ed.; University Press: Cambridge, UK, 1992; ISBN 10: 0521383307.
30. Lasemia, N.; Pachera, U.; Zhigilei, L.V.; Bomatí-Miguela, O.; Lahozd, R.; Kautek, W. Pulsed laser ablation and incubation of nickel, iron and tungsten in liquids and air. *Appl. Surf. Sci.* **2018**, *433*, 772–779. [CrossRef]
31. Artyukov, I.A.; Zayarniy, D.A.; Ionin, A.A.; Kudryashov, S.I.; Makarov, S.V.; Saltuganov, P.N. Relaxation Phenomena in Electronic and Lattice Subsystems on Iron Surface during Its Ablation by Ultrashort Laser Pulses. *JETP Lett.* **2014**, *99*, 51–55. [CrossRef]
32. Fernandez-Pañella, A.; Ogitsu, T.; Engelhorn, K.; Correa, A.A.; Barbrel, B.; Hamel, S.; Prendergast, D.G.; Pemmaraju, D.; Beckwith, M.A.; Bae, L.J.; et al. Reduction of electron-phonon coupling in warm dense iron. *Phys. Rev. B* **2020**, *101*, 184309. [CrossRef]
33. Amendola, V.; Pilot, R.; Frasconi, M.; Maragò, O.M.; Iatì, M.A. Surface plasmon resonance in gold nanoparticles: A review. *J. Phys. Condens. Matter* **2017**, *29*, 203002. [CrossRef] [PubMed]
34. Yamashita, T.; Hayes, P. Analysis of XPS spectra of Fe^{2+} and Fe^{3+} ions in oxide materials. *Appl. Surf. Sci.* **2008**, *254*, 2441–2449. [CrossRef]
35. Shokrollahi, H. A review of the magnetic properties, synthesis methods and applications of maghemite. *J. Magn. Magn. Mater.* **2017**, *426*, 74–81. [CrossRef]
36. Biesinger, M.C.; Payne, B.P.; Grosvenor, A.P.; Lau, L.W.M.; Gerson, A.R.; Smart, R.S. Resolving surface chemical states in XPS analysis of first row transition metals, oxides and hydroxides: Cr, Mn, Fe, Co and Ni. *Appl. Surf. Sci.* **2011**, *257*, 2717–2730. [CrossRef]
37. Pannu, C.; Bala, M.; Khan, S.A.; Srivastava, S.K.; Kabiraj, D.; Avasthi, D.K. Synthesis and characterization of Au–Fe alloy nanoparticles embedded in a silica matrix by atom beam sputtering. *RSC Adv.* **2015**, *5*, 92080–92088. [CrossRef]
38. Liu, M.; Zhou, W.; Wang, T.; Wang, D.; Liu, L.; Ye, J. High Performance Au-Cu Alloy for Enhanced Visible-light Water Splitting Driven by Coinage Metals. *Chem. Commun.* **2016**, *52*, 4694–4697. [CrossRef] [PubMed]
39. Liu, H.L.; Wu, J.H.; Min, J.H.; Kim, Y.K. Synthesis of monosized magnetic-optical AuFe alloy nanoparticles. *J. Appl. Phys.* **2008**, *103*, 07D529. [CrossRef]
40. Amendola, V.; Meneghetti, M.; Bakr, O.M.; Riello, P.; Polizzi, S.; Anjum, D.H.; Fiameni, S.; Arosio, P.; Orlando, T.; Fernandez, C.J.; et al. Coexistence of plasmonic and magnetic properties in $Au_{89}Fe_{11}$ nanoalloys. *Nanoscale* **2013**, *5*, 5611–5619. [CrossRef]

Disclaimer/Publisher's Note: The statements, opinions and data contained in all publications are solely those of the individual author(s) and contributor(s) and not of MDPI and/or the editor(s). MDPI and/or the editor(s) disclaim responsibility for any injury to people or property resulting from any ideas, methods, instructions or products referred to in the content.

Article

A Visualization Experiment on Icing Characteristics of a Saline Water Droplet on the Surface of an Aluminum Plate

Yingwei Zhang, Xinpeng Zhou, Weihan Shi, Jiarui Chi, Yan Li * and Wenfeng Guo *

College of Engineering, Northeast Agricultural University, Harbin 150030, China; zhangyingweineau@163.com (Y.Z.); neauzxp@163.com (X.Z.); a13114593518@163.com (W.S.); 17357711921@163.com (J.C.)
* Correspondence: liyanneau@163.com (Y.L.); guowenfengmail@163.com (W.G.)

Abstract: When the offshore device, such as an offshore wind turbine, works in winter, ice accretion often occurs on the blade surface, which affects the working performance. To explore the icing characteristics on a microscale, the freezing characteristics of a water droplet with salinity were tested in the present study. A self-developed icing device was used to record the icing process of a water droplet, and a water droplet with a volume of 5 µL was tested under different salinities and temperatures. The effects of salinity and temperature on the profile of the iced water droplet, such as the height and contact diameter, were analyzed. As the temperature was constant, along with the increase in salinity, the height of the iced water droplet first increased and then decreased, and the contact diameter decreased. The maximum height of the iced water droplet was 1.21 mm, and the minimum contact diameter was 3.67 mm. With the increase in salinity, the icing time of the water droplet increased, yet a minor effect occurred under low temperatures such as −18 °C. Based on the experimental results, the profile of the iced water droplet was fitted using the polynomial method, with a coefficient of determination (R^2) higher than 0.99. Then the mathematical model of the volume of the iced water droplet was established. The volume of the iced water droplet decreased along with temperature and increased along with salinity. The largest volume was 4.1 mm^3. The research findings provide a foundation for exploring the offshore device icing characteristics in depth.

Keywords: offshore wind turbine; icing; water droplet; salinity; visualization test

1. Introduction

Wind energy is one of the most popular renewable energies around the world and is mainly used in the field of wind power [1]. The wind turbine realizes the conversion of wind energy into electric power without any pollution [2]. However, most high-quality wind energy resources are mainly located in the high-altitude and high-latitude regions [3]. In these regions, the temperatures were low in the winter. Therefore, when wind turbines work in such an environment, ice accretion often occurs on the blade surface. Under this condition, the aerodynamic profile of the blade is destroyed by ice. Then the aerodynamic characteristics of the blade degrade, and the power efficiency of the wind turbine decreases [4,5]. Additionally, icing results in an increase in mass and the destruction of the balance of the rotor, which affects the stability and lifespan of wind turbines. Under some extreme conditions, a shedding ice event also threatens the nearby buildings and staff.

Therefore, exploring the icing characteristics of wind turbines is necessary to ensure stability and power efficiency. With the rapid development of wind farms, onshore high-quality wind energy resources are gradually becoming scarcer. Thus, more and more offshore wind farms are being developed all over the world. For offshore wind turbines, an icing event is prone to presenting on the blade surface because the humidity in the offshore region is far higher than that in the onshore region. In comparison with liquid water icing, wind turbine icing has its own characteristics. Many scholars have conducted research on wind turbine icing on the macroscale [6,7]. These research findings introduced

Citation: Zhang, Y.; Zhou, X.; Shi, W.; Chi, J.; Li, Y.; Guo, W. A Visualization Experiment on Icing Characteristics of a Saline Water Droplet on the Surface of an Aluminum Plate. *Coatings* **2024**, *14*, 155. https://doi.org/10.3390/coatings14020155

Academic Editors: Ludmila B. Boinovich and Dariusz Bartkowski

Received: 5 December 2023
Revised: 19 January 2024
Accepted: 19 January 2024
Published: 23 January 2024

Copyright: © 2024 by the authors. Licensee MDPI, Basel, Switzerland. This article is an open access article distributed under the terms and conditions of the Creative Commons Attribution (CC BY) license (https://creativecommons.org/licenses/by/4.0/).

macro-icing characteristics, including ice amount, ice shape, ice distribution, ice type, and so on. Nevertheless, the selection of icing conditions in this research was decided according to the working conditions of onshore wind turbines [8,9]. The water droplets used in the air flow were pure water. In contrast, the working conditions of offshore wind turbines are significantly different from those of onshore wind turbines [10,11]. In addition to the difference in humidity, the composition of water droplets is also different. For the onshore wind turbine icing, the water droplets in the air are pure water vapor from freshwater. In contrast, for the offshore wind turbine icing, the water droplets in the airflow contain salts from sea water. For this reason, the icing characteristic between the onshore wind turbine and the offshore one is different. Therefore, it is necessary to explore offshore wind turbine icing, especially. At present, few experimental conditions have been selected based on the working conditions of offshore wind turbines [12,13]. Moreover, the research methods include the analytical method, simulation method, and experimental method [14,15]. In addition, the wind turbine icing also has micro-icing characteristics. The formation of ice on the macroscale stems from the water droplets flowing in the air and impacting the cold blade surface. Therefore, the micro-icing characteristics of water droplets also need to be explored, which could disclose the mechanism of wind turbine icing. In the previous research findings [16–20], the freezing process of a pure water droplet has been conducted on a metal surface under different icing conditions, such as temperature, water droplet volume, surface characteristics of the substrate, and so on. In comparison, the freezing characteristics of water droplets with salinity have been less studied [21–23], especially for the ingredients of sea water. Scholars examined the freezing characteristics of salty water droplets on the superhydrophobic surface, such as freezing time, contact angle, ice nucleation propagation, and so on.

In summary, macro- and micro-icing characteristics are both important to wind turbine icing. Previous explorations in these two aspects both provide valuable experience and research methods for carrying out the present study. In the present study, the freezing process and characteristics of a water droplet with the ingredient of sea water were tested. An experimental system with a freezing part and an image acquisition part was developed. The freezing process of the water droplet was recorded and processed. The effect of salinity on the icing characteristics of the water droplet was analyzed, including the height, contact diameter, and icing time of the water droplet. And the profile of a frozen water droplet was regressed by the polynomial method. The research findings provide a basis for exploring the macro-icing characteristics of an offshore wind turbine.

2. Experimental Scheme and System

In the present study, the freezing characteristics of water droplets under different salinities were tested. The experimental scheme was listed in Table 1.

Table 1. Experimental scheme.

Items	Values
Volume of water droplet	5 µL
Substrate temperatures	−6 °C, −12 °C, −18 °C
Salinities of water droplets	0 g/L, 6 g/L, 12 g/L, 18 g/L, 26.7 g/L
Substrate material	Aluminum

As listed in Table 1, the volume of the water droplet was 5 µL, and three kinds of temperatures were selected in the present study. The temperatures were decided according to the ice types of the wind turbines. When the temperature is high, the ice type is glaze ice. On the contrary, the ice type is rime ice in low-temperature conditions. In medium temperatures, the ice type is mixed ice [24]. For the salinity, five kinds were determined, which were measured and selected according to the salinity of the Bohai Sea in China. The Bohai Sea is located in the north of China. The lowest temperature was approximated at −20 °C. To make real sea water as much as possible, sea salt and pure water were used according to the measurement results. Additionally, aluminum was selected as a precooling substrate for contact with a water droplet. The reason for the selection is that aluminum has the characteristics of high and isotropic heat conductivity, which is widely used by many scholars. In addition, some small-scale wind turbine blades are made of aluminum material.

For carrying out the icing tests of a water droplet based on the experimental scheme, an icing system was designed and built in the present study. The diagrammatic sketch is shown in Figure 1.

As shown in Figure 1, the experimental system is mainly composed of a freezing device, an industrial camera, a DC power supply, a chiller, a micro-pump, a computer, and so on. The freezing device was a closed chamber that was made of acrylic material because of its better transparency for capturing images. It was used to realize the freezing process of water droplets. The Peltier element (12710) was used as the freezing component in the freezing device. It has the advantages of a small scale, no noise, no vibration, a stable temperature, and low power consumption. In the present study, three pieces were used. One piece was used to cool the substrate, and the other two were used to cool the space over the substrate in the freezing device. The cooling substrate was a core component that was adhered between an aluminum block for dissipating heat and a copper base for the cooling substrate. When it works, one side is used to freeze the copper base, and the other side, which generates heat, is cooled by the aluminum block. The chiller (JZ-5200, JIZHI Electromechanical Co., Ltd. Guangzhou in China) circulates the cooling water flowing through the aluminum block. The aluminum substrate for water droplet icing was located on the copper base. The water droplet was injected out of a syringe by a micro-pump (LSP02-2A, LONGER Co., Ltd., Baoding, China). The moving resolution of it is 0.03125 µm. The industrial camera (XG1205GC-T, MindVision Technology Co., Ltd., Shenzhen, China) was used to record the freezing process of the water droplet. The visual range of the lens is 4 mm × 3.2 mm, and the image resolution is 4096 × 3072. After the test, the image of the iced water droplet profile was processed by image processing software (Photoshop 2023).

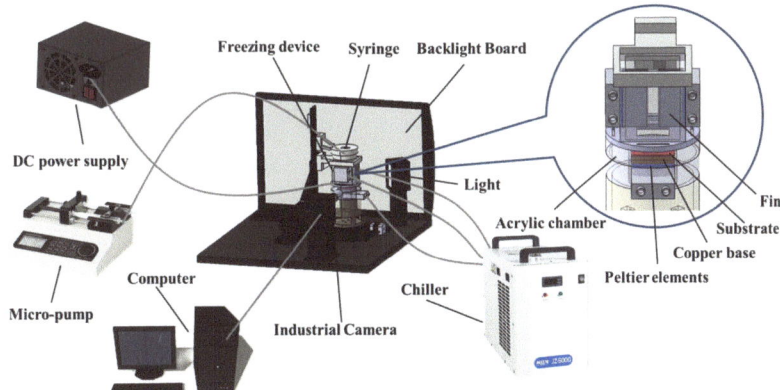

Figure 1. Schematic diagram of experimental system.

3. Experimental Results

According to the experimental scheme, icing tests of water droplets on the aluminum plate surface were carried out. An aluminum plate (1060, CHINALCO SOUTHWEST ALUMINUM GROUP, Chongqing, China) with a size of 40 mm × 40 mm × 1 mm was used as the substrate. The thickness of the aluminum plate is 2 mm, and the surface roughness Ra is 0.17 µm. For exploring the freezing process of sea water, sea salt was used to make a saline solution in the present study. The key ingredients in the saline solution are listed in Table 2.

Table 2. Key ingredients of sea salt.

Ingredients	Concentration (mg/L)
Na^+	9500~10,500
Cl^-	16,500~18,500
SO_4^{2-}	2000~2600
Mg^{2+}	950~1400
K^+	280~380
Ca^{2+}	300~420

In the present study, a thermocouple temperature sensor (TT-K-36) was used to monitor the substrate temperature. Before experiments, the temperatures over the substrate were measured when the substrate temperatures were −6 °C, −12 °C, and −18 °C, respectively. The measurements show that the temperature over the substrate was 2 °C~4 °C lower than the one on the substrate surface. It satisfied the requirement of a low-temperature environment. The water droplet was injected from the syringe and fell onto the aluminum plate surface with an approximate velocity of 1 m/s. Before each test, the aluminum plate was cleaned with deionized water in an ultrasonic cleaning machine. Then it was dried by hot air. After that, it was placed in the freezing device and precooled to the experimental temperature. The icing results under different salinities and temperatures are shown in Figure 2.

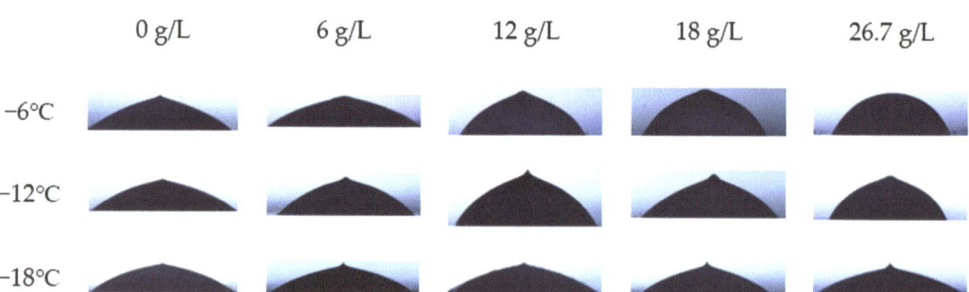

Figure 2. Results of iced water droplets under different salinities and temperatures.

As shown in Figure 2, when the temperature was high, such as −6 °C, the salinity had a significant effect on the profile of the iced water droplet. Under low-salinity conditions, the contact angle of the iced water droplet was low and generated a tip on the top after freezing. On the contrary, the contact angle increased along with salinity, and the tip disappeared. In contrast, when the temperature was low, such as −18 °C, the effect of salinity became insignificant. The contact angles were small, and there were tips under all salinity conditions. This phenomenon resulted from the brine rejection effect [25]. In the process of freezing the water droplet with salinity, the salt ions were rejected from the frozen part into the unfrozen part. It led to an increase in salinity in the unfrozen part [26–28]. This effect increased the concentration and surface tension and gradually decreased the freezing point of the unfrozen solution. Meanwhile, the icing time of the water droplet was delayed, the volume expansion was slight, and there might have been a little unfrozen salty water at the top of the water droplet. In this case, tip disappearance was observed at last. In the present study, when the temperature was high, such as −6 °C, the freezing process was slow, and there was enough time to reject salt ions from the frozen part into an unfrozen one in a water droplet, which resulted in tip disappearance. In contrast, with the decrease in temperature, such as −12 °C and −18 °C, the freezing time shortened, and the salt ions in the frozen part could not diffuse into the unfrozen part instantly. For this reason, the freezing rate was high, and the volume expansion effect was obvious at the tip of the water droplet, which led to the tip generation at last. However, even at a low temperature, the brine rejection effect still existed in the freezing process.

Based on the icing test, the images of iced water droplets were processed, and the profiles are shown in Figure 3.

As shown in Figure 3, according to the processing results, the profiles of iced water droplets can be quantitatively analyzed, which will be discussed in the next sections.

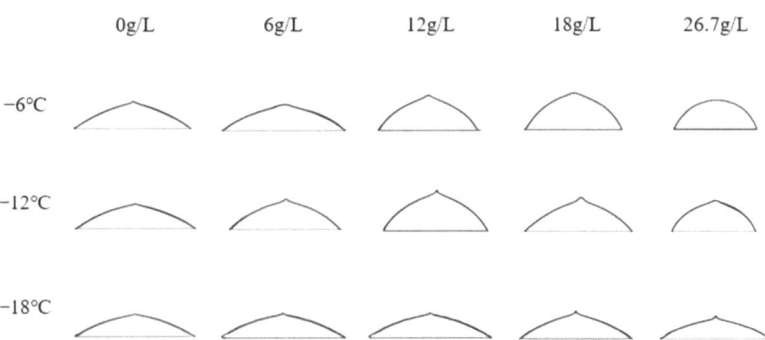

Figure 3. Profiles of iced water droplets under different salinities and temperatures.

4. Discussion and Analysis

Based on the icing tests and image processing results, the effect of salinity on the profile of iced water droplets was analyzed, including the height h and contact diameter d of the iced water droplet, which are shown in Figure 4.

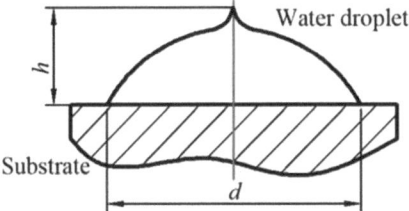

Figure 4. Schematic profile of an iced water droplet.

4.1. Effect of Salinity on Height of Iced Water Droplet

The variation in height of the iced water droplet with salinity is shown in Figure 5. As shown in Figure 5, salinity has a significant effect on the height of the iced water droplet. When the temperature was constant, the height of the iced water droplet increased first. The reason for this result is that salinity affects the surface tension of water droplets. The surface tension of salty water increases along with salinity [29]. Therefore, in comparison with the water droplet without salinity, the one with salinity had a higher surface tension to make the surface of the water droplet contract tightly, which resulted in an increase in the height of water droplet. However, as shown in Figure 5, when the salinity is too high, such as 26.7 g/L, the increase in height was not obvious and even somewhat decreased. As discussed in Section 2, the brine rejection effect decreases the freezing point of a water droplet in the freezing process. Then the icing time increased and the water droplet flew peripherally, which resulted in a decrease in height and tip disappearance. Additionally, at a high temperature, such as −6 °C, the variation in height is a little different from the ones at low temperatures of −12 °C and −18 °C. When the salinity was low, the height decreased first and then increased. The reason for this difference is that low salinity has little effect on the surface tension of water droplets. After contact with the substrate, the water droplet did not freeze in a short time and flew peripherally under a high temperature.

It resulted in a decrease in the height of the iced water droplet. With the increase in the salinity, the rule of variation was the same for the ones at low temperatures of −12 °C and −18 °C.

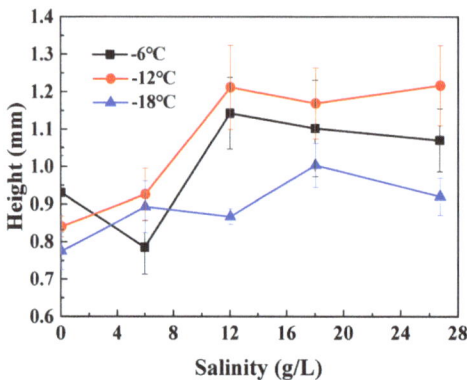

Figure 5. Variation in the height of iced water droplet with salinity.

In addition, from Figure 5, as the salinity was constant, the height of the iced water droplet at −12 °C was the highest, the one at −18 °C was the lowest, and the one at −6 °C was medium. In comparison with the conditions between −6 °C and −12 ° the lower the temperature was, the higher the height of the iced water droplet was. The reason was that the icing time of the water droplet decreased along with temperature. In this case, the flowability of the water droplet decreased with the decrease in temperature. The water droplet was frozen in a shorter time, which resulted in an increase in the height of the water droplet. In contrast, when the temperature decreased further, such as −18 °C, the height decreased again. The reason for this result is that the thickness of frost generating on the cold blade surface increased with the decrease in temperature [30]. In the tests, it was found that there was frost generated on the cold blade surface. At low temperatures, such as −18 °C, the thickness of the frost was higher than the that in the conditions of −6 °C and −12 °C. In this case, the thick frost layer isolated the cold substrate surface absolutely. Under this condition, the water droplet could not make contact with the substrate surface directly but rather the frost layer. After that, the heat transfer between the substrate and water droplet was conducted via the frost layer. For this reason, the frost began melting into water, which merged into a water droplet. The melted water film on the substrate surface intensified the flowability of the water droplet and decreased the concentration of salt in the water droplet, which increased the the contact diameter, decreased the height, and shortened the icing time of the water droplet. In contrast, when the temperature was high, such as −6 °C or −12 °C, the frost layer was thin, and the effect of melted water film on the flowability of the water droplet was weak. That is why the height of the iced water droplet was higher than the one at a low temperature.

4.2. Effect of Salinity on Contact Diameter of Iced Water Droplet

The variation in contact diameter with salinity is shown in Figure 6. As shown in Figure 6, both salinity and temperature have significant effects on the contact diameter of the iced water droplet. Under non-salinity conditions, the effect of temperature on the contact diameter is slight. The contact diameters under different temperature conditions were approximate. With the increase in salinity, the discrepancies among different temperatures became obvious. When the temperature was high, such as −6 °C and −12 °C, the contact diameter decreased with the increase in salinity. However, as the temperature was low, such as −18 °C, the contact diameter varied little with the increase in salinity.

The reasons for the variations are that the icing process of the water droplet was affected by frost generation and heat transfer rate together. At a temperature of −18 °C, the thickness of the frost layer generated on the substrate surface and the heat transfer rate between the water droplet and substrate were both high. As discussed in Section 4.1, the thick frost layer increased the flowability of the water droplet after melting. That is why the contact diameter was larger than the ones under the temperatures of −6 °C and −12 °C. Additionally, the high heat transfer rate under low-temperature conditions resulted in a decrease in the flowability of the water droplet after impacting on the substrate surface, which led to slight variation under different salinities.

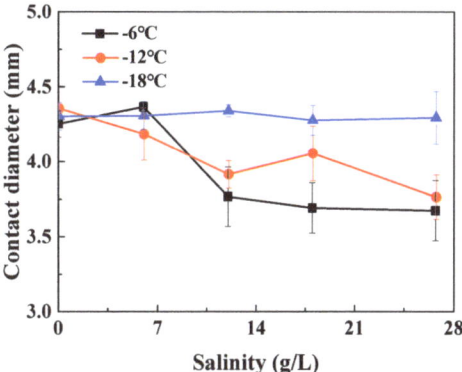

Figure 6. Variation of the contact diameter with salinity.

In contrast, at a temperature of −6 °C or −12 °C, the flowability of the water droplet and the heat transfer rate were low in comparison with those at a temperature of −18 °C, which resulted in a decrease in the contact diameter. Nevertheless, the surface tension of the water droplet increased along with salinity, as discussed above. Then the contact diameter decreased. The higher the salinity, the smaller the contact diameter. Especially at a temperature of −6 °C, the contact diameter under low salinity increased first and then decreased. It shows that the surface tension has a smaller effect on contact diameter under low-salinity conditions. The decrease in freezing point had a significant effect, which resulted in an increase in contact diameter. In the condition of high salinity, surface tension played a key role in decreasing the contact diameter.

4.3. Effect of Salinity on Icing Time of Water Droplet

Similarly, the effect of salinity on the icing time of the water droplet was also analyzed, as shown in Figure 7.

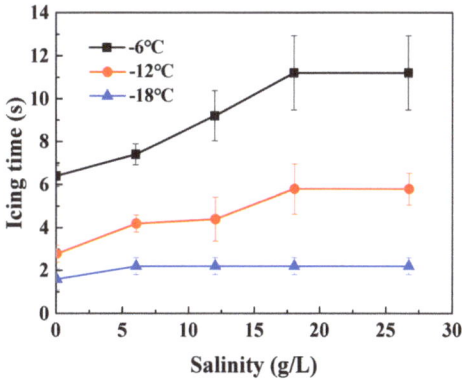

Figure 7. Variation of the icing time of water droplet with salinity.

As shown in Figure 7, the icing time of the water droplet increases along with salinity as the temperature is constant. The reason for this is that salinity decreases the freezing point of the water droplet. The higher the salinity, the lower the freezing point. Therefore, the icing time increased along with salinity. From Figure 7, when the temperature is high, such as −6 °C and −12 °C, the icing time increases dramatically at first. As the salinity is higher than 18 g/L, the salinity has little effect on the icing time. It validates that with the increase in salinity; the freezing point of the water droplet decreased dramatically first and then slightly. It is also the same as previous research findings [31]. However, under low-temperature conditions, such as −18 °C, salinity has little effect on the icing time of the water droplet. The icing time of the water droplet with salinity was just a little higher than that of the water droplet without salinity. It shows that a low temperature increases the heat transfer rate and shortens the icing time of water droplets.

4.4. Effect of Salinity on Profile of Iced Water Droplet

Based on the experimental results of the water droplet with salinity, the profiles of iced water droplets under different icing conditions were processed, which are shown in Figure 3. For analyzing the effect of salinity on the profile quantitatively, the processed profile of an iced water droplet is located in a coordinate, the contact interface between the water droplet and substrate was selected as an abscissa or x axis, and the leftmost point of the contact diameter was selected as the origin.

According to the image of the iced water droplet, some key points along the profile of the iced water droplet were selected to characterize the profile. For these points, the abscissa values were isometric along the contact diameter. In this way, the profile of an iced water droplet can be characterized quantitatively. Therefore, the profiles of iced water droplets under different salinities and temperatures are shown in Figures 8–10.

As shown in Figures 8–10, the profiles of iced water droplets were reconstructed in the coordinates. Then, based on the profile parameters, the profile was fitted by the polynomial method. The power of each fitting profile was decided according to the coefficient of determination R^2. In the present study, the value of R^2 under each condition should be higher than 0.99. According to the criteria, the mathematical expressions of the fitting profiles are listed in Table 3. The fitting results coincide well with the experimental data.

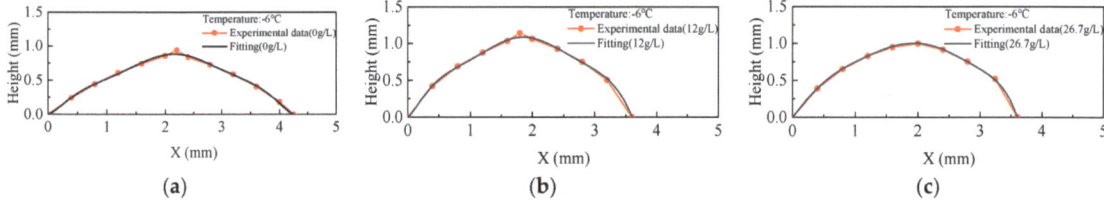

Figure 8. Fitting profiles of iced water droplets at the temperature of −6 °C: (**a**) 0 g/L; (**b**) 12 g/L; (**c**) 26.7 g/L.

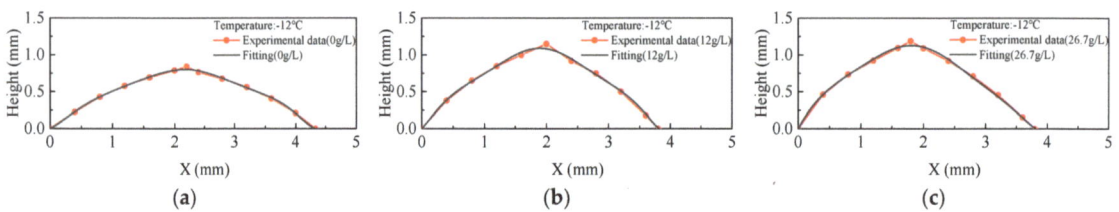

Figure 9. Fitting profiles of iced water droplets at the temperature of −12 °C: (**a**) 0 g/L; (**b**) 12 g/L; (**c**) 26.7 g/L.

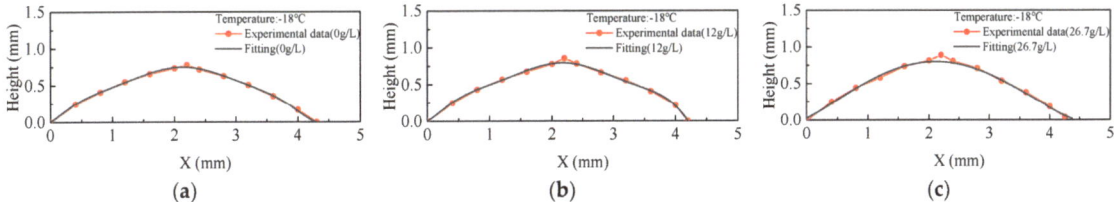

Figure 10. Fitting profiles of iced water droplets at the temperature of −18 °C: (**a**) 0 g/L; (**b**) 12 g/L; (**c**) 26.7 g/L.

Table 3. Mathematical expressions of profile fitted by polynomial method.

Temperature	Salinity	Polynomial Equation	Coefficient of Determination
−6 °C	0 g/L	$y = -0.1839x^2 + 0.7892x - 0.0351$	$R^2 = 0.9914$
	12 g/L	$y = 0.0211x^4 - 0.1387x^3 - 0.017x^2 + 0.9263x + 0.0203$	$R^2 = 0.9939$
	26.7 g/L	$y = 0.0191x^3 - 0.3626x^2 + 1.1348x - 0.011$	$R^2 = 0.9911$
−12 °C	0 g/L	$y = -0.163x^2 + 0.7083x - 0.0174$	$R^2 = 0.9954$
	12 g/L	$y = 0.0273x^4 - 0.1947x^3 + 0.1434x^2 + 0.7851x + 0.017$	$R^2 = 0.9939$
	26.7 g/L	$y = 0.0291x^4 - 0.1821x^3 + 0.0125x^2 + 0.994x + 0.0216$	$R^2 = 0.9956$
−18 °C	0 g/L	$y = -0.1542x^2 + 0.6599x - 0.003$	$R^2 = 0.9948$
	12 g/L	$y = -0.1575x^2 + 0.6843x - 0.0063$	$R^2 = 0.9961$
	26.7 g/L	$y = -0.1681x^2 + 0.717x - 0.0105$	$R^2 = 0.9918$

x (mm) is the position coordinate along contact diameter; y (mm) is the height of iced water droplet; R^2 is the coefficient of determination.

According to the mathematical expressions in Table 2, the volumes of iced water droplets were calculated in the present study for analyzing the effect of salinity on the volume of iced water droplets quantitatively. The schematic diagram of calculation method is shown in Figure 11.

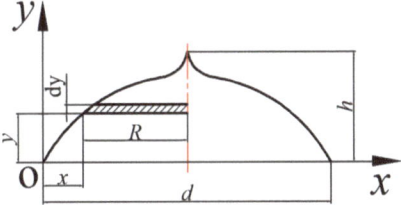

Figure 11. Model of the volume of an iced water droplet.

As shown in Figure 11, an infinitesimal element **dy** was selected at any point along the profile of an iced water droplet. The coordinates at this point are considered (x, y), and the radius of the iced water droplet at this point is R, which is equal to the difference between the half-contact diameter $d/2$ and the coordinate x. Therefore, the infinitesimal volume of the iced water droplet is expressed in Equation (1).

$$dV = \pi(\frac{d}{2} - x)^2 \cdot dy \tag{1}$$

Then the whole volume of the iced water droplet is expressed as Equation (2).

$$V = \int_0^h \pi(\frac{d}{2} - x)^2 \cdot dy \tag{2}$$

Combining Equation (2) with Table 3, the volume of the iced water droplet can be calculated theoretically. The formation of dy and the scope of x under different conditions are listed in Table 4.

Table 4. Formation of dy and scope of x under different conditions.

Temperature	Salinity (g/L)	dy/dx	d/2 (mm)	x
−6 °C	0	$dy/dx = -0.3678x + 0.7892$	2.1	[0.0449, 4.2465]
	12	$dy/dx = 0.0844x^3 - 0.4161x^2 - 0.034x + 0.9263$	1.99	[−0.0219, 3.9549]
	26.7	$dy/dx = 0.0573x^2 - 0.3626x^2 + 1.1348x - 0.011$	1.96	[0.0097, 3.9394]
−12 °C	0	$dy/dx = -0.326x + 0.7083$	2.15	[0.0247, 4.3207]
	12	$dy/dx = 0.1092x^3 - 0.5841x^2 + 0.2868x + 0.7851$	1.99	[−0.0217, 3.9545]
	26.7	$dy/dx = 0.1164x^3 - 0.5463x^2 + 0.025x + 0.994$	1.99	[−0.0217, 3.9669]
−18 °C	0	$dy/dx = -0.3084x + 0.6599$	2.14	[0.0046, 4.275]
	12	$dy/dx = -0.315x + 0.6843$	2.16	[0.0092, 4.3355]
	26.7	$dy/dx = -0.3362x + 0.717$	2.12	[0.0147, 4.2506]

The variations in volumes with salinity under different temperatures are shown in Figure 12.

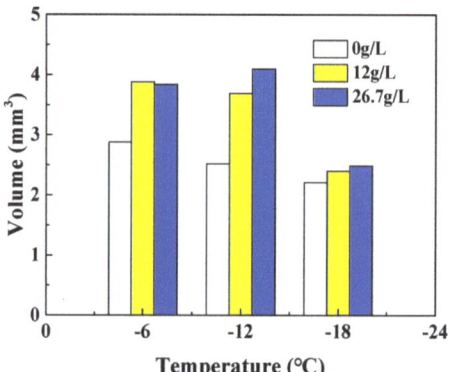

Figure 12. Variations in the volumes of iced water droplets with salinity and temperature.

As shown in Figure 12, both the salinity and the temperature have effects on the volume of iced water droplets. When the temperature was constant, the volume of an iced water droplet with salinity was higher than one without salinity. In contrast, for a salty water droplet, the effect of salinity on the volume was slight. However, as the salinity was constant, the volume decreased along with the temperature. For the low-temperature conditions, such as −18 °C, the volumes among all salinity conditions were approximate. It validates that the salinity had little effect on the volume of water droplet in the process of a phase change.

5. Conclusions

In the present study, the freezing characteristics of a water droplet with salinity were studied, and the effect of salinity on the profile of an iced water droplet was analyzed. Some conclusions were obtained and listed as follows:
1. The effects of salinity and temperature on the height of the iced water droplet have been explored. As the temperature was constant, the height of an iced water droplet increased dramatically first because of the increase in surface tension. In contrast, under high-salinity conditions, the increment decreased. In addition, as the temperature was low, the height decreased because there was thicker frost generated on the cold

surface, which led to an increase in the flowability of a water droplet after melting due to heat transfer.
2. The effects of salinity and temperature on the contact diameter of an iced water droplet have also been analyzed. Generally, the contact diameter of an iced water droplet decreased along with the increase in salinity because of the high surface tension of the salty solution. However, for low-temperature conditions, salinity had little effect on the contact diameter because of the increase in frost generation, which increases the flowability of water droplets after melting during heat transfer between the substrate and the water droplet.
3. The profile of an iced water droplet was obtained from the processing image. Based on the data, the fitting profile of an iced water droplet was also given by the polynomial method in a mathematical expression. The power of each polynomial expression under different conditions was decided according to the coefficient of determination. The fitting results coincide well with the experimental data. Based on the fitting results, the volumes of iced water droplets were also calculated by a mathematical model established in the present study. It validates that the volume decreased along with temperature and increased along with salinity.

Author Contributions: Conceptualization, Y.L. and W.G.; methodology, Y.Z.; software, X.Z.; validation, X.Z., W.S. and J.C.; formal analysis, W.G.; investigation, Y.Z.; resources, Y.L. and Y.Z.; data curation, X.Z. and W.G.; writing—original draft preparation, Y.Z.; writing—review and editing, Y.L. and W.G.; visualization, Y.Z.; supervision, Y.L.; project administration, W.G.; funding acquisition, W.G. All authors have read and agreed to the published version of the manuscript.

Funding: This research was Funded by the Key Laboratory of Icing and Anti/De-icing of CARDC (Grant No. IADL 20220109).

Institutional Review Board Statement: Not applicable.

Informed Consent Statement: Not applicable.

Data Availability Statement: The data presented in this study are available on request from the corresponding author.

Conflicts of Interest: The authors declare no conflicts of interest.

References

1. Nwaigwe, K.N. Assessment of wind energy technology adoption, application and utilization: A critical review. *Int. J. Environ. Sci. Technol.* **2022**, *19*, 4525–4536. [CrossRef]
2. Balat, M. A Review of Modern Wind Turbine Technology. *Energy Sources Part A* **2009**, *31*, 1561–1572. [CrossRef]
3. Farina, A.; Anctil, A. Material consumption and environmental impact of wind turbines in the USA and globally. *Resour. Conserv. Recycl.* **2022**, *176*, 105938. [CrossRef]
4. Dalili, N.; Edrisy, A.; Carriveau, R. A review of surface engineering issues critical to wind turbine performance. *Renew. Sustain. Energy Rev.* **2009**, *13*, 428–438. [CrossRef]
5. Liu, Z.; Zhang, Y.; Li, Y. Superhydrophobic coating for blade surface ice-phobic properties of wind turbines: A review. *Prog. Org. Coat.* **2024**, *187*, 108145. [CrossRef]
6. Gao, L.; Liu, Y.; Hu, H. An experimental investigation of dynamic ice accretion process on a wind turbine airfoil model considering various icing conditions. *Int. J. Heat Mass Transf.* **2019**, *133*, 930–939. [CrossRef]
7. Yang, X.; Bai, X.; Cao, H. Influence analysis of rime icing on aerodynamic performance and output power of offshore floating wind turbine. *Ocean Eng.* **2022**, *258*, 111725. [CrossRef]
8. Shu, L.; Liang, J.; Hu, Q.; Jiang, X.; Ren, X.; Qiu, G. Study on small wind turbine icing and its performance. *Cold Reg. Sci. Technol.* **2017**, *134*, 11–19. [CrossRef]
9. Yi, X.; Wang, K.C.; Ma, H.L.; Zhu, G. Computation of icing and its effect of horizontal axis wind turbine. *Acta Energiae Sol. Sin.* **2014**, *35*, 1052–1058. (In Chinese)
10. Li, J.; Wang, G.; Li, Z.; Yang, S.; Chong, W.T.; Xiang, X. A review on development of offshore wind energy conversion system. *Int. J. Energy Res.* **2020**, *44*, 9283–9297. [CrossRef]
11. Cao, H.Q.; Bai, X.; Ma, X.D.; Yin, Q.; Yang, X.Y. Numerical Simulation of Icing on Nrel 5-MW Reference Offshore Wind Turbine Blades Under Different Icing Conditions. *China Ocean Eng.* **2022**, *36*, 767–780. [CrossRef]
12. Mu, Z.; Li, Y.; Guo, W.; Shen, H.; Tagawa, K. An Experimental Study on Adhesion Strength of Offshore Atmospheric Icing on a Wind Turbine Blade Airfoil. *Coatings* **2023**, *13*, 164. [CrossRef]

13. Mu, Z.; Guo, W.; Li, Y.; Tagawa, K. Wind tunnel test of ice accretion on blade airfoil for wind turbine under offshore atmospheric condition. *Renew. Energy* **2023**, *209*, 42–52. [CrossRef]
14. Anderson, D. Methods for scaling icing test conditions. In Proceedings of the 33rd Aerospace Sciences Meeting and Exhibit, Reno, NV, USA, 9–12 January 1995.
15. Jin, J.Y.; Virk, M.S. Study of ice accretion along symmetric and asymmetric airfoils. *J. Wind Eng. Ind. Aerodyn.* **2018**, *179*, 240–249. [CrossRef]
16. Zhu, Z.; Zhang, X.; Zhao, Y.; Huang, X.; Yang, C. Freezing characteristics of deposited water droplets on hydrophilic and hydrophobic cold surfaces. *Int. J. Therm. Sci.* **2022**, *171*, 107241. [CrossRef]
17. Tembely, M.; Dolatabadi, A. A comprehensive model for predicting droplet freezing features on a cold substrate. *J. Fluid Mech.* **2019**, *859*, 566–585. [CrossRef]
18. Zhang, X. Research on Freezing and Impact Processes of Supercooled Water Droplet and Their Coupling Characteristics. Ph.D. Thesis, Tsinghua University, Beijing, China, 2019. (In Chinese)
19. Enríquez, O.R.; Marín, G.; Winkels, K.G.; Snoeijer, J.H. Freezing singularities in water drops. *Phys. Fluids* **2012**, *24*, 091102. [CrossRef]
20. Fang, W.-Z.; Zhu, F.; Tao, W.-Q.; Yang, C. How different freezing morphologies of impacting droplets form. *J. Colloid Interface Sci.* **2021**, *584*, 403–410. [CrossRef]
21. Boinovich, L.B.; Emelyanenko, A.M. Recent progress in understanding the anti-icing behavior of materials. *Adv. Colloid Interface Sci.* **2024**, *323*, 103057. [CrossRef]
22. Carpenter, K.; Bahadur, V. Saltwater icephobicity: Influence of surface chemistry on saltwater icing. *Sci. Rep.* **2015**, *5*, 17563. [CrossRef]
23. Boinovich, L.B.; Emelyanenko, A.M.; Emelyanenko, K.A.; Maslakov, K.I. Anti-icing properties of a superhydrophobic surface in a salt environment: An unexpected increase in freezing delay times for weak brine droplets. *Phys. Chem. Chem. Phys.* **2016**, *18*, 3131–3136. [CrossRef] [PubMed]
24. *ISO 12494*; Atmospheric Icing of Structures. International Organization for Standardization: Geneva, Switzerland, 2017.
25. Singha, S.K.; Das, P.K.; Maiti, B. Influence of Salinity on the Mechanism of Surface Icing: Implication to the Disappearing Freezing Singularity. *Langmuir* **2018**, *34*, 9064–9071. [CrossRef] [PubMed]
26. Bauerecker, S.; Ulbig, P.; Buch, V.; Vrbka, L.; Jungwirth, P. Monitoring Ice Nucleation in Pure and Salty Water via High-Speed Imaging and Computer. *J. Phys. Chem. C* **2008**, *112*, 7631–7636. [CrossRef]
27. Vrbka, L.; Jungwirth, P. Brine Rejection from Freezing Salt Solutions: A Molecular Dynamics Study. *Phys. Rev. Lett.* **2005**, *95*, 148501. [CrossRef] [PubMed]
28. Carignano, M.; Baskaran, E.; Shepson, P.; Szleifer, I. Molecular dynamics simulation of ice growth from supercooled pure water and from salt solution. *Ann. Glaciol.* **2006**, *44*, 113–117. [CrossRef]
29. Zhang, Z.; Jin, L.; Yuan, Z.; He, S. The seawater surface tension coefficient representation with different salinities. *J. Exp. Fluid Mech.* **2017**, *31*, 45–50. (In Chinese)
30. Şahin, A.Z. An experimental study on the initiation and growth of frost formation on a horizontal plate. *Exp. Heat Transf.* **1994**, *7*, 101–119. [CrossRef]
31. Huang, W. The Experimental Determination and Prediction of Freezing Point of Salt-Water Systems. Master's Thesis, Xinjiang University, Urumqi, China, 2015. (In Chinese)

Disclaimer/Publisher's Note: The statements, opinions and data contained in all publications are solely those of the individual author(s) and contributor(s) and not of MDPI and/or the editor(s). MDPI and/or the editor(s) disclaim responsibility for any injury to people or property resulting from any ideas, methods, instructions or products referred to in the content.

Article

Exploring the Influence of the Deposition Parameters on the Properties of NiTi Shape Memory Alloy Films with High Nickel Content

André V. Fontes, Patrícia Freitas Rodrigues, Daniela Santo and Ana Sofia Ramos *

CEMMPRE-Department of Mechanical Engineering, University of Coimbra, 3030-788 Coimbra, Portugal; valentim.universidade@gmail.com (A.V.F.); pf.rodrigues@uc.pt (P.F.R.); daniela93santo@gmail.com (D.S.)
* Correspondence: sofia.ramos@dem.uc.pt

Abstract: NiTi shape memory alloy films were prepared by magnetron sputtering using a compound NiTi target and varying deposition parameters, such as power density, pressure, and deposition time. To promote crystallization, the films were heat treated at a temperature of 400 °C for 1 h. For the characterization, scanning electron microscopy, energy dispersive X-ray spectroscopy, atomic force microscopy, synchrotron X-ray diffraction, and nanoindentation techniques were used on both as-deposited and heat-treated films. Apart from the morphology and hardness of the as-deposited films that depend on the deposition pressure, the power applied to the target and the deposition pressure did not seem to significantly influence the characteristics of the NiTi films studied. After heat treatment, austenitic (B2) crystalline superelastic films with exceptionally high nickel content (~60 at.%) and vein-line cross-section morphology were produced. The crystallization of the films resulted in an increase in hardness, Young's modulus, and elastic recovery.

Keywords: shape memory alloys; NiTi; Ni-rich; microelectromechanical systems (MEMS); superelasticity; magnetron sputtering

Citation: Fontes, A.V.; Freitas Rodrigues, P.; Santo, D.; Ramos, A.S. Exploring the Influence of the Deposition Parameters on the Properties of NiTi Shape Memory Alloy Films with High Nickel Content. *Coatings* **2024**, *14*, 138. https://doi.org/10.3390/coatings14010138

Academic Editor: Elena Villa

Received: 19 December 2023
Revised: 12 January 2024
Accepted: 19 January 2024
Published: 20 January 2024

Copyright: © 2024 by the authors. Licensee MDPI, Basel, Switzerland. This article is an open access article distributed under the terms and conditions of the Creative Commons Attribution (CC BY) license (https://creativecommons.org/licenses/by/4.0/).

1. Introduction

Over recent decades, technological advancements have significantly transitioned from single macro-scale devices to intricate micro-scale solutions. This shift is exemplified in the emergence of microelectromechanical systems (MEMS), which integrate both electronic components and moving parts at a microscopic scale. This evolution in technology drives a continuous demand for innovative mechanical and functional solutions underpinned by scientific advancements. In this dynamic landscape, shape memory alloys (SMAs), especially noted for their unique properties, have become pivotal in advancing MEMS technology [1,2].

SMAs, particularly renowned for their shape memory effect (SME) and superelasticity effect (SE), exhibit remarkable functional properties. These materials, when subjected to specific external stimuli, such as stress or temperature variations, demonstrate extraordinary responses. For example, in response to the superelasticity effect, these materials can undergo significant deformation without permanent damage. Similarly, the shape memory effect allows the original shape of these materials to be reverted upon heating above their austenitic transformation temperature. The manifestation of these effects is intricately linked to the precise balance between the material's processing techniques and its chemical composition. The shape memory effect or superelasticity effect can only be achieved if the alloy has a martensite or austenite phase, respectively. Even minor variations in the chemical composition, namely in Ni content, can lead to significant changes in the material's morphology and properties, profoundly influencing end-use performance [3–7].

Among deposition techniques such as pulsed laser deposition, electron beam evaporation, ion beam deposition, plasma spray technique, ion plating, and flash evaporation, magnetron sputtering has proven to be the most successful and widely used technique for preparing NiTi-based films [1]. Sputtering is a physical vapor deposition technique that works by applying a high-voltage electric field on a target, accelerating ions towards the target and ejecting its atoms, which eventually collide with the substrate, creating a film. For this process, an inert gas, usually Argon (Ar), is used due to its good compromise between ion size and cost. Low Ar gas pressures (<0.05 Pa) lead to high-energy atoms reaching the substrate, causing atomic peening and increasing compressive stresses in the film. Intermediate pressures (from 0.05 Pa to 0.5 Pa) create rather dense films with few defects. Films produced with higher argon pressures (>1 Pa) are less dense, more brittle, and more porous and exhibit a columnar morphology [1,8,9]. John A. Thornton [10] succeeded in relating the substrate temperature and Ar deposition pressure with the films' surface and cross-section microstructure.

As-deposited films sputtered from NiTi alloy target tend to exhibit amorphous characteristics [8,9]. Therefore, the equilibrium austenitic phase can only be achieved after crystallization, which is possible either by depositing the films at high substrate temperatures [7] or by post-annealing treatment [8]. In the literature, there are several works regarding the impact of the deposition temperature on the final phasic composition of films produced using an alloy target. When the films are deposited using substrate temperatures around 425 °C, the NiTi peak becomes evident in the X-ray diffraction (XRD) diffractograms [6,11]. Using high-power impulse magnetron sputtering (HiPIMS), X. Bai et al. [12] obtained in situ crystalline NiTi thin films at a low substrate temperature (230 °C). Although less usual, it is possible to use separate targets and heat the substrates during the deposition to obtain in situ crystalline NiTi films. In this case, the reported substrate temperature is 450 °C [6]. Alternatively, whatever the sputtering approach, the deposition stage can be separated from the heat-treatment stage. In this case, the films are typically heat-treated at temperatures above 400 °C [13]. For near equiatomic monolithic films, temperatures of 600 °C are used to obtain well-defined NiTi XRD peaks [7,14]. The heat-treatment temperature is critical since, if low, it might not be enough to fully crystallize the films. On the other hand, a temperature higher than necessary can promote the precipitation of undesired phases, such as $NiTi_2$, Ni_3Ti, and oxides, which can influence the films' phase transformation temperatures, as well as the mechanical properties [7,15]. In addition, the mechanical properties are also influenced by grain growth during the heat treatment.

The study of the hardness and Young's modulus is important to characterize the shape memory alloy films. The mechanical properties play a decisive role in the identification of SMAs' functional properties, namely in the superelasticity that is related to the stress-induced martensitic transformation [8]. The hardness and Young's modulus can be evaluated using depth-sensing indentation [16]. According to the literature, a maximum indentation depth close to 10% of the films' thickness should guarantee the absence of the substrate's influence on the hardness measurements [17,18].

Despite the potential of NiTi SMAs, there is a gap in the literature regarding their detailed characterization, particularly in high-nickel content NiTi alloy films (Ni ranging from 55 to 60 wt.%). This study aims to bridge this gap by exploring the correlation between chemical composition, topographical/morphological characteristics, structure, and mechanical properties. Through a methodical approach involving the production of NiTi films via sputtering and their subsequent characterization using a range of analytical techniques, this research seeks to deepen the understanding of NiTi-based films. The insights gained could not only enhance the properties of these films but also potentially revolutionize their application in various fields, especially in MEMS.

2. Materials and Methods

2.1. Deposition Technique

NiTi films were deposited onto mirror-finished monocrystalline silicon substrates via d.c. magnetron sputtering from a NiTi compound target (99.9% pure) measuring 150 mm × 150 mm × 6 mm. The target had a near equiatomic chemical composition (Ti–49.9 at.% Ni).

The silicon substrates were subjected to ultrasound cleaning in acetone and alcohol baths for 5 min each. After being dried with hot air, the Si substrates were fixed onto a copper substrate holder, promoting heat dissipation and, thus, keeping substrates at a considerably low temperature during deposition.

The equipment used for the magnetron sputtering depositions is a semi-industrial German apparatus from Hartec. The NiTi films were produced according to varying deposition parameters, such as the power applied to the target, the deposition pressure, and the deposition time, while the other parameters were kept constant. For all the films, the target-substrate distance, the substrates' rotation speed, and the substrate's bias were 75 mm, 23 rpm, and −50 V, respectively. Before deposition, the sputtering chamber was evacuated until a vacuum pressure below 5×10^{-4} Pa was reached.

Table 1 summarizes the deposition parameters of the NiTi films under observation, together with the thickness evaluated using profilometry. The first film can be considered as the reference, while in the following depositions, a specific parameter was varied. The nomenclature followed for the NiTi films is F_pressure[Pa]_power[W]. The film deposited during a longer time as a "t" at the end. The deposition time of the F_0.3_1700 film was adjusted in order to obtain a thickness similar to the reference film.

Table 1. Deposition parameters of the NiTi films produced using magnetron sputtering.

Film	Target	Pressure [Pa]	Power Density [Wmm^{-2}]	Dep Time [min]	Thickness [μm]
F_0.3_1000	NiTi	0.3	4.44×10^{-2} (1000 W)	60	3.2
F_0.3_1700	NiTi	0.3	7.56×10^{-2} (1700 W)	28	2.7
F_0.5_1000	NiTi	0.5	4.44×10^{-2} (1000 W)	60	3.4
F_0.3_1000t	NiTi	0.3	4.44×10^{-2} (1000 W)	120	5.1

Since the as-deposited NiTi films produced using sputtering tend to be amorphous, all the films were subjected to heat treatment at 400 °C for 1 h in a horizontal furnace under a hydrogenated argon pressure of around 0.5 Pa. The nomenclature used for the heat-treated films is the same as that for the as-deposited films, followed by underscore HT. The as-deposited and the heat-treated films were characterized using several techniques.

2.2. Characterization Techniques

The surface and cross-section of the films were analyzed by scanning electron microscopy (SEM). The equipment used for the SEM analyses was a field-emission gun microscope (Zeiss, Merlin, Oberkochen, Germany) equipped with energy dispersive spectroscopy (EDS). Accelerating voltages of 2 and 10 kV were used for imaging and EDS, respectively. To evaluate the chemical composition of the films, EDS measurements were carried out on four different areas (110 μm × 80 μm), randomly selected from the central region of each film.

The surface topography of the films was thoroughly examined using atomic force microscopy (AFM) in tapping mode. AFM micrographs were taken over 6 × 6 μm^2 and 3 × 3 μm^2, and 2D and 3D profiles of each sample were generated (Veeco, Innova, New York, NY, USA). The average roughness (Sa) and the root mean square roughness (Sq) were obtained through the roughness subroutine of the AFM apparatus from four independent measurements.

In this work, synchrotron radiation X-ray diffraction (SR-XRD) experiments were carried out for phase indexation. The equipment used for the SR-XRD analysis is located in the P07 High-Energy Materials Science (HEMS) beamline of Petra III/DESY (Deutsches Elektronen-Synchrotron) in Hamburg, Germany. The measurements were carried out at room temperature using a wavelength of 0.1467 Å (87 keV), a beam spot of 200 × 200 μm^2, and a two-dimensional (2D) detector PERKIN ELMER XRD 50 1621 was placed at 1.00 m from the samples. The raw 2D images were treated using the Fit2D program [19] to calculate the individual XRD patterns by integration from 0° to 360° (azimuthal angles).

The hardness and Young's modulus of the films were evaluated using nanoindentation (Micro Materials NanoTest, Wrexham, UK). The nanoindentation equipment had a Berkovich indenter, and the experiments were performed in load control mode using a maximum load of 5 mN. The maximum load was selected to make sure that the influence of the substrate on the hardness was avoided by guaranteeing that the maximum depth was below 10% of the films' thickness. A minimum of 30 indentations were performed for each film. The results were treated according to the Oliver and Pharr [16] method, including thermal drift correction. Fused quartz was used as a reference material to determine the Berkovich tip area function.

3. Results and Discussion

3.1. As-Deposited Films

Table 2 shows the chemical composition of the films obtained through EDS. All the films present minor amounts of oxygen (below 4.0 at.%), and this content is similar for all the films under study. Therefore, to have a more precise idea of the films' stoichiometry, EDS results are presented after quantifying only titanium and nickel. The Ni and Ti contents presented in Table 2 correspond to the average of the four areas analyzed for each film.

Table 2. Chemical composition of the NiTi films.

Film	Ni [at.%]	Ti [at.%]
F_0.3_1000	60.7 ± 0.4	39.3 ± 0.4
F_0.3_1700	61.4 ± 0.1	38.6 ± 0.1
F_0.5_1000	60.3 ± 0.3	39.7 ± 0.3
F_0.3_1000t	59.4 ± 0.4	40.6 ± 0.4

As can be observed, the films are enriched in nickel, with Ni atomic percentages above 60%, a percentage much higher than that of the target from which they were prepared. This result is in accordance with the available literature and is attributed to the lower sputtering yield of Ti compared with Ni [7,20]. In fact, to obtain equiatomic chemical compositions, it is common to use a Ti-rich NiTi target [6,21], two separate targets (NiTi and Ti) [20], or even a NiTi target with Ti foils/slices superimposed [12,22]. Since the Ni-rich chemical composition is required to reach the objective of obtaining superelastic NiTi films, in this work, an equiatomic target was used to intentionally obtain films with high Ni content.

Compared with the reference film (F_0.3_1000), the increase in the power applied to the NiTi target led to a minor increase in the Ni content, while the increase in the argon pressure resulted in a slightly lower Ni%. At the beginning of the sputtering process, more nickel atoms are ejected from the target, but as the target's surface becomes depleted in Ni, more titanium atoms are ejected, and in some cases, a steady state where the films' chemical composition is close to the target's chemical composition can be reached. In this study, the steady state was not reached, although the film deposited during more time (F_0.3_1000t) had somewhat more Ti.

SEM analyses were performed to characterize both the surface and cross-section morphology of the as-deposited films (Figure 1). Except for the film deposited at a higher pressure, a cauliflower shape surface morphology could be observed for all films, where the larger features should correspond to the top of the columns and, inside these features, some grains could be distinguished. The grain size of the F_0.3_1000 and F_0.3_1700 films seems similar. This cauliflower-type morphology is characteristic of metallic films deposited by magnetron sputtering, which usually results in columnar growth [23]. The SEM surface images of the F_0.3_1000 and F_0.3_1700 films, shown in Figures 1a and 1c, respectively, show smaller surface features compared with the F_0.3_1000t film (Figure 1g). As a consequence of the columnar growth, the film deposited during a longer time exhibits larger features, corresponding to the top of wider columns. In the SEM image corresponding to the F_0.5_1000 film (Figure 1e), the cauliflower morphology is not noticeable, as this film has a surface morphology that seems smoother.

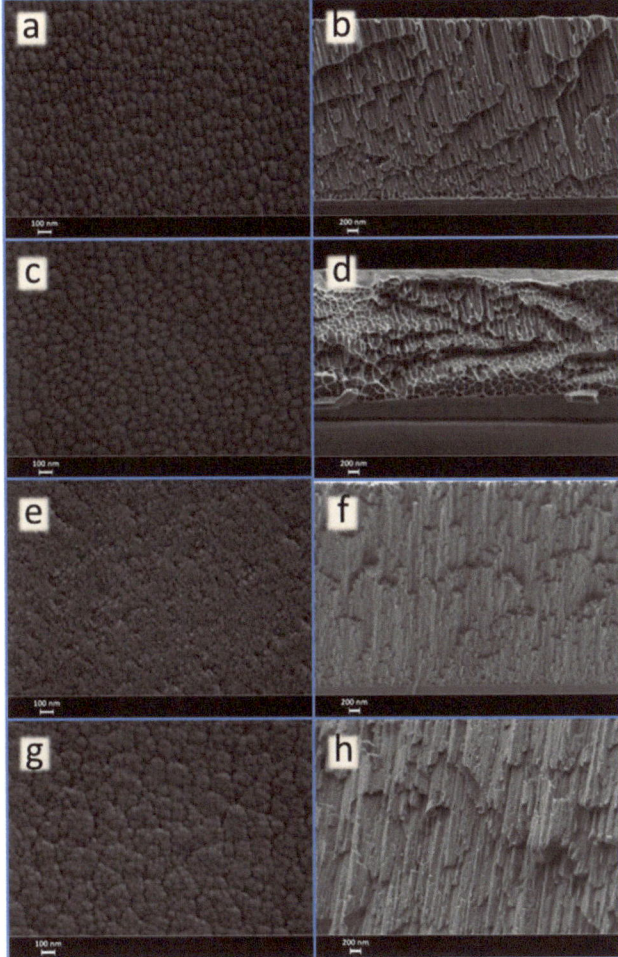

Figure 1. SEM surface micrographs of as-deposited films: (**a**) F_0.3_1000, (**c**) F_0.3_1700, (**e**) F_0.5_1000, and (**g**) F_0.3_1000t. SEM cross-section micrographs of as-deposited films: (**b**) F_0.3_1000, (**d**) F_0.3_1700, (**f**) F_0.5_1000, and (**h**) F_0.3_1000t.

Regarding the cross-section morphologies, also shown in Figure 1, all the films presented rather compact morphology, without pores or discontinuities. The film deposited at the highest pressure (Figure 1f) has a dense columnar morphology, according to the Thornton [10] model, and a higher thickness than the reference film (F_0.3_1000). Higher pressure means more Ar+ ions bombarding the target(s) surface and more collisions of the ejected atoms when traveling towards the substrates. In the present case, the first effect prevailed, and an increase in thickness was observed for the film deposited using a higher pressure. The increase in thickness could also indicate a less dense film.

The F_0.3_1000, F_0.3_1700, and F_0.3_1000t films' cross-section presented an interesting morphology; in particular, the film deposited applying a higher power to the NiTi target. In the SEM cross-section image of the F_0.3_1700 film (Figure 1d), vein-like features, which are characteristic of metallic glass thin films produced by sputtering, can be observed [24,25]. The increase in the power applied to the target should be responsible for the more notorious vein-like morphology. This morphology points to a more ductile behavior and could suggest the presence of a film with a higher elastic recovery.

The influence of the deposition time can be seen when comparing films F_0.3_1000 and F_0.3_1000t under the same magnification (Figures 1b and 1h, respectively). Besides the obvious higher thickness, the F_0.3_1000t film has wider columns, as expected, as time increases due to the shape of the columns.

Figure 2 shows representative AFM images of the NiTi films deposited onto Si substrates using different parameters. The films reveal rather uniform surface topographies, with grain size in the nanometer range.

The topographic images of the films are in accordance with the SEM surface micrographs, except for the film deposited at a higher pressure. The roughness values measured using AFM and resulting from an average of two different zones for each film (minimum of 4 measurements per film) are compiled in Table 3. The films under study present low roughness, and the power applied to the NiTi target, as well as the pressure, does not seem to influence the Sa and Sq values. Although the SEM surface image of the F_0.5_1000 film suggests a smoother surface, the AFM results do not corroborate it. As expected, the film deposited during a longer time has higher roughness compared with the other films due to the larger features observed at the surface of this film and that correspond to the top of the columns.

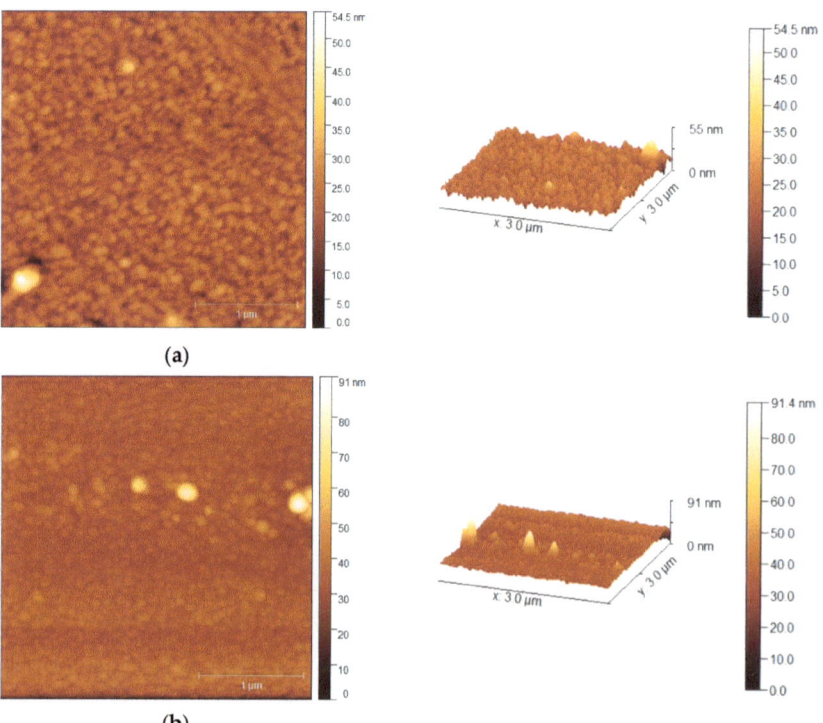

Figure 2. Cont.

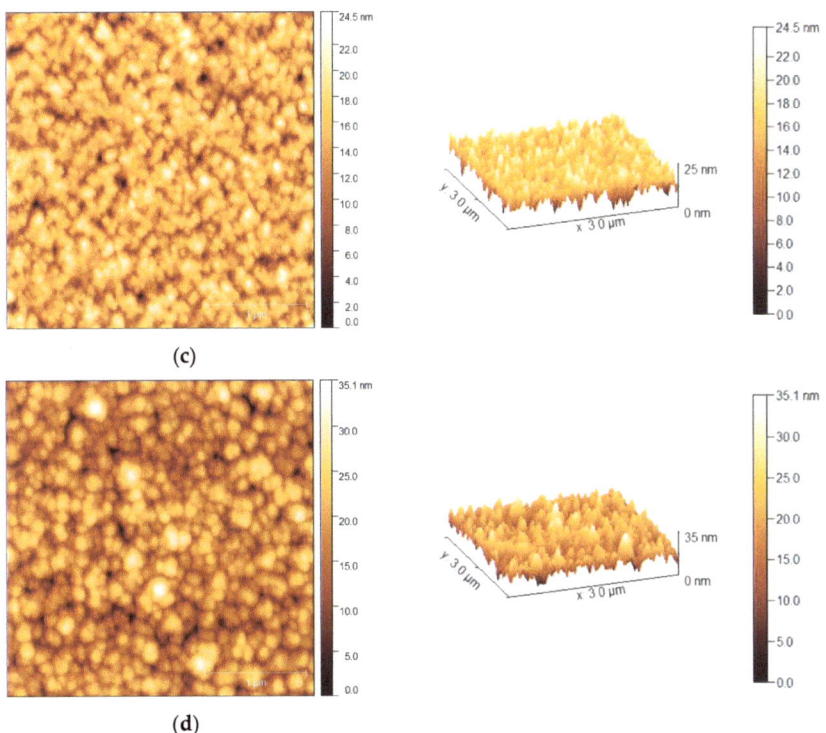

Figure 2. 2D and 3D AFM images (3 μm × 3 μm) of as-deposited films: (**a**) F_0.3_1000, (**b**) F_0.3_1700, (**c**) F_0.5_1000, and (**d**) F_0.3_1000t.

Table 3. Sa and Sq values of the NiTi films measured using AFM.

Film	F_0.3_1000	F_0.3_1700	F_0.5_1000	F_0.3_1000t
Sa (nm)	2.5 ± 0.1	2.6 ± 0.4	2.6 ± 0.1	3.8 ± 0.4
Sq (nm)	3.4 ± 0.1	4.0 ± 0.8	3.4 ± 0.3	4.9 ± 0.6

Analyzing the XRD diffractogram of Figure 3, the presence of a single broad peak for the reference film is evident (F_0.3_1000). This observation suggests the absence of crystalline phases, leading to the conclusion that this film is amorphous. All the other films show similar diffractograms, confirming the amorphous nature of the as-deposited films under study. This finding is in line with the available literature.

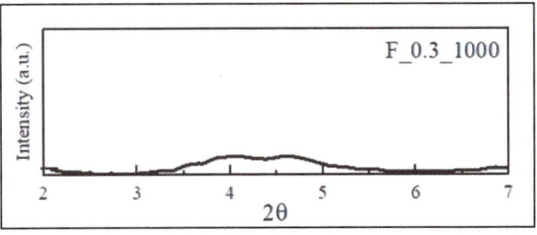

Figure 3. Synchrotron radiation XRD diffractograms—F_0.3_1000.

3.2. Heat-Treated Films

As described in the Materials and Methods section, the films were subjected to heat treatment at 400 °C for 1 h, being thereafter characterized. After heat treatment, the surface and cross-section of the films were analyzed by SEM. Comparing the SEM surface micrographs of Figure 4a,c with the surface SEM micrographs of the as-deposited films (Figure 1), it can be perceived that the features observed in F_0.3_1000 and F_0.3_1700 films are slightly larger, but the surface of the heat-treated films seems more compact when the voids between the columns are less defined. During heat treatment, grain growth should occur, explaining the morphological differences observed.

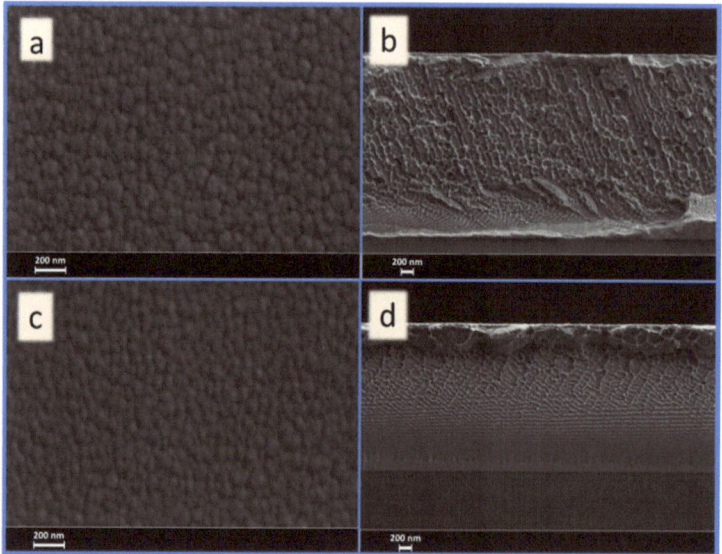

Figure 4. SEM surface micrographs of heat-treated films: (**a**) F_0.3_1000_HT and (**c**) F_0.3_1700_HT; SEM cross-section micrographs of heat-treated films: (**b**) F_0.3_1000_HT and (**d**) F_0.3_1700_HT.

Cross-section SEM images of the heat-treated films are also shown in Figure 4. When analyzing the cross-section of the F_0.3_1000 and F_0.3_1700 films (Figure 4b,d), there are changes from the as-deposited to the heat-treated morphology. Although still with a vein-like appearance, the heat-treated F_0.3_1000 and F_0.3_1700 films transited to a more compact morphology, confirmed by the decrease in thickness from 3.2 to 2.9 µm and from 2.7 to 2.3 µm, respectively. A thin layer of oxide seems to have formed at the surface of both films.

After heat treatment, SR-XRD analyses were conducted to ensure that the films had crystalized, forming intermetallic phases, in particular, the desired B2 phase (austenitic phase). SR-XRD was used to study the structural characteristics of the NiTi films because it is a powerful technique that uses high-energy X-rays in transmission mode with a small beam size. SR-XRD allows for detailed structural analysis of the films, which is fundamental for materials such as NiTi that may display minor phases with reduced dimensions (precipitates) [26]. Since all the films are enriched in nickel, besides the B2 phase, the indexation of Ni-rich crystalline phases is possible. Films that were initially amorphous (Figure 3) were found to crystallize, revealing numerous XRD peaks, as can be observed in the diffractograms of Figure 5. Among these, several were indexed as B2, while a few can be identified as Ni-rich precipitates [27–29].

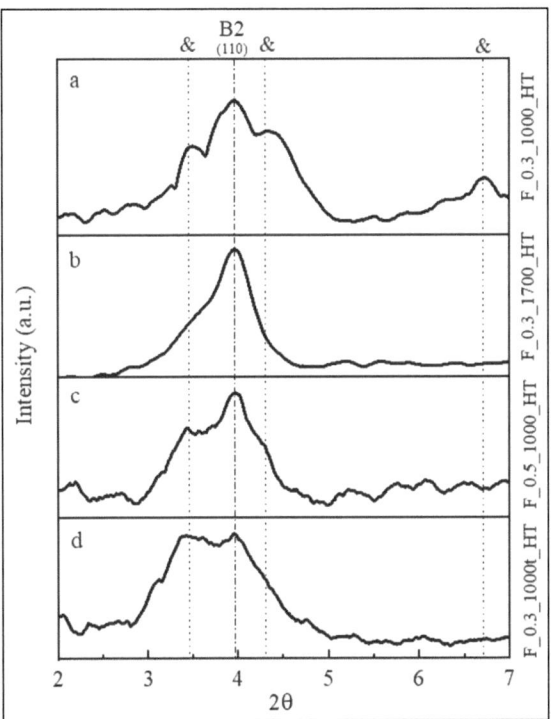

Figure 5. Synchrotron radiation XRD diffractograms: (**a**) F_0.3_1000_HT, (**b**) F_0.3_1700_HT, (**c**) F_0.5_1000_HT, and (**d**) F_0.3_1000t_HT.

The information from the XRD diffractograms is valuable because it proves that the chosen heat-treatment temperature (400 °C) is enough to obtain the B2 phase, the main driving factor for the existence of a superelastic NiTi film. The austenite formed presents the (110) crystallographic plane for all the films. Furthermore, along with the austenite phase, the heat-treatment temperature promoted the formation of Ni-rich precipitates (highlighted by & in Figure 5), which can be interesting for improving the mechanical properties of the films, as shown in the mechanical behavior section [30].

3.3. Mechanical Behavior

Table 4 shows the results of the nanoindentation experiments for the as-deposited films. In this table, the films' thickness is also included to more easily confirm that there is no influence of the substrate on the measured hardness. Actually, as can be seen in Table 4, the maximum indentation depths are always below 10% of the films' thickness, assuring that there is no influence of the substrate on the hardness values.

Table 4. Mechanical properties of the as-deposited films.

Film	Thickness [μm]	Max Depth [nm]	Plastic Depth [nm]	Hardness [GPa]	Er [GPa]	E [GPa]	ERP
F_0.3_1000	3.2	162.1 ± 5.8	133.9 ± 6.2	7.1 ± 0.5	137 ± 8	143 ± 8	0.211 ± 0.02
F_0.3_1700	2.7	159.8 ± 6.3	131.6 ± 6.7	7.3 ± 0.6	138 ± 5	145 ± 5	0.215 ± 0.02
F_0.5_1000	3.4	193.9 ± 9.4	166.1 ± 9.9	5.0 ± 0.5	116 ± 7	119 ± 7	0.168 ± 0.02

Er—reduced Young's modulus; E—Youngs modulus; ERP—Elastic recovery parameter.

For the F_0.3_1000 and F_0.3_1700 films, a hardness of around 7 GPa and Young's modulus close to 140 GPa was obtained, meaning that the power applied to the NiTi target does not significantly influence the films' mechanical behavior. Indeed, these films have a similar nickel content (EDS results) and similar morphology based on the SEM surface and cross-section analyses (Figure 1a–d). For the F_0.5_1000 film, a lower hardness was obtained (5.0 GPa). The increase of the pressure also results in a decrease in the Young's modulus. This film presents a significantly different morphology (Figure 1e,f) from the other films and that is unusual for sputtered metallic films, but based on the SEM images of Figure 1, it is difficult to infer whether the increase in the deposition pressure promotes less dense films. However, due to the interelectrode collisions, at higher pressures, the adatoms arrive at the substrate with lower mobility, and the films tend to be less dense, which can explain the lower hardness and Young's modulus of the F_0.5_1000 film. The thickness of the F_0.5_1000 film is higher than the thickness of the F_0.3_1000 film, which is also in accordance with a less dense film for a higher deposition pressure. Compared with the film prepared using higher pressure, the elastic recovery parameter (ERP) of the other films is higher, with ERP values close to 0.21. This result is in accordance with the as-deposited films' cross-section morphology that points to a more ductile behavior of the F_0.3_1000 and F_0.3_1700 films (see Figure 1). The ERP, defined as the ratio of recoverable indentation depth to maximum indentation depth, is a dimensionless parameter that highlights the elastoplastic behavior of the material; as ERP increases, the elastic work increases, while as it decreases, the plastic work increases [31].

Heat treating the films promoted changes in their mechanical properties, as demonstrated in Table 5. When analyzing the nanoindentation results after heat treatment at 400 °C, it is evident that the hardness and Young's modulus increased for all films. The hardness increase is due to the formation of hard intermetallic phases but also because heat treatment promotes compaction of the films, as confirmed by SEM. In fact, the comparison between both surface and cross-section images of the films (Figures 1 and 4) reveals more compact morphology after heat treatment.

Table 5. Mechanical properties of the heat-treated films.

Film	Max Depth [nm]	Plastic Depth [nm]	Hardness [GPa]	Er [GPa]	E [GPa]	ERP
F_0.3_1000_HT	154.8 ± 6.9	126.7 ± 6,8	7.7 ± 0.7	143 ± 9	151 ± 9	0.222 ± 0.02
F_0.3_1700_HT	150.6 ± 4.6	122.6 ± 5.0	8.1 ± 0.6	147 ± 6	156 ± 6	0.229 ± 0.02
F_0.5_1000_HT	160.6 ± 5.1	133.0 ± 5.8	7.1 ± 0.5	140 ± 5	147 ± 5	0.208 ± 0.02

In order to highlight the differences in hardness, Figure 6 shows the hardness of the as-deposited and heat-treated films. The more pronounced rise in hardness upon heat treatment is observed for the high-pressure film. After heat treatment, the F_0.3_1000_HT and F_0.3_1700_HT films still have similar hardness values, again in accordance with their similar morphology (Figure 4), while the F_0.5_1000_HT film as a slightly lower hardness. The high-pressure film has a slightly lower Ni content, and thus, the presence of Ni-rich precipitates might be slightly less pronounced, which can explain the lower hardness of this film.

Regarding Young's modulus, there are no significant differences, and all the heat-treated films have values higher than those found in the literature for equiatomic B2-NiTi films (~95–100 GPa [11,32]). The current films are proven to have formed B2-NiTi as the major phase, which should have brought Young's modulus closer to the literature values. The higher Young's modulus values are probably related to the nanoindentation measurements and can be explained by the influence of the silicon substrate. Even though the maximum depths are lower than 10% of the films' thickness, when it comes to Young' modulus, the substrate's influence is still possible. Tests conducted using the same nanoindentation equipment, also with a maximum load of 5 mN, revealed a reduced Young's modulus for monocrystalline silicon around 175 GPa. Therefore, if there is an influence of the substrate, it should result in a higher Young's modulus than expected for NiTi. It should also be noted that the measured hardness values are also higher than those expected for superelastic NiTi films. According to the work of Ni et al. [32], the hardness of a superelastic NiTi film deposited onto an aluminum alloy substrate is 4.7 GPa, and Young's modulus is 95 GPa. These mechanical properties were obtained by nanoindentation using loads ranging from 5 to 200 mN. Interestingly, the H/Er ratio reported by Ni et al. [32] (~0.050) is similar to the H/Er values of the heat-treated films under study, which are between 0.051 and 0.054. Comparing Tables 4 and 5, it is possible to conclude that the heat-treated films always have higher ERP than the corresponding as-deposited film, indicating an increase in the elastic work, which is consistent with a superelastic behavior and thus, together with the presence of austenite phase, suggests that the Ni-rich films under study are superelastic.

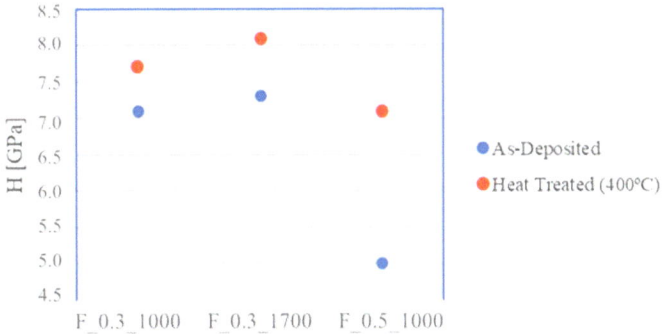

Figure 6. Hardness of as-deposited and heat-treated films.

4. Conclusions

Ni-rich films were produced by magnetron sputtering from a NiTi target using different deposition parameters. The films had to be heat-treated to obtain the NiTi shape memory alloy. The characterization of the as-deposited and heat-treated films allowed for the following conclusions to be taken:

(i) The increase in the power applied to the NiTi target results in films with a slightly higher Ni content, while the increase in pressure had the opposite effect.

(ii) The as-deposited films present a cauliflower-like surface morphology (except for the film deposited at a higher pressure), while the cross-section morphology of most films is columnar with traces of vein-like features.

(iii) The power applied to the NiTi target and the pressure do not seem to influence the roughness, with all the films presenting Sa and Sq values below 4 and 5 nm, respectively.

(iv) After deposition, all the films are amorphous, as is usual in NiTi sputtered films without substrate heating.

(v) Heat treatment is responsible for grain growth, thickness reduction, and more compact films. The vein-like cross-section morphology is dominant in the F_0.3_1000_HT and F_0.3_1700_HT films.

(vi) The 400 °C heat treatment temperature is enough to promote the crystallization of the films with the formation of austenitic B2-NiTi as a major phase together with Ni-rich precipitates.

(vii) Due to the formation of intermetallic phases, the heat treatment results in an increase in hardness in the films. Heat treatment also increases the Young's modulus of the films. The H/Er ratio of the heat-treated films is similar to that found in the literature on superelastic NiTi films.

(viii) The elastic recovery of the Ni-rich austenitic films is higher in comparison with the corresponding as-deposited amorphous films.

Considering the conclusions above and the versatility of the sputtering technique, the high Ni content NiTi films produced show potential for MEMS applications.

Author Contributions: Conceptualization, P.F.R. and A.S.R.; methodology, A.V.F., D.S., P.F.R. and A.S.R.; formal analysis, P.F.R. and A.S.R.; investigation, A.V.F., P.F.R. and A.S.R.; writing—original draft preparation, A.V.F.; writing—review and editing, D.S., P.F.R. and A.S.R.; supervision, P.F.R. and A.S.R. All authors have read and agreed to the published version of the manuscript.

Funding: This research was funded by national funds through FCT—Fundação para a Ciência e a Tecnologia, under projects UIDB/EMS/00285/2020 and LA/P/0112/2020.

Institutional Review Board Statement: Not applicable.

Informed Consent Statement: Not applicable.

Data Availability Statement: Data are contained within the article.

Acknowledgments: We acknowledge DESY (Hamburg, Germany), a member of the Helmholtz Association HGF, for the provision of the experimental facilities. Parts of this research were carried out at PETRA III, and we would like to thank Norbert Schell and Emad Maawad for their assistance in using Hereon—X-ray Diffraction/Small Angle X-ray Scattering (EH1). Beamtime was allocated for proposal I-20221253 EC.

Conflicts of Interest: The authors declare no conflicts of interest.

References

1. Khanlari, K.; Ramezani, M.; Kelly, P. 60NiTi: A Review of Recent Research Findings, Potential for Structural and Mechanical Applications, and Areas of Continued Investigations. *Trans. Indian Inst. Met.* **2018**, *71*, 781–799. [CrossRef]
2. Xu, G.X.; Zheng, L.J.; Zhang, F.X.; Zhang, H. Influence of solution heat treatment on the microstructural evolution and mechanical behavior of 60NiTi. *J. Alloys Compd.* **2019**, *775*, 698–706. [CrossRef]
3. Zhang, F.; Zheng, L.; Wang, F.; Zhang, H. Effects of Nb additions on the precipitate morphology and hardening behavior of Ni-rich Ni55Ti45 alloys. *J. Alloys Compd.* **2018**, *735*, 2453–2461. [CrossRef]
4. Khanlari, K.; Shi, Q.; Li, K.; Hu, K.; Tan, C.; Zhang, W.; Cao, P.; Achouri, I.E.; Liu, X. Fabrication of Ni-Rich 58NiTi and 60NiTi from Elementally Blended Ni and Ti Powders by a Laser Powder Bed Fusion Technique: Their Printing, Homogenization and Densification. *Int. J. Mol. Sci.* **2022**, *23*, 9495. [CrossRef] [PubMed]
5. Li, Y.; Dai, J.; Song, Y. Research Progress of First Principles Studies on Oxidation Behaviors of Ti-Al Alloys and Alloying Influence. *Metals* **2021**, *11*, 985. [CrossRef]
6. Tillmann, W.; Momeni, S. In-situ annealing of NiTi thin films at different temperatures. *Sens. Actuators A Phys.* **2015**, *221*, 9–14. [CrossRef]
7. Sharma, S.K.; Mohan, S. Influence of annealing on structural, morphological, compositional and surface properties of magnetron sputtered nickel–titanium thin films. *Appl. Surf. Sci.* **2013**, *282*, 492–498. [CrossRef]
8. Otsuka, K.; Ren, X. Physical metallurgy of Ti–Ni-based shape memory alloys. *Prog. Mater. Sci.* **2005**, *50*, 511–678. [CrossRef]
9. Young, M.L.; Wagner, M.-X.; Frenzel, J.; Schmahl, W.W.; Eggeler, G. Phase volume fractions and strain measurements in an ultrafine-grained NiTi shape-memory alloy during tensile loading. *Acta Mater.* **2010**, *58*, 2344–2354. [CrossRef]
10. Thornton, J.A. The microstructure of sputter-deposited coatings. *J. Vac. Sci. Technol. A Vac. Surf. Film.* **1986**, *4*, 3059–3065. [CrossRef]
11. Tillmann, W.; Momeni, S. Deposition of superelastic composite NiTi based films. *Vacuum* **2014**, *104*, 41–46. [CrossRef]
12. Bai, X.; Cai, Q.; Xie, W.; Zeng, Y.; Chu, C.; Zhang, X. In-situ crystalline TiNi thin films deposited by HiPIMS at a low substrate temperature. *Surf. Coat. Technol.* **2023**, *455*, 129196. [CrossRef]
13. Kumar, A.; Singh, D.; Kaur, D. Grain size effect on structural, electrical and mechanical properties of NiTi thin films deposited by magnetron co-sputtering. *Surf. Coat. Technol.* **2009**, *203*, 1596–1603. [CrossRef]
14. Fu, Y.; Huang, W.; Du, H.; Huang, X.; Tan, J.; Gao, X. 26 Characterisation of TiNi Shape-Memory Alloy Thin Films for MEMS Applications. *Surf. Coat. Technol.* **2001**, *145*, 107–112. [CrossRef]
15. Marupalli, B.C.G.; Behera, A.; Aich, S. A Critical Review on Nickel–Titanium Thin-Film Shape Memory Alloy Fabricated by Magnetron Sputtering and Influence of Process Parameters. *Trans. Indian Inst. Met.* **2021**, *74*, 2521–2540. [CrossRef]
16. Oliver, W.C.; Pharr, G.M. Measurement of hardness and elastic modulus by instrumented indentation: Advances in understanding and refinements to methodology. *J. Mater. Res.* **2004**, *19*, 3–20. [CrossRef]
17. Logothetidis, S.; Charitidis, C. Elastic properties of hydrogen-free amorphous carbon thin films and their relation with carbon–carbon bonding. *Thin Solid Films* **1999**, *353*, 208–213. [CrossRef]
18. Fasaki, I.; Koutoulaki, A.; Kompitsas, M.; Charitidis, C. Structural, electrical and mechanical properties of NiO thin films grown by pulsed laser deposition. *Appl. Surf. Sci.* **2010**, *257*, 429–433. [CrossRef]

19. Hammersley, A.P.; Svensson, S.O.; Hanfland, M.; Fitch, A.N.; Hausermann, D. Two-dimensional detector software: From real detector to idealised image or two-theta scan. *High Press. Res.* **1996**, *14*, 235–248. [CrossRef]
20. Shih, C.-L.; Lai, B.-K.; Kahn, H.; Phillips, S.M.; Heuer, A.H. A robust co-sputtering fabrication procedure for TiNi shape memory alloys for MEMS. *J. Microelectromech. Syst.* **2001**, *10*, 69–79. [CrossRef]
21. Tillmann, W.; Momeni, S. Comparison of NiTi thin films sputtered from separate elemental targets and Ti-rich alloy targets. *J. Mater. Process. Technol.* **2015**, *220*, 184–190. [CrossRef]
22. Miyazaki, S.; Ishida, A. Martensitic transformation and shape memory behavior in sputter-deposited TiNi-base thin films. *Mater. Sci. Eng. A* **1999**, *273–275*, 106–133. [CrossRef]
23. Gachon, J.-C.; Rogachev, A.; Grigoryan, H.; Illarionova, E.; Kuntz, J.-J.; Kovalev, D.; Nosyrev, A.; Sachkova, N.; Tsygankov, P. On the mechanism of heterogeneous reaction and phase formation in Ti/Al multilayer nanofilms. *Acta Mater.* **2005**, *53*, 1225–1231. [CrossRef]
24. Apreutesei, M.; Steyer, P.; Billard, A.; Joly-Pottuz, L.; Esnouf, C. Zr–Cu thin film metallic glasses: An assessment of the thermal stability and phases' transformation mechanisms. *J. Alloys Compd.* **2015**, *619*, 284–292. [CrossRef]
25. Chang, J.-C.; Lee, J.-W.; Lou, B.-S.; Li, C.-L.; Chu, J.P. Effects of tungsten contents on the microstructure, mechanical and anticorrosion properties of Zr–W–Ti thin film metallic glasses. *Thin Solid Films* **2015**, *584*, 253–256. [CrossRef]
26. Rodrigues, P.F.; Fernandes, F.B.; Magalhães, R.; Camacho, E.; Lopes, A.; Paula, A.; Basu, R.; Schell, N. Thermo-mechanical characterization of NiTi orthodontic archwires with graded actuating forces. *J. Mech. Behav. Biomed. Mater.* **2020**, *107*, 103747. [CrossRef]
27. Hou, H.; Hamilton, R.F.; Horn, M.W. Crystallization of nanoscale NiTi alloy thin films using rapid thermal annealing. *J. Vac. Sci. Technol. B Nanotechnol. Microelectron. Mater. Process. Meas. Phenom.* **2016**, *34*, 06KK01. [CrossRef]
28. Somsen, C.; Zähres, H.; Kästner, J.; Wassermann, E.; Kakeshita, T.; Saburi, T. Influence of thermal annealing on the martensitic transitions in Ni–Ti shape memory alloys. *Mater. Sci. Eng. A* **1999**, *273–275*, 310–314. [CrossRef]
29. Otsuka, K.; Wayman, C.M. Mechanism of Shape Memory Effect and Superelasticity. In *Shape Memory Materials*, 1st ed.; Cambridge University Press: New York, NY, USA, 1998; ISBN 0-521-44487-X.
30. Yu, H.; Qiu, Y.; Young, M.L. Influence of Ni4Ti3 precipitate on pseudoelasticity of austenitic NiTi shape memory alloys deformed at high strain rate. *Mater. Sci. Eng. A* **2021**, *804*, 140753. [CrossRef]
31. Di Egidio, G. Evaluation by nanoindentation of the influence of heat treatments and the consequent induced microstructure on the mechanical response of the heat-treated L-PBF AlSi10Mg alloy. *Metall. Ital.* **2022**, *6*, 8–16.
32. Ni, W.; Cheng, Y.-T.; Lukitsch, M.; Weiner, A.M.; Lev, L.C.; Grummon, D.S. Novel layered tribological coatings using a superelastic NiTi interlayer. *Wear* **2005**, *259*, 842–848. [CrossRef]

Disclaimer/Publisher's Note: The statements, opinions and data contained in all publications are solely those of the individual author(s) and contributor(s) and not of MDPI and/or the editor(s). MDPI and/or the editor(s) disclaim responsibility for any injury to people or property resulting from any ideas, methods, instructions or products referred to in the content.

Article

The Roughness Effect on the Preparation of Durable Superhydrophobic Silver-Coated Copper Foam for Efficient Oil/Water Separation

Aikaterini Baxevani, Fani Stergioudi * and Stefanos Skolianos

Physical Metallurgy Laboratory, School of Mechanical Engineering, Aristotle University of Thessaloniki, GR-54124 Thessaloniki, Greece; ampaxeva@meng.auth.gr (A.B.); skol@meng.auth.gr (S.S.)
* Correspondence: fstergio@auth.gr

Abstract: In recent decades, there has been a significant interest in superhydrophobic coatings owing to their exceptional properties. In this research work, a superhydrophobic coating was developed on copper foams with a different roughness via immersion in $AgNO_3$ and stearic acid solutions. The resulting foams exhibited water contact angles of 180°. Notably, surface roughness of the substrate influenced the development of silver dendrites and stearic acid morphologies, leading to different structures on rough and smooth copper foams. Separation efficiency was maintained above 94% for various pollutants, suggesting good stability and durability, irrespective of the substrate's roughness. Conversely, absorption capacity was influenced by surface roughness of the substrate, with smooth copper foams demonstrating higher absorption values, primarily due to its uniform porosity and microstructure, which allowed for efficient retention of pollutants. Both copper foams exhibited excellent thermal and chemical stability and maintained their hydrophobic properties even after a 40 h exposure to harsh conditions. Mechanical durability of modified copper foams was tested by dragging and in ultrasounds, exhibiting promising results. The samples with the smooth substrate demonstrated improved coating stability.

Keywords: superhydrophobicity; copper foam; durability; stability; oil/water separation; roughness; absorption capacity

Citation: Baxevani, A.; Stergioudi, F.; Skolianos, S. The Roughness Effect on the Preparation of Durable Superhydrophobic Silver-Coated Copper Foam for Efficient Oil/Water Separation. *Coatings* **2023**, *13*, 1851. https://doi.org/10.3390/coatings13111851

Academic Editor: Alexandra Muñoz-Bonilla

Received: 22 September 2023
Revised: 21 October 2023
Accepted: 24 October 2023
Published: 27 October 2023

Copyright: © 2023 by the authors. Licensee MDPI, Basel, Switzerland. This article is an open access article distributed under the terms and conditions of the Creative Commons Attribution (CC BY) license (https://creativecommons.org/licenses/by/4.0/).

1. Introduction

In the past few decades, there has been a notable interest in superhydrophobic coatings due to their remarkable characteristics and wide range of applications, including anti-icing, anti-fogging, self-cleaning capability and the ability to separate oil and water, which contribute significantly to their widespread usage [1–3]. Water contamination due to oil exploration, transportation, refining and effluents from various industries leads to severe environmental and economic consequences [4,5]. Conventional oil–water separation strategies including air flotation, in situ burning, oil-absorbing materials, oil skimmers, flocculation, gravity settling, centrifugation, gas flotation and electrochemical techniques suffer from limitations such as low selectivity, prolonged separation time, complex separation processes and inefficiency in large-scale applications [6–8]. Hence, it is crucial to develop a low-cost and efficient strategy to achieve complete separation of oil/water mixtures.

Taking inspiration from the exceptional water-repellent properties exhibited with natural elements like lotus leaves, desert beetles and water striders, artificial superhydrophobic surfaces have garnered significant interest [9–12]. Lotus leaves have emerged as a symbol of superhydrophobicity and self-cleaning surfaces, giving rise to what is known as the 'lotus effect'. While several other plants boast superhydrophobic surfaces with nearly identical contact angles, the lotus stands out for its superior stability and the impeccable quality of its water-repellent properties. This exceptional performance is attributed to the unique shape and density of the papillae, which result in an exceptionally reduced contact area between the leaf surface and water droplets [10,13,14]. The wettability of a solid surface is influenced by two crucial factors: surface chemical compositions and micro-nanostructures [15,16]. Over time, various methods have been developed to prepare superhydrophobic materials including hydrothermal techniques, chemical etching, electrochemical deposition, the sol-gel method, spray coating, self-assembly and laser etching [15–18]. These preparation approaches primarily rely on these strategies: roughening hydrophobic surfaces and modifying rough surfaces with substances possessing low surface energy. The development of a straightforward, one-step technology for acquiring high-performance and stable superhydrophobic materials is immensely desirable [19,20].

In recent years, there has been significant interest in three-dimensional (3D) porous materials, such as organic sponges, metallic foams as well as aerogels, which offer large specific surface areas plus porosity and are considered ideal substrates for developing super-wettability materials for water/oil separation [21–24]. Copper foam is a favorable choice as a substrate for oil/water separation since it is highly amenable to various surface modifications and treatments, owning to its electrical conductivity, which allows for the integration of various functional coatings and nanomaterials. Moreover, copper foam is known for its mechanical robustness. It can withstand the handling and manipulation required during the fabrication process and can maintain its structural integrity in practical applications. Additionally, copper foams are widely available and relatively affordable, which can contribute to cost-effective fabrication methods for oil/water separation materials.

To achieve superhydrophobicity, the surface of the copper foam can be micro/nanostructured and then modified using thiols or silanes or fatty acids. This can be achieved using several methods such as chemical etching, laser ablation or chemical treatments. Xin et al. [25] created superhydrophobic and superhydrophilic sides on copper foam using laser ablation and chemical modification, involving 1H,1H,2H,2H-perfluorodecyltriethoxysilane (FAS) and graphene oxide. Zhang et al. [26] created a surface layer of HKUST-1, which was grown in situ on the copper foam and then modified with 1-Hexadecanethiol, resulting in superhydrophobicity. Song et al. [27] created superhydrophobic copper meshes by immersing them in an $AgNO_3$ solution, followed by 1-dodecanethiol immersion, which formed a superhydrophobic surface on the copper. Zhang et al. [28] adopted a procedure that involved chemical etching in $FeCl_3$/HCl solutions to create rough structures on copper foam and subsequent modification with four sulfhydryl compounds to lower surface energies. In Xu et al.'s study [29], copper foams were cleaned and immersed in an ethanolic stearic acid solution to create superhydrophobic surfaces. Zhou et al. [30] employed a process involving dopamine, $AgNO_3$ reduction and n-dodecyl mercaptan to create superhydrophobic copper foam. In Chen et al.'s study [31], copper foam was anodized in NaOH, modified with APTES and carbon nanotubes (CNTs) and immersed in silanes to achieve superhydrophobicity.

Hydrophobic nanoparticles or nanostructures incorporated into the coating that can create a rough surface structure, to increase water repellency, are also reported in the literature. Gao et al. [32] electrodeposited CeO_2 nanostructures on copper foam, followed by immersion in an n-dodecyl mercaptan solution to achieve superhydrophobicity. Li et al. [33] fabricated superhydrophobic surfaces by growing $Cu(OH)_2$ nanowires and CuS nanostructures on copper foam using electrodeposition, followed by chemical modification. Zhu et al. [34] created superhydrophobic copper foams coated with various patterned nanostructures, including $Cu(OH)_2$ nanoneedles, ZnO nanocones and ZnO nanorods. In Rong et al.'s study [35], copper foam was immersed in a mixed solution containing $(NH_4)_2S_2O_8$ and Na_2HPO_4, resulting in the growth of $Cu_3(PO_4)_2 \cdot H_2O$ nanosheets on the surface. In Zhu et al.'s study [36], copper foams were coated with a composite material through a spray-on method using hydroxyl-functionalized multi-wall carbon nanotubes and water-based melamine formaldehyde. More complicated methods such as in Liu et al.'s study [37], who constructed superhydrophobic Janus separation materials using commercially available copper foam and silane agents through a controlled hydrophobic molecular vapor deposition process, are also reported.

The provided techniques for creating superhydrophobic copper foams indeed show promise for separating water and oil mixtures. However, there are some disadvantages that need to be addressed. Many of these techniques involve multiple steps and processes, which can be time-consuming. Some methods require specialized equipment, which may not be readily available or cost-effective for widespread adoption. The utilization of expensive materials or reagents is also an essential consideration for practical applications. Most importantly, there is a lack of understanding of interaction mechanisms.

Though there is an abundance of research related to the properties of hydrophobic coatings, when applied to surfaces, a notable void in the literature pertains to hydrophobic metal foams. As mentioned, there is a need for a better understanding of the interaction mechanisms on superhydrophobic copper foams. Additionally, the majority of studies focus on the roughness of the coating and its impact on hydrophobicity. Nonetheless, there exists a noticeable dearth of the literature concerning substrate roughness and its influence on the morphology of the hydrophobic coating.

In this research work, the alteration of the superhydrophobic coating's structure was thoroughly investigated, focusing solely on variations in the substrate's roughness. The production process of the superhydrophobic coating remains identical and is extensively detailed in our previous work [38]. Another crucial aspect examined was the mechanical resilience of the developed hydrophobic coatings. Numerous research studies are related to the growth of superhydrophobic coatings, but the lack of mechanical durability along with satisfying chemical and thermal stability hinder their widespread application.

In this research work, an effective superhydrophobic coating was successfully developed on copper foams with varying degrees of roughness, via immersion in $AgNO_3$ and stearic acid solutions. Silver nitrate displays a swift reaction with copper and even a small amount is sufficient to adequately coat the surface. Consequently, the overlay process was simple, fast and cost-effective. Silver nitrate was employed to enhance the surface micro-roughness, a pivotal element in hydrophobic coating fabrication [39,40]. In addition, fatty acids like stearic acid, palmitic acid, lauric acid and oleic acid are renowned for their capacity to form coordinating bonds with metal nanoparticles and thus for improving the performance of superhydrophobic materials [41]. To test the coating's chemical and thermal stability, contact angle measurements were conducted after its exposure in low and high temperatures as well as in acidic and alkaline aqueous mixtures. To assess the mechanical resilience of the altered copper foams, samples were subjected to abrasion with 600-grit SiC paper with various weights on them. Samples also underwent ultrasonic treatment and retained their hydrophobic properties. In terms of absorption capacity, the superhydrophobic copper foams with a smooth substrate exhibited a greater ability to absorb. Separation efficiency remained above 94% for a range of contaminants after 10 separation cycles, with both samples indicating excellent stability and durability. The durability of the superhydrophobic copper foam, as well as the cost-effectiveness of the chosen materials and processes, should be considered as an asset in potential use in filtration applications.

2. Materials and Methods

2.1. Materials

Silver nitrate was obtained from Panreac (Barcelona, Spain) and stearic acid ($C_{18}H_{36}O_2$) was a product of Merck (Darmstadt, Germany). Rough copper metal foams were provided by Metafoam Technologies (Brossard, Canada) and smooth copper foams were purchased from Hui Rui Siwang (Hengshui City, China). Both copper foams have 94% porosity and are in the form of 10 cm × 10 cm × 1.6 mm sheets. Ethanol was purchased from Central-chem (Bratislava, Slovakia). Sodium hydroxide (NaOH) and hydrochloric acid (HCl), used in chemical stability tests, are products of Honeywell (Seelze, Germany) and Panreac (Barcelona, Spain), respectively.

2.2. Preparation of Superhydrophobic Copper Foam

Copper foam sheets with different struts' roughness (rough and smooth), measuring 15 mm × 15 mm × 1.6 mm, underwent ultrasonic cleaning using acetone and anhydrous ethanol. Smooth and rough substrates were produced via electrodeposition and sintering methods, respectively. Pretreated copper foam samples were submerged in a 20 mm ethanolic solution of silver nitrate ($AgNO_3$) at 50 °C for 20 min to enhance the micro/nanoroughness of the foams. Subsequently, they were immersed in a 15 mm ethanolic solution of stearic acid for 50 min to reduce surface energy. The optimum parameters were found and analyzed in previous research work [38]. The achieved water contact angle (WCA) of the modified copper foams was 180°.

2.3. Characterization and Testing

Surface morphology of the superhydrophobic copper foams and substrates' roughness measurements were examined through a scanning electron microscope (Phenom ProX desktop SEM, Thermo Fisher Scientific, Eindhoven, The Netherlands). The microstructure and phases formed on superhydrophobic films were characterized using an energy dispersive spectrometer (EDS). The foam substrates' roughness measurements were examined through 3D EDS Tomography, which is a technique of a scanning electron microscope (Phenom ProX desktop SEM, Thermo Fisher Scientific). Static water contact angle measurements were conducted to evaluate wettability. Specifically, 8 µL water droplets were deposited onto the surfaces under examination in standard ambient conditions. A laboratory incubator was used for the thermal treatment of the coated samples (CLIMACELL incubator, MMM Group, Munich, Germany). Oil viscosities were measured using a rheometer (TA Discovery Hybrid Rheometer HR30). The calculation of the separation efficiency (s) involves dividing the weight of oil collected in the filtrate tank after separation (m_1) by the weight of oil added before separation (m_0). Separation efficiency was calculated as follows (Equation (1)):

$$\text{Separation efficiency} = \frac{m_1}{m_0} \times 100\% \qquad (1)$$

Absorption capacity was calculated through Equation (2):

$$\text{Absorption capacity} = \frac{W_1 - W_2}{W_1} \qquad (2)$$

where W_1 is the weight of the foam before the absorption and W_2 is the weight of the foam after the absorption.

3. Results and Discussion

3.1. Analysis of the Developed Morphologies on Coated Copper Foams

Surface roughness plays a crucial role in the shape of developed structures. Observed morphologies of silver dendrites, in this research work, can be divided into two categories: tree- and moss-like on rough and smooth copper foam, respectively [42,43]. Figure 1a,b show a schematic illustration of these morphologies. Figure 1c,d depict the mean value of roughness of the surface irregularities for rough and smooth copper foams, respectively. At least 10 measurements were conducted with each sample. R_z represents the mean magnitude of the heights of the five highest peaks and the depths of the five deepest valleys found within the specified measurement distance. R_a denotes the mean value obtained by calculating the absolute heights of the surface profile and taking their arithmetic average over the specified measurement distance. Rough copper foam has higher values of inherent roughness than the smooth substrate. Figure 1e,f depict SEM images of the copper foams used. The inherent roughness plays a crucial role in the shape of developed structures. Tree-like silver dendrites and flower-like structures grow on rougher copper foams, while nanowire morphologies and moss-like Ag dendrites are commonly found at smoother samples (see Figure 2).

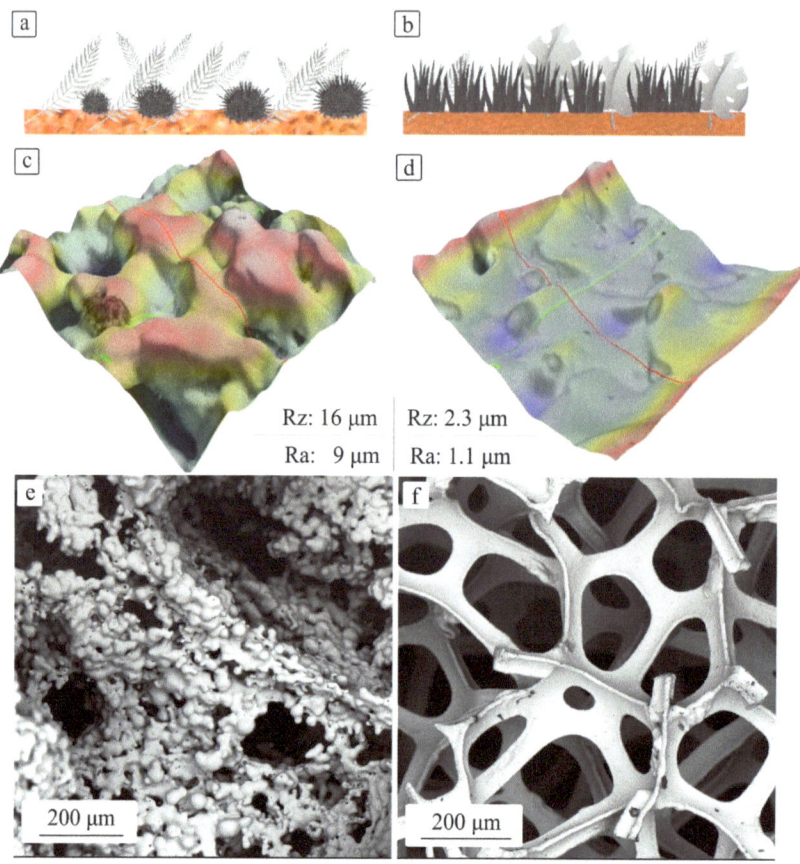

Figure 1. Schematic illustration of developed morphologies (**a**) on rough copper foam and (**b**) on smooth copper foam. Strut roughness of (**c**) rough copper foam and (**d**) smooth copper foam. SEM images of (**e**) rough copper foam and (**f**) smooth copper foam.

The emergence of Ag dendrites can be elucidated through a straightforward reaction mechanism. In this galvanic reaction, the reduction reaction $Ag^+ + e^- = Ag$ occurs exclusively on the metal surface, initially on copper (Cu) and subsequently on the newly formed silver (Ag) [44]. The necessary electrons are supplied with the oxidation reaction $Cu = Cu^{2+} + 2e^-$, which takes place on the Cu surface, leading to the release of Cu^{2+} ions into the solution. Electron transport within the metal enables the reduction of Ag^+ ions on the Ag surface. The spontaneous reduction of silver ions using copper can be described with the following reaction: $2Ag^+ + Cu \rightarrow 2Ag + Cu^{2+}$ [45]. By immersing copper substrates in a silver nitrate solution, a series of thin silver films were deposited onto the substrates. Due to the minimal roughness of smooth copper foam, the reaction occurs at a slow pace, resulting in the formation of tiny voids within the film that are nearly imperceptible. However, when the roughness of the strut increases (rough copper foam), the reaction accelerates, leading to the emergence of larger fractal-like silver structures. As a result, the size of the voids between these structures also expands. (See Figure 1a,b).

The formation of silver dendrites has been the subject of extensive research studies. Silver dendrites are known for their ability to generate extensive specific surface areas [46]. Dendrite growth can be understood through various models, including deposition, diffusion and aggregation (DDA); diffusion-limited aggregation (DLA); oriented attachment (OA); and cluster–cluster aggregation (CCA) [47]. The anisotropic crystal growth of dendrites occurs when kinetic factors dominate over thermodynamic factors, resulting in a non-equilibrium condition. The diffusion-limited aggregation model describes the fortuitous aggregation and asymmetric growth of nanoparticles, leading to the formation of fractal structures when the growth rate is limited by the diffusion rate of solute atoms (also known as random walkers) to the reaction interface [48]. Conversely, oriented attachment involves the spontaneous self-assembly and alignment of adjacent particles, resulting in a shared crystallographic orientation and the joining of these particles at a planar interface. The Cluster–Cluster Agglomeration (CCA) model, a dynamic cluster model, utilizes random motions to generate larger clusters by repeating the diffusion-limited aggregation (DLA) process [49].

Figure 2. SEM images of superhydrophobic coating developed at 20 min immersion time in AgNO$_3$ and 50 min in stearic acid solution, on rough copper foam (**a**–**c**), and on smooth copper foam (**d**–**f**).

In general, a fully formed main structure characterizes the dendritic silver nanostructure, showcasing distinct branches, stems and leaves. Certain secondary branches of the dendrites steadily extend and evolve into fresh trunks. These dendrite formations possessing such a shape are referred to as secondary branch structures. It is important to note that explaining the formation of Ag dendrites by considering only one of these mechanisms oversimplifies the actual phenomena. Therefore, it is common to invoke multiple mechanisms simultaneously to better understand the growth process of Ag dendrites [50].

Dendritic fractals can have one, two or multiple branches, extending from worm-like structures as primary dendrite arms as schematically shown in Figure 3. In the case of silver, the slightly elongated structures act as primary dendrite arms, serving as a central trunk for the growth of secondary dendritic structures through branching [51]. The transformation from branched structures to dendritic structures in silver is facilitated with the continuous supply of a new portion of the AgNO$_3$ solution to the surface of the primary dendrite arm [52].

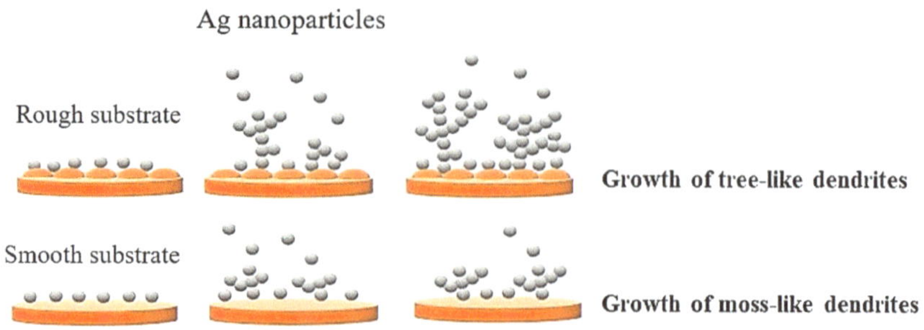

Figure 3. Schematic illustration of dendrites' growing mechanism on rough substrate and smooth substrate.

As the reaction time progresses, the concentration of silver ions diminishes, leading to a gradual reduction in the length of symmetrically shaped dendrites. Moreover, the dendrite structure undergoes degeneration, transitioning into a higher-order branched structure and the gaps between dendrites become narrower [43,53].

In the case of multilayer films of stearic acid, it is generally accepted that the first monolayer forms a chemical bond with the surface, while subsequent monolayers stack on top. A fatty acid becomes complexed with the surface metal atoms and forms the first monolayer alongside neighboring molecules [54–57]. The ordered arrangement of long-chain organic amphiphiles in monolayers is usually created using two main techniques: the Langmuir–Blodgett (LB) deposition method or through the process of spontaneous adsorption from a solution, commonly known as self-assembly (SA). There is a great similarity between monolayers produced through the self-assembly (SA) method and those created using the Langmuir–Blodgett (LB) method [58–61]. The adsorption process of stearic acid onto the silver surface is highly energetically favorable, making it relatively effortless to form complete monolayers by simply bringing the silver substrate into contact with the acid solution [47,48].

3.2. Oil–Water Separation

The oil–water separation performance of superhydrophobic copper foams with a different substrate roughness was conducted at 25 °C in a small tank (Figure 4a) with a volume ratio of $V_{oil}:V_{water}$ = 1:3. Nine different types of oil with different viscosities (Figure 4b) were used. Separation efficiency was maintained above 94% after 10 cycles of filtration (Figure 4c). The significance of viscosity lies in its influence on the separation efficiency and absorption capacity of superhydrophobic copper foams. The efficiency of separating high viscous oils is lower compared to less viscous oils, whereas this fact is not observed in terms of absorption (Figure 4d). All pollutants have lower density than water. The findings from the separation efficiency results, as depicted in Figure 4c, indicate that roughness holds little significance as a contributing factor, with the primary determinant of efficiency being the crucial role played by viscosity. Conversely, it should be noted that when it comes to absorption capacity, roughness is a determining factor. Figure 4d depicts the absorption capacity values of superhydrophobic copper foams with the smooth and rough substrate, respectively. Smooth copper foam displays significantly greater absorption values in contrast to the foam with the rough base. Due to the greater uniformity of the pores, it is able to maintain a satisfactory level of oil evenly over its entire surface, unlike foam with a rough substrate. Furthermore, the microstructure of the coating on the smooth foam exhibits greater uniformity, without sizable voids, consequently leading to improved retention of contaminants.

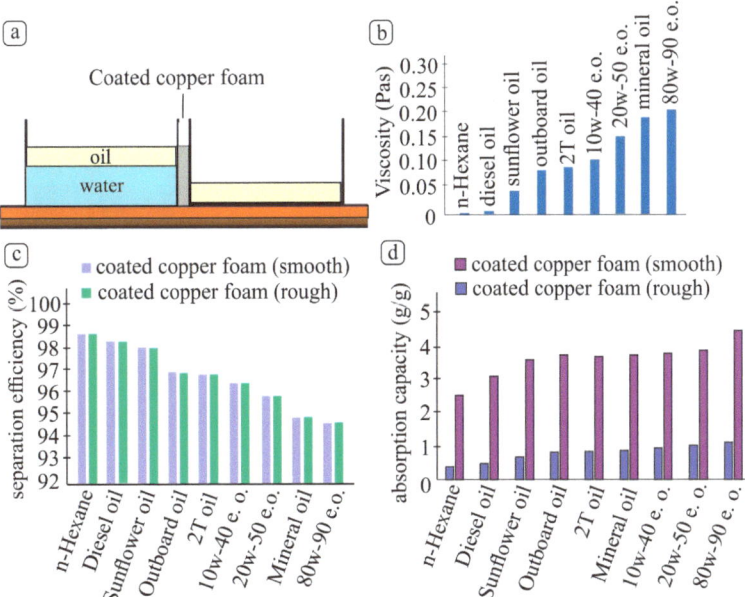

Figure 4. (**a**) Tank for oil–water separation, (**b**) oil viscosities used, (**c**) separation efficiency for different types of oil after 10 filtration cycles, (**d**) absorption capacity for various oils.

3.3. Thermal and Chemical Stability

To assess the environmental stability and durability of superhydrophobic copper foams, a comprehensive investigation was conducted to examine their response to exposure to different solutions and temperatures. Superhydrophobic copper foam samples were exposed to 100 °C in a laboratory incubator and to −15 °C in a cooling chamber, for 1, 4, 9, 16 and 40 h, respectively, in order to evaluate their stability under thermal conditions. Figure 5a illustrates WCA measurements for rough and smooth copper foam, taken after thermal loads, and the results show that their hydrophobic properties were maintained. It is noteworthy that in the smooth foam, the WCA values remain consistently high at around 140°, even after 40 h of exposure at both temperatures (100 °C and −15 °C), while the rough foam's contact angle decreased to 120° at 100 °C and 110° at −15 °C, after the same duration. Chemical stability of superhydrophobic copper foams was evaluated in acidic and alkaline solutions. The foam samples were immersed in HCl and NaOH solutions for 1, 4, 9, 16 and 40 h, correspondingly. Figure 5b illustrates the WCA values obtained after subjecting the samples to these chemical treatments. There is a consistent stabilization of the contact angle observed in both acidic and alkaline environments for both foams. Nonetheless, it is worth noting that the contact angles are notably higher in the foam with the smooth substrate. All the foams maintain their hydrophobic properties, even after 40 h of immersion.

Figure 6a–d depict the morphology of superhydrophobic smooth copper foams, after 40 h of exposure at 100 °C and −15 °C. Nanowire morphologies of stearic acid create aggregates, a fact that leads to an uneven distribution of the structure across the surface, and the leaf-like structure has undergone partial degradation. SEM images of the modified foams' morphologies following 40 h in the HCl and NaOH solutions are presented in Figure 6e–h. Regarding the morphologies when exposed to the acidic solution (Figure 6e,f), an extensive structural degradation was observed. Initial morphology of the coating's components was disrupted, leading to the formation of aggregates. Copper foam remains superhydrophobic as the substrate remains coated and no exfoliation of the coating was noted. Submerging the coated foam in an alkaline solution has a minor impact on its morphology, but WCA reduction is probably due to the decrease in the quantity of its constituents on the foam substrate (Figure 6g,h).

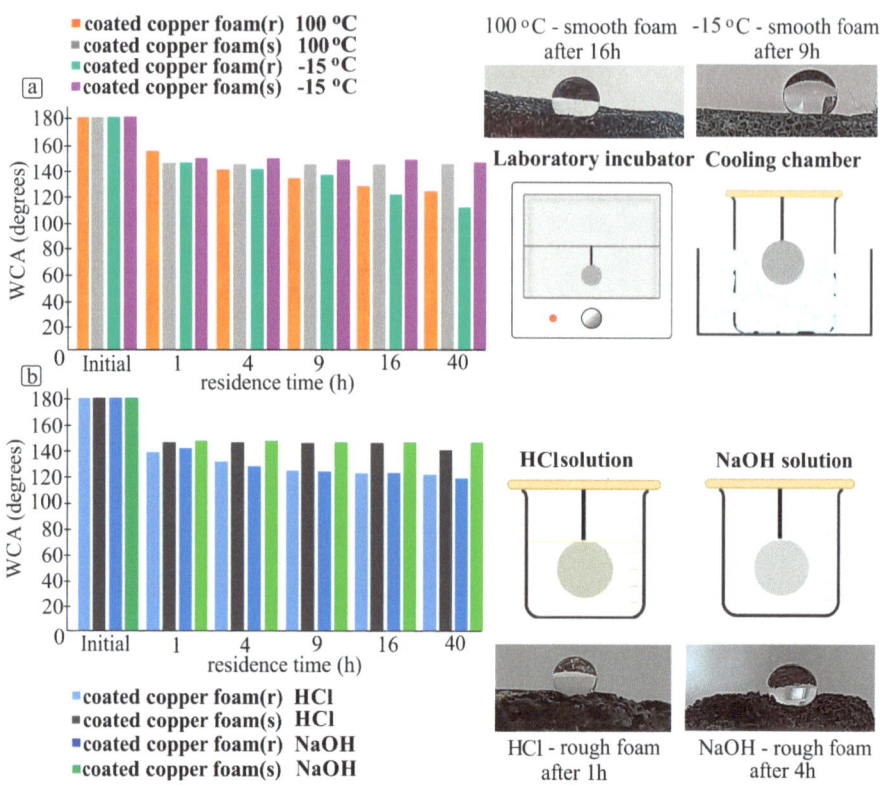

Figure 5. Contact angle measurements and indicative water droplet pictures (**a**) after thermal treatment at 100 °C and −15 °C and (**b**) after chemical treatment in HCl and NaOH solutions.

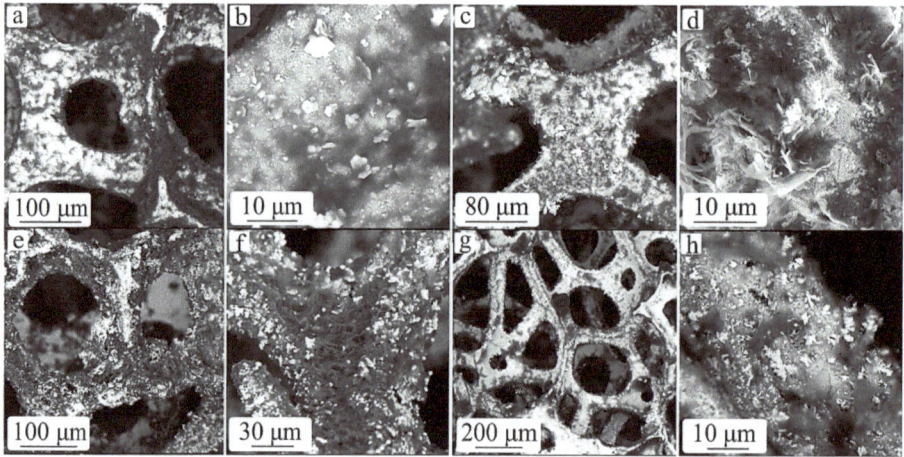

Figure 6. SEM images of smooth superhydrophobic copper foams, (**a**,**b**) after thermal treatment at 100 °C and (**c**,**d**) at −15 °C, and (**e**,**f**) after chemical treatment in HCl and (**g**,**h**) NaOH solutions, for 40 h.

Figure 7a,b illustrate the coating's structures on rough copper foam, after exposure for 40 h at 100 °C. Tree-like silver dendrites appear to have undergone slight melting and consolidation, while micro-flowers' petals undergo partial destruction and form aggregates. Regarding the exposure of the sample to −15 °C for 40 h (Figure 7c,d), there is a destruction as well as a detachment of dendritic morphologies and an extended degradation of micro-clusters. Dendritic branches separated from the stems and fused together, while micro-flowers were detached and completely lost their initial shape, after 40 h of residence in the HCl solution (Figure 7e,f). Figure 7g,h show both structures after 40 h of residence in the NaOH solution. The dendritic structures have undergone fragmentation in some places and aggregation in others, whilst micro-clusters have been thoroughly demolished and dispersed across different locations on the foam.

Figure 7. SEM images of rough superhydrophobic copper foams, (**a,b**) after thermal treatment at 100 °C and (**c,d**) at −15 °C, and after chemical treatment in (**e,f**) HCl solutions and (**g,h**) NaOH solutions, for 40 h.

3.4. Stability in Water and Sodium Chloride Solutions

To assess the viability of utilizing superhydrophobic foams for water purification applications involving the removal of pollutants such as oils, a comprehensive examination of their stability in water and the sodium chloride solution (3.5% w/v) was conducted. Figure 8a depicts copper foam's WCA measurements as a function of residence time in water and the sodium chloride solution. Both modified copper foams maintained their hydrophobic properties even after 40 h of immersion in both solutions. In comparison to immersion in water, the reduction in angle values is greater when the foams are submerged in the sodium chloride solution. During the initial hours of exposure, a decline in the angle is detected in both foams, whether they are immersed in water or a sodium chloride solution. Following approximately 10 h of immersion in the solutions, the angle in both foams reaches a stable state, maintaining a constant value even after 40 h in water and the NaCl solution. The foams retain their hydrophobic properties, as evidenced with the angle remaining close to 140°.

SEM images of silver-coated copper foams' morphologies after immersion in solutions for 40 h (Figure 8b–d) show that nanowire and dendritic structures on smooth copper foam have undergone reduction and partial degradation, while micro-flowers and tree-like morphologies on rough copper foam disintegrate and one structure adheres to another (Figure 8e–g). After exposure to the sodium chloride solution, deposits from the NaCl components are observed on the foam (Figure 8h–j). Figure 8k–m depict that silver dendrites broken down into small particles and micro-flowers altered in shape from their original state.

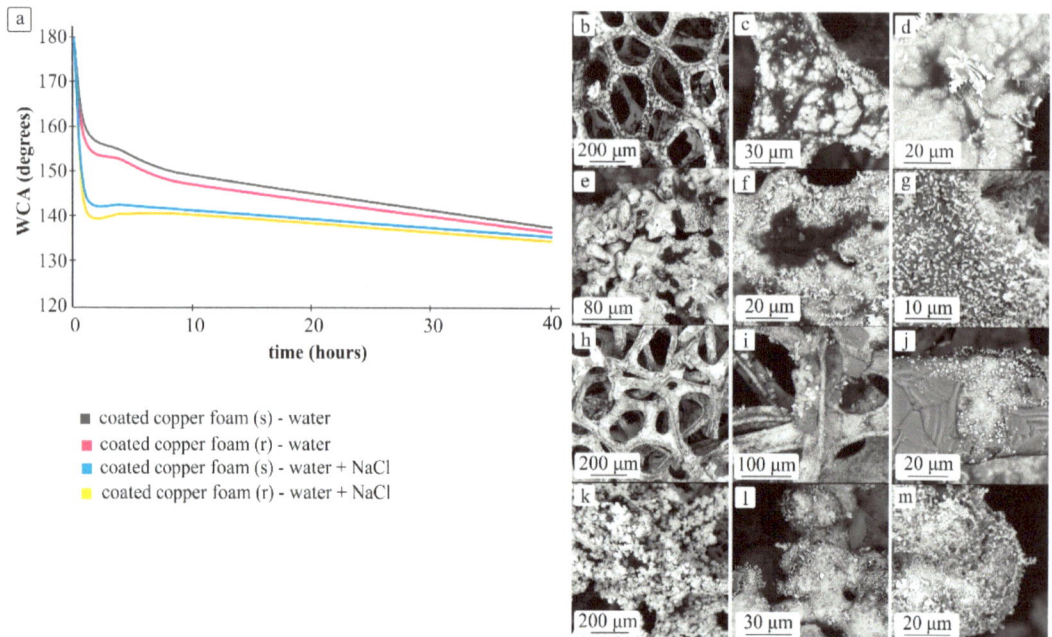

Figure 8. (**a**) Contact angle measurements after immersion in water and sodium chloride solution (SEM images of coated copper foams), (**b**–**d**) (smooth) after immersed in water, (**e**–**g**) (rough) after immersed in water, (**h**–**j**) (smooth) after immersed in sodium chloride solution, (**k**–**m**) (rough) after immersed in sodium chloride solution, for 40 h.

3.5. Mechanical Durability

Abrasion tests were carried out to evaluate the mechanical durability of the modified copper foams. Superhydrophobic copper foams were placed on a 600-grit SiC paper, using weights of 35 gr and 90 gr, and measurements of WCA were performed during multiple cycles of abrasion of a 20 cm length each (Figure 9a). After five cycles, a significant decrease in the contact angle is noted, but in the subsequent cycles, there is a substantial stabilization. This stability persists, maintaining the hydrophobicity in both foams for 35 gr and 90 gr. Hence, the substrate's surface roughness does not impact the mechanical stability of the coating during this test. Even after 40 cycles of abrasion, the samples maintained their hydrophobic properties. The weights applied are significantly larger in comparison to the weights of the foams, which amount to 0.2 gr for the rough foam and 0.07 gr for the smooth foam. It is evident that, for instance, 90 gr is 1285 times greater than the weight of the smooth foam. Consequently, both samples exhibit exceptional mechanical stability.

The stability of the superhydrophobic foams was also investigated using an ultrasonic device. The samples were positioned in the apparatus for various time durations. Samples exhibited hydrophobicity even after 90 min (Figure 9b). The foam with the smooth substrate exhibits slightly higher WCA values, demonstrating enhanced coating stability, due to the coating's uniformity.

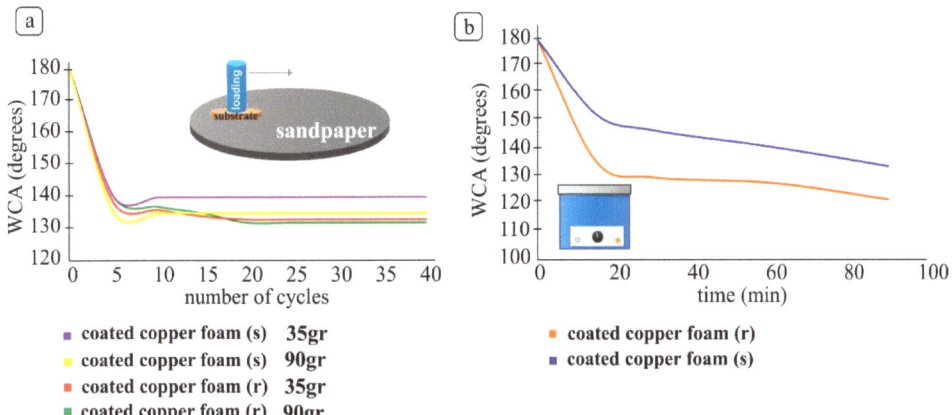

Figure 9. Contact angle measurements (**a**) after several cycles of abrasion on SiC paper with several weights on modified copper foams and (**b**) after various time durations of treatment in ultrasonic device.

4. Conclusions

A superhydrophobic coating was effectively developed on copper foams with different levels of roughness by immersing them in AgNO$_3$ and stearic acid solutions. The goal was to attain a long-lasting superhydrophobic copper foam featuring a water contact angle of 180 degrees, coupled with outstanding chemical and thermal resilience, as well as superior separation capability. Due to this fact, roughness of the substrate was evaluated with respect to the foam properties. The following conclusions can be drawn:

(1) As the strut's roughness increases, stearic acid's nanosheets and self-assembled Ag clusters grow larger in size and length. Particularly, the stearic acid nanosheets form and start to vertically expand, resembling chrysanthemum petals.

(2) The modified copper foams exhibited significant chemical and thermal stability. Specifically, WCA of the foam with the smooth substrate consistently stays high at approximately 140° even after 40 h of exposure at both temperatures (100 °C and −15 °C). In contrast, after the same time duration, the contact angle of the foam having a rough substrate decreases to 120° at 100 °C and 110° at −15 °C. A stable contact angle is consistently observed in both acidic and alkaline environments for both types of foams. However, contact angles for the samples with the smooth substrate have greater values. All the foams retain their hydrophobic characteristics, even following a 40 h immersion.

(3) Mechanical durability of modified copper foams was tested with dragging and in ultrasounds, exhibiting outstanding results. The samples with the smooth substrate show higher WCA values, indicating improved coating stability attributed to the uniformity of the coating.

(4) The separation efficiency of both foams remained above 94% for pollutants with several viscosities, indicating excellent stability and durability. Regarding absorption, the foam with the smooth substrate proved to be more efficient, retaining larger quantities of pollutants. The uniform porosity and coating's microstructure lacking significant voids leads to retaining the oil across its entire surface, a fact that improves its ability to retain contaminants.

Author Contributions: Conceptualization, F.S.; Methodology, A.B.; Validation, A.B.; Formal analysis, A.B.; Investigation, A.B.; Resources, F.S.; Data curation, A.B.; Writing—original draft, A.B.; Writing—review & editing, Visualization, A.B. and F.S.; Supervision, F.S.; Project administration, S.S. Funding acquisition, S.S. All authors have read and agreed to the published version of the manuscript.

Funding: This research was funded by the Research Committee, Aristotle University of Thessaloniki, grand number 89251.

Institutional Review Board Statement: Not applicable.

Informed Consent Statement: Not applicable.

Data Availability Statement: Data is contained within the article.

Conflicts of Interest: The authors declare no conflict of interest.

References

1. Chanwoo, P.; Taegun, K.; Yong, K.; Min, W.L.; Seongpil, A.; Sam, Y. Supersonically sprayed transparent flexible multifunctional composites for self-cleaning, anti-icing, anti-fogging, and anti-bacterial applications. *Compos. B* **2021**, *222*, 109070.
2. Yuxing, B.; Haiping, Z.; Yuanyuan, S.; Hui, Z.; Jesse, Z. Recent Progresses of Superhydrophobic Coatings in Different. Application Fields: An Overview. *Coatings* **2021**, *11*, 116.
3. Darband, G.B.; Aliofkhazraei, M.; Khorsand, S.; Sokhanvar, S.; Kaboli, A. Science and Engineering of Superhydrophobic Surfaces: Review of Corrosion Resistance, Chemical and Mechanical Stability. *Arab. J. Chem.* **2020**, *13*, 1763–1802. [CrossRef]
4. Rasouli, S.; Rezaei, N.; Hamedi, H.; Zendehboudi, S.; Duan, X. Superhydrophobic and superoleophilic membranes for oil-water separation application: A comprehensive review. *Mater. Des.* **2021**, *204*, 109599. [CrossRef]
5. Bhushan, B. Bioinspired oil–water separation approaches for oil spill clean-up and water purification. *Philos. Trans.* **2019**, *377*, 20190120. [CrossRef] [PubMed]
6. Gupta, R.K.; Dunderdale, G.J.; England, M.W.; Hozumi, A. Oil/water separation techniques: A review of recent progresses and future directions. *J. Mater. Chem.* **2017**, *5*, 16025–16058. [CrossRef]
7. Yue, H.; Yanji, Z.; Huaiyuan, W.; Chijia, W.; Hongwei, L.; Xiguang, Z.; Ruixia, Y.; Yiming, Z. Facile preparation of superhydrophobic metal foam for durable and high efficient continuous oil-water separation. *J. Chem. Eng.* **2017**, *322*, 157–166.
8. Yangyang, C.; Shengke, Y.; Qian, Z.; Dan, Z.; Chunyan, Y.; Zongzhou, W.; Runze, W.; Rong, S.; Wenke, W.; Yaqian, Z. Effects of Surface Microstructures on Superhydrophobic Properties and Oil-Water Separation Efficiency. *Coatings* **2019**, *9*, 69.
9. Collins, C.M.; Safiuddin, M. Lotus-Leaf-Inspired Biomimetic Coatings: Different Types, Key Properties, and Applications in Infrastructures. *Infrastructures* **2022**, *7*, 46. [CrossRef]
10. Ensikat, H.J.; Ditsche-Kuru, P.; Neinhuis, C.; Barthlott, W. Superhydrophobicity in perfection: The outstanding properties of the lotus leaf. *Beilstein J. Nanotechnol.* **2011**, *2*, 152–161. [CrossRef]
11. Parvate, S.; Chattopadhyay, S. Complex Polymeric Microstructures with Programmable Architecture via Pickering Emulsion-Templated In Situ Polymerization. *Langmuir* **2022**, *38*, 1406–1421. [CrossRef]
12. Mengru, J.; Qianli, X.; Zikang, C. A Review: Natural Superhydrophobic Surfaces and Applications. *J. Biomater. Nanobiotechnol.* **2020**, *11*, 110–149.
13. Aguilar-Morales, A.I.; Alamri, S.; Voisiat, B.; Kunze, T.; Lasagni, A.F. The Role of the Surface Nano-Roughness on the Wettability Performance of Microstructured Metallic Surface Using Direct Laser Interference Patterning. *Materials* **2019**, *12*, 2737. [CrossRef]
14. Goswami, A.; Pillai, S.C.; McGranaghan, G. Surface modifications to enhance dropwise condensation. *Surf. Interfaces* **2021**, *5*, 101143. [CrossRef]
15. Dingliang, D.; Jianhao, Q.; Lu, Z.; Hong, M.; Jianfeng, Y. Amino-functionalized Ti-metal-organic framework decorated BiOI sphere for simultaneous elimination of Cr(VI) and tetracycline. *J. Colloid Interface Sci.* **2022**, *607*, 933–941.
16. Tao, L.; Hebin, L.; Wenxuan, C.; Yankang, D.; Qingli, Q.; Wenjing, M.; Ranhua, X.; Chaobo, H. Blow-spun nanofibrous composite Self-cleaning membrane for enhanced purification of oily wastewater. *J. Colloid Interface Sci.* **2022**, *608*, 2860–2869.
17. Nia, L.; Zhua, C.; Zhanga, S.; Caia, P.; Airoudj, A.; Vonna, L.; Hajjar-Garreaub, S.; Chemtob, A. Light-induced crystallization-driven formation of hierarchically ordered superhydrophobic sol-gel coatings. *Prog. Org. Coat.* **2019**, *135*, 255–262. [CrossRef]
18. Yu, Z.; Zhentao, Z.; Junling, Y.; Yunkai, Y.; Huafu, Z. A Review of Recent Advances in Superhydrophobic Surfaces and Their Applications in Drag Reduction and Heat Transfer. *Nanomaterials* **2022**, *12*, 44. [CrossRef]
19. Sotoudeh; Mousavi, S.M.; Karimi, N.; Lee, B.J.; Abolfazli-Esfahani, J.; Manshadi, M.K.D. Natural and synthetic superhydrophobic surfaces: A review of the fundamentals, structures, and applications. *Alex. Eng. J.* **2023**, *68*, 587–609. [CrossRef]
20. Shuangshuang, X.; Qing, W.; Ning, W. Chemical Fabrication Strategies for Achieving Bioinspired Superhydrophobic Surfaces with Micro and Nanostructures: A Review. *Adv. Eng. Mater.* **2021**, *23*, 2001083.
21. Avrămescu, R.E.; Ghica, M.V.; Dinu-Pîrvu, C.; Prisada, R.; Popa, L. Superhydrophobic Natural and Artificial Surfaces—A Structural Approach. *Materials* **2018**, *11*, 866. [CrossRef] [PubMed]
22. Liping, D.; Meng, C.; Huiying, L.; Haochen, H.; Xia, L.; Yanqing, W. 3D multiscale sponges with plant-inspired controllable superhydrophobic coating for oil spill cleanup. *Prog. Org. Coat.* **2021**, *151*, 106075. [CrossRef]
23. Yihao, G.; Fangqin, C.; Zihe, P. Superwetting Polymeric Three Dimensional (3D) Porous Materials for Oil/Water Separation: A Review. *Polymers* **2019**, *11*, 806. [CrossRef]
24. Tao, L.; Hebin, L.; Wenxuan, C.; Yankang, D.; Qingli, Q.; Wenjing, M.; Ranhua, X.; Chaobo, H. Blow-spun Nanofibrous Membrane for Simultaneous Treatment of Emulsified Oil/WaterMixtures, Dyes, and Bacteria. *Langmuir* **2022**, *38*, 15729–15739.

25. Guoqiang, X.; Congyi, W.; Weinan, L.; Miaozheng, W.; Yu, H.; Youmin, R. Fabrication of super-wetting copper foam based on laser ablation for selective and efficient oil-water separation. *Surf. Coat. Technol.* **2021**, *424*, 127650.
26. Wanqing, Z.; Shaohua, W.; Wenlong, T.; Kang, H.; Cheng-xing, C.; Yalei, Z.; Yuping, Z.; Zheng, W.; Shouren, Z.; Lingbo, Q. Fabrication of a superhydrophobic surface using a simple in situ growth method of HKUST-1/copper foam with hexadecanethiol modification. *New J. Chem.* **2020**, *44*, 7065.
27. Song, Y.; Liu, Y.; Zhan, B.; Kaya, C.; Stegmaier, T.; Han, Z.; Ren, L. Fabrication of Bioinspired Structured Superhydrophobic and Superoleophilic Copper Mesh for Efficient Oil-water Separation. *J. Bionic Eng.* **2017**, *14*, 497–505. [CrossRef]
28. Yu-Ping, Z.; Jing-Hua, Y.; Ling-Li, L.; Cheng-Xing, C.; Ying, L.; Shan-Qin, L.; Xiao-Mao, Z.; Ling-Bo, Q. Facile Fabrication of Superhydrophobic Copper-Foam and Electrospinning Polystyrene Fiber for Combinational Oil–Water Separation. *Polymers* **2019**, *11*, 97. [CrossRef]
29. Jia, X.; Jinliang, X.; Yang, C.; Xianbing, J.; Yuying, Y. Fabrication of non-flaking, superhydrophobic surfaces using a one-step solution-immersion process on copper foams. *Appl. Surf. Sci.* **2013**, *286*, 220–227. [CrossRef]
30. Wei, Z.; Guangji, L.; Liying, W.; Zhifeng, C.; Yinlei, L. A facile method for the fabrication of a superhydrophobic polydopamine-coated copper foam for oil/water separation. *Appl. Surf. Sci.* **2017**, *413*, 140–148.
31. Zehao, C.; Jihao, Z.; Ting, Z.; Qing, T.; Yunjun, N.; Shouping, X.; Jiang, C.; Xiufang, W.; Pihui, P. Superhydrophobic copper foam bed with extended permeation channels for water-in-oil emulsion separation with high efficiency and flux. *J. Environ. Chem. Eng.* **2023**, *11*, 109018. [CrossRef]
32. Ruixi, G.; Xiang, L.; Tian, C.Z.; Like, Q.; Ying, L.; Shaojun, Y. Superhydrophobic Copper Foam Modified with n-Dodecyl Mercaptan-CeO$_2$ Nanosheets for Efficient Oil/Water Separation and Oil Spill Cleanup. *Ind. Eng. Chem. Res.* **2020**, *59*, 21510–21521.
33. Ji, L.; Ruixi, G.; Yuan, W.; Tian, C.Z.; Shaojun, Y. Superhydrophobic palmitic acid modified Cu(OH)$_2$/CuS nanocomposite-coated copper foam for efficient separation of oily wastewater. *Colloids Surf. A Physicochem. Eng.* **2022**, *637*, 128249. [CrossRef]
34. Haiyan, Z.; Lin, G.; Xinquan, Y.; Caihua, L.; Youfa, Z. Durability evaluation of superhydrophobic copper foams for long-term oil-water separation. *Appl. Surf. Sci.* **2017**, *407*, 145–155. [CrossRef]
35. Jian, R.; Tao, Z.; Fengxian, Q.; Jicheng, X.; Yao, Z.; Dongya, Y.; Yuting, D. Design and preparation of efficient, stable and superhydrophobic copper foam membrane for selective oil absorption and consecutive oil–water separation. *Mater. Des.* **2018**, *142*, 83–92. [CrossRef]
36. Haiyan, Z.; Doudou, L.; Mingjuan, C.; Xinquan, Y.; Youfa, Z. Sprayed superamphiphilic copper foams for long term recoverable oil-water separation. *Surf. Coat. Technol.* **2018**, *334*, 394–401. [CrossRef]
37. Chunhua, L.; Yun, P.; Conglin, H.; Yuzhen, N.; Jiaoping, S.; Yibao, L. Bioinspired Superhydrophobic/Superhydrophilic Janus Copper Foam for On-Demand Oil/Water Separation. *ACS Appl. Mater. Interfaces* **2022**, *14*, 11981–11988.
38. Baxevani, A.; Stergioudi, F.; Patsatzis, N.; Malletzidou, L.; Vourlias, G.; Skolianos, S. Preparation and Characterization of Stable Superhydrophobic Copper Foams Suitable for Treatment of Oily Wastewater. *Coatings* **2023**, *13*, 355. [CrossRef]
39. Stergioudi, F.; Baxevani, A.; Mavropoulos, A.; Skordaris, G. Deposition of Super-Hydrophobic Silver Film on Copper Substrate and Evaluation of Its Corrosion Properties. *Coatings* **2021**, *11*, 1299. [CrossRef]
40. Xiulan, L.; Xiaohong, H.; Yao, L.; Zhongxiang, B.; Chenchen, L.; Xiaobo, L.; Kun, J. In-situ growth of silver nanoparticles on sulfonated polyarylene ether nitrile nanofibers as super-wetting antibacterial oil/water separation membranes. *J. Membr. Sci.* **2023**, *675*, 121539.
41. Dong, C.; Zhang, X.; Cai, H.; Cao, C.; Zhou, K.; Wang, X.; Xiao, X. Synthesis of stearic acid-stabilized silver nanoparticles in aqueous solution. *Adv. Powder Technol.* **2016**, *27*, 2416–2423. [CrossRef]
42. Liu, H.; Cheng, X.B.; Jin, Z.; Zhang, R.; Wang, G.; Chen, L.Q.; Liu, Q.B.; Huang, J.Q.; Zhang, Q. Recent advances in understanding dendrite growth on alkali metal anodes. *Energy Chem.* **2019**, *1*, 100003. [CrossRef]
43. Cai, W.F.; Pu, K.B.; Ma, Q.; Wang, Y.H. Insight into the fabrication and perspective of dendritic Ag nanostructures. *J. Exp. Nanosci.* **2017**, *12*, 319–337. [CrossRef]
44. Khaskhoussi, A.; Calabrese, L.; Patané, S.; Proverbio, E. Effect of Chemical Surface Texturing on the Superhydrophobic Behavior of Micro–Nano-Roughened AA6082 Surfaces. *Materials* **2021**, *14*, 7161. [CrossRef] [PubMed]
45. Yong, J.; Chen, F.; Yang, Q.; Huoa, J.; Hou, X. Superoleophobic surfaces. *Chem. Soc. Rev.* **2017**, *46*, 4113–4376. [CrossRef]
46. Safaee, A.; Sarkar, D.K.; Farzaneh, M. Superhydrophobic properties of silver-coated films on copper surface by galvanic exchange reaction. *Appl. Surf. Sci.* **2008**, *254*, 2493–2498. [CrossRef]
47. Bahadori, S.R.; Mei, L.; Athavale, A.; Chiu, Y.; Pickering, C.S.; Hao, Y. New Insight into Single-Crystal Silver Dendrite Formation and Growth Mechanisms. *Cryst. Growth Des.* **2020**, *20*, 7291–7299. [CrossRef]
48. Chen, R.; Nguyen, Q.N.; Xia, Y. Oriented Attachment: A Unique Mechanism for the Colloidal Synthesis of Metal Nanostructures. *ChemNanoMat* **2022**, *8*, e202100474. [CrossRef]
49. Chu, J.; Zhao, Y.; Li, S.H.; Li, W.W.; Chen, X.Y.; Huang, Y.X.; Chen, Y.P.; Qu, W.G.; Yu, H.Q.; Xu, A.W.; et al. A highly-ordered and uniform sunflower-like dendritic silver nanocomplex array as reproducible SERS substrate. *RSC Adv.* **2015**, *5*, 3860. [CrossRef]
50. Ge, D.; Yao, J.; Ding, J.; Babangida, A.A.; Zhu, C.; Ni, C.; Zhao, C.; Qian, P.; Zhang, L. Growth mechanism of silver dendrites on porous silicon by single-step electrochemical synthesis method. *Appl. Phys.* **2022**, *128*, 908. [CrossRef]
51. Boles, M.A.; Engel, M.; Talapin, D.V. Self-Assembly of Colloidal Nanocrystals: From Intricate Structures to Functional Materials. *Chem. Rev.* **2016**, *116*, 11220–11289. [CrossRef] [PubMed]

52. Dinga, H.P.; Xina, G.Q.; Chena, K.C.; Zhanga, M.; Liub, Q.; Haoa, J.; Liu, H.G. Silver dendritic nanostructures formed at the solid/liquid interface via electroless deposition. *Colloids Surf. A Physicochem. Eng.* **2010**, *353*, 166–171. [CrossRef]
53. Gawert, C.; Bähr, R. Automatic Determination of Secondary Dendrite Arm Spacing in AlSi-Cast Microstructures. *Materials* **2021**, *14*, 2827. [CrossRef]
54. Ahmed, I.; Haque, A.; Bhattacharyya, S.; Patra, P.; Plaisier, J.R.; Perissinotto, F.; Bal, J.K. Vitamin C/Stearic Acid Hybrid Monolayer Adsorption at Air-Water and Air-Solid Interfaces. *ACS Omega* **2018**, *3*, 15789–15798. [CrossRef] [PubMed]
55. Norazura, A.M.H.; Sivaruby, K.; Mat Dian, N.L.H. Blended palm fractions as confectionery fats: A preliminary study. *J. Oil Palm Res.* **2021**, *33*, 360–380. [CrossRef]
56. Xu, C.L.; Wang, Y.Z. Self-assembly of stearic acid into nano flowers induces the tunable surface wettability of polyimide film. *Mater. Des.* **2018**, *138*, 30–38. [CrossRef]
57. Chao, Y.; Yaoguang, W.; Hongyan, F.; Sainan, Y.; Yingming, Z.; Hairong, Y.; Wei, J.; Bin, L. A stable eco-friendly superhydrophobic/superoleophilic copper mesh fabricated by one-step immersion for efficient oil/water separation. *Surf. Coat. Technol.* **2019**, *359*, 108–116.
58. Bike, M.J.B.; Benessoubo, K.D.; Eko, M.C.; Tekoumbo, T.L.C.; Elimbi, A.; Kamga, R. Adsorption mechanisms of pigments and free fatty acids in the discoloration of shea butter and palm oil by an acid-activated Cameroonian smectite. *Sci. Afri.* **2020**, *9*, e00498.
59. Patti, A.; Lecocq, H.; Serghei, A.; Acierno, D.; Cassagnau, P. The universal usefulness of stearic acid as surface modifier: Applications to the polymer formulations and composite processing. *J. Ind. Eng. Chem.* **2021**, *96*, 1–33. [CrossRef]
60. Syed Arshad, H.; Bapi, D.; Bhattacharjee, D.; Mehta, N. Unique supramolecular assembly through Langmuir Blodgett (LB) technique. *Heliyon* **2018**, *4*, e01038. [CrossRef]
61. Swierczewski, M.; Bürgi, T. Langmuir and Langmuir–Blodgett Films of Gold and Silver Nanoparticles. *Langmuir* **2023**, *39*, 2135–2151. [CrossRef] [PubMed]

Disclaimer/Publisher's Note: The statements, opinions and data contained in all publications are solely those of the individual author(s) and contributor(s) and not of MDPI and/or the editor(s). MDPI and/or the editor(s) disclaim responsibility for any injury to people or property resulting from any ideas, methods, instructions or products referred to in the content.

Article

Anisotropy of the Tribological Performance of Periodically Oxidated Laser-Induced Periodic Surface Structures

Pavels Onufrijevs [1,*], Liga Grase [2], Juozas Padgurskas [3], Mindaugas Rukanskis [3], Ramona Durena [1], Dieter Willer [4], Mairis Iesalnieks [2], Janis Lungevics [5], Jevgenijs Kaupuzs [1], Raimundas Rukuiža [3,*], Rita Kriūkienė [6], Yuliya Hanesch [4] and Magdalena Speicher [4]

1. Institute of Technical Physics, Faculty of Materials Science and Applied Chemistry, Riga Technical University, P. Valdena 7, LV-1048 Riga, Latvia; ramona.durena@rtu.lv (R.D.); kaupuzs@latnet.lv (J.K.)
2. Institute of Materials and Surface Engineering, Faculty of Materials Science and Applied Chemistry, Riga Technical University, P. Valdena 7, LV-1048 Riga, Latvia; liga.grase@rtu.lv (L.G.); mairis.iesalnieks@rtu.lv (M.I.)
3. Department of Mechanical, Energy and Biotechnology Engineering, Faculty of Engineering, Vytautas Magnus University, 11 Studentų Str., LT-53362 Kaunas, Lithuania; juozas.padgurskas@vdu.lt (J.P.); rukanskis@gmail.com (M.R.)
4. Materials Testing Institute, University of Stuttgart, Pfaffenwaldring 32, 70569 Stuttgart, Germany; dieter.willer@mpa.uni-stuttgart.de (D.W.); yuliya.hanesch@mpa.uni-stuttgart.de (Y.H.); magdalena.speicher@mpa.uni-stuttgart.de (M.S.)
5. Department of Mechanical Engineering and Mechatronics, Riga Technical University, Kipsalas 6b, LV-1048 Riga, Latvia; janis.lungevics@rtu.lv
6. Lithuanian Energy Institute, Breslaujos 3, LT-44403 Kaunas, Lithuania; rita.kriukiene@lei.lt
* Correspondence: onufrijevs@latnet.lv (P.O.); raimundas.rukuiza@vdu.lt (R.R.)

Abstract: Laser-induced periodic surface structures (LIPSS) enable advanced surface functionalization with broad applications in various fields such as micro- and nanoelectronics, medicine, microbiology, tribology, anti-icing systems, and more. This study demonstrates the possibility of achieving anisotropy in the tribological behavior of C45-grade steel structured by nanosecond laser radiation using the LIPSS method. The lateral surface of the steel roller was irradiated with a pulsed Nd:YAG laser at an optimum intensity I = 870 MW/cm^2 for the formation of LIPSS. Two sets of samples were formed with LIPSS that were perpendicular and parallel to the roller's rotational motion direction. The Raman intensity maps revealed that the LIPSS structure consisted of periodically arranged oxides at the top of hills. At the same time, the valleys of the LIPSS structures were almost not oxidized. These results correlated well with scanning electron microscopy energy dispersive X-ray spectroscopy mapping and atomic force microscopy measurements. A comparison of Raman and X-ray photoelectron spectroscopy spectra revealed that both the magnetite phase and traces of the hematite phase were present on the surface of the samples. Tribological tests were performed in two cycles with periodic changes in the normal clamping force and sliding speed. It was found that the LIPSS structures which were formed perpendicularly to the sliding direction on the roller had a significantly greater impact on the friction processes. Structures oriented perpendicular to the direction of motion had a positive influence on reducing the energy consumption of a friction process as well as increasing the wear resistance compared to LIPSS formed parallel to the direction of motion or ones having a non-texturized surface. Laser texturing to produce LIPSS perpendicular to the direction of motion could be recommended for friction pairs operating under low-load conditions.

Keywords: Nd:YAG laser; LIPSS; anisotropy; tribology; wear resistance; energy consumption

1. Introduction

Laser texturing is a method which involves the remelting or removal of material from a surface to create more durable and shear-resistant textures compared to other methods that supplement the material to the surface [1]. Recently, laser texturing has also focused on cold region applications such as the development of ice-phobic surfaces for road signs, wind turbine blades, building constructions, airplane constructions, etc., preventing ice build-ups on them [2–6]. Laser-induced surface texturing has many advantages, such as its flexibility, high-speed texturing, well-controlled surface characteristics, applicability for the metal industry, and being environmentally friendly [7,8].

Among the techniques that can be used for patterning surfaces is the formation of laser-induced periodic surface structures (LIPSS, also known as ripples) using a laser beam with linearly polarized radiation [6,9,10]. LIPSS can be classified by their spatial frequency: low spatial frequency LIPSS (LSFL) and high spatial frequency LIPSS (HSFL). In the first case, the period Λ of the LSFL is of the order of the used laser wavelength ($\Lambda \sim \lambda$) and its orientation is perpendicular to the pulse polarization [11]. In the second case, the period Λ of the HSFL is much smaller than that of the laser wavelength $\Lambda \ll \lambda$ and is generated with an orientation parallel to that of the light polarization [12]. The LIPSS technique can be applied to almost any material, including metals [13], semiconductors [14], superconductors [15], polymers [16,17], dielectrics [18], or 2D nanomaterials [19].

The mechanism of the LIPSS formation remains the subject of debate, and generally, it can be divided into two kind of theories according to the materials and laser parameters: (1) electromagnetic theories which describe the deposition of the optical energy into the solid; (2) matter reorganization theories, which are based on the redistribution of matter at the surface layer [20,21]. One of the most widely accepted mechanisms for the formation of LSFL structures is the action of surface plasmon polaritons on rough metal surfaces [22]. The initial roughness of the material is crucial for producing scattering which may lead to the generation of surface plasmon polaritons that interfere with the incident light and modulate the finally absorbed fluence "imprints" in the material, while selectively ablating the parallel periodic structures [6].

Numerous studies have revealed an increase in friction using LIPSS fabricated through ultra-short pulse laser processing on various substrate materials [23]. Nevertheless, the formation of LIPSS in environmental atmosphere or specific gas atmosphere has an influence not only on structured material topography, wettability [2–5,7,24,25], and tribological properties [2,4,23], but also on its chemical properties [10,26]. There are long-lasting debates in the LIPSS community regarding the input of topography and surface chemistry for surface functionalization [10]. Side effects such as metal oxidation processes at higher laser intensities could take place during the laser irradiation of metals in an air environment [26,27]. The resulting superficial oxide layer would play an important role in tribological performances [28,29].

Recent publications have shown that a laser-induced oxide layer led to a reduction in friction and wear resistance [28,30]. Bonse et al. [29] studied the chemical effects and tribological performances of a titanium alloy (Ti6Al4V) while forming various types of femtosecond-laser-generated surface structures, such as LIPSS, grooves, and spikes. However, changes in the friction coefficient for LIPSS in both sliding directions for a 1 N load applied on the ball in linear motion have not been revealed. Nevertheless, anisotropic micro-texturing has previously been used for friction control depending on the direction [1]. Gachot et al. [31–33] studied dry friction of stainless steel surfaces patterned in a micro-length scale obtained with laser irradiation. The study showed that the use of direct laser interference patterning (DLIP) structures reduced the coefficient of friction after running-in. Zhang and Kyriakos [34] conducted a study highlighting the properties of nanoscale ripples obtained by oblique Ar+ ion beam irradiation on silicon surfaces. Their research revealed that these ripples exhibit scale-dependent nanomechanical behavior and display anisotropic friction characteristics. The team provided insightful explanations for the observed experimental trends by delving into the underlying physical mechanisms governing deformation behavior during indentation loading and considering the impact of adhesion forces at the sliding contact interface on the overall frictional forces. This research sheds light on the intricate nature of nanoscale systems and contributes to our understanding of surface engineering and nanotribology.

In this study, we demonstrated the attainment of anisotropic tribological behavior in a liquid lubricant medium for C45-grade steel through the implementation of nanosecond laser radiation and the LIPSS method. The primary objective of our investigation was to gain insights into the effects of normal load and sliding speed on frictional losses and wear resistance, with a particular emphasis on understanding these influences in relation to the direction of the LIPSS. We analyzed these parameters to understand the interplay between surface structuring, tribological performance, and the resulting wear characteristics. The findings of this study contribute to the broader field of surface engineering and provide valuable information for enhancing the design and performance of lubricated systems.

2. Materials and Methods

2.1. Sample Preparation and Laser Processing

Hollow cylinders (rollers) made of C45-grade steel (HRC 40–43), consisting of 0.45% C, 0.4% Si, 0.6% Mn, 0.02% P, 0.03% S, and Fe, were mechanically polished with diamond paste followed by a GOI polishing paste to obtain a mirror-like surface with an average roughness value of 29 nm. The samples were washed in isopropyl alcohol in an ultrasonic bath and dried under dry airflow.

The lateral surface of a cylinder with an outer diameter $Ø_o = 35$ mm and inner diameter $Ø_I = 16$ mm was irradiated in air with a flash-lamp pumped Q-switched pulsed Nd:YAG laser model NL301G, produced by Ekspla (Lithuania) with the following parameters: wavelength of the laser $\lambda = 1064$ nm; laser pulse intensity $I = 870$ MW/cm^2 (no changes to structural properties were observed at a laser intensity lower than 870 MW/cm^2), pulse duration $\tau = 6$ ns, repetition rate of $\nu = 10$ Hz, beam profile "Hat-Top" [35], and beam diameter of $Ø = 0.5$ mm. A hatch spacing of 0.4 mm was used and the rotating speed of the cylinder was 1.0 mm/s. The irradiation of the samples was carried out in an air environment at room temperature and ambient pressure.

2.2. Samples Characterisation Prior to Friction and Wear Tests

The surfaces of the samples were analyzed by atomic force microscopy (AFM) with a Smena (NT-MDT) instrument in semi-contact mode using a golden silicone probe NSG03 (NT-MDT) to characterize the surface roughness and topography of the samples. Areas of 15×15 μm in size were scanned by AFM using a scan velocity of 12 μm/s and step size of 78 nm.

A field emission scanning electron microscope (FE-SEM) Zeiss Auriga Crossbeam and energy dispersive X-ray (EDX) analysis detector were used to characterize the surface structures and perform chemical analyses through elemental mapping.

Micro-Raman shift measurements and mapping were carried out with a Renishaw In-ViaV727 spectrometer with a backscattering geometry at room temperature. The phonon excitation was induced with a green laser (Ar$^+$, $\lambda = 514.5$ nm, grating–1200 mm^{-1}), the exposure time for one accumulation was 3 s, and each spectrum consisted of three accumulations. The integral sum of the Raman spectrum intensity from 77 to 1000 cm^{-1} was used to create a Raman map (with a spatial lateral resolution of 0.25 μm and axial resolution <1 μm) to display the changes in the chemical composition of different parts of the sample.

The effect of the surface modifications was analyzed by X-ray photoelectron spectroscopy (XPS, Escalab Xi+, Thermo Scientific, Waltham, MA, USA) with an Al K-alpha X-ray source with spatial resolution (spot size) of 500 μm, and without further surface cleaning with an ion gun. The sample was attached to the sample holder specially designed for thick samples (Thermo Scientific) using carbon tape. The advantageous carbon peak at 284.8 eV was used as a calibration point. Peak fitting was performed using the Avantage 5.9925 software.

2.3. Friction and Wear Test

The block, as a counter-body of friction tests, was made of construction C45-grade steel, and a molybdenum coating was applied with the electro spark method using an industrial EFI-10M machine (Moldova) that operated with a current intensity of 0.7–2.0 A under the conditions of an unprotected medium.

The friction and wear tests were performed on a modernized SMC-2 friction machine using the block-on-roll test (testing scheme shown in Figure 1), which corresponded to the block-on-ring test scheme of the ASTM D2714 standard.

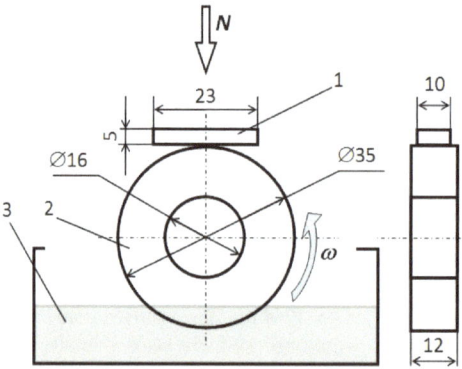

Figure 1. Principal scheme of the block-on-roll test: 1–block; 2–cylindrical shape steel sample; 3–oil bath; N–normal load; ω–roller rotation direction.

During the tribological test, the normal load F and roller rotation speed varied in different cycles, as shown in Figure 2. In total, two test cycles were performed and the friction distance of each cycle was 20,000 m. During each cycle, the tests were performed under three different loads F of 300, 450, and 600 N, and the rollers were rotated at three different speeds of 400, 600, and 800 rpm. The tribological tests were performed in an oil bath. The test roller was partly submerged in commercial 15 W-40 mineral engine oil produced by SCT Lubricants to ensure mixed-boundary lubrication conditions. The roughness of the counter-body blocks with a molybdenum coating prepared for the tribological tests was Ra = 0.41 μm. As mentioned in Section 2.1, the working surface of the roller was mechanically polished before being irradiated with a laser.

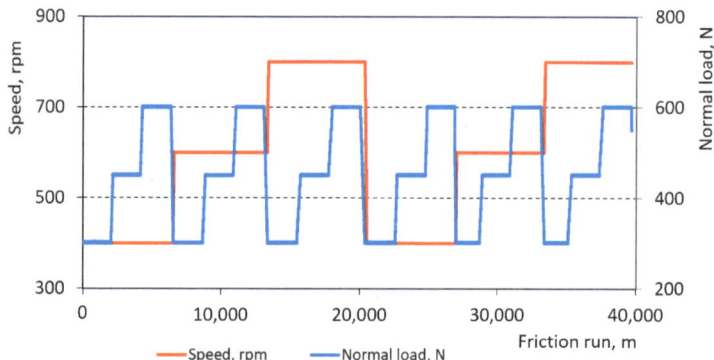

Figure 2. Diagram of the block-on-roll test performed in two cycles of revolutions per minute (each cycle: n = 400, 600, and 800 rpm) (red line) and six cycles of load force (each cycle: F = 300, 450, and 600 N) (blue line).

Surface roughness was measured using a MahrSurf GD 25 stylus profilometer (Mahr GmbH, Goettingen, Germany) with a stylus tip with a radius of 2 μm and a measurement length of 3 mm. The wear of the segments and rollers was determined after friction tests by weighing with a ABJ 120-4 M electronic scale (Kern & Sohn GmbH, Balingen, Germany) with an accuracy of 0.1 mg.

3. Results and Discussion

3.1. Surface Morphology and Chemical Composition

The images of the obtained samples (a,b) and optical images (c,d) of LIPSS formed with the Nd:YAG laser on the ring-shaped C45-grade steel are shown in Figure 3. Due to the interference of the formed structures, one (Figure 3a) or two (Figure 3b) stripes of rainbow-like colors could be observed depending on the LIPSS direction when the samples were exposed to light [36]. The period (Λ) between such structures (Figure 3c,d) was typically close to that of the laser wavelength, which corresponded to an applied laser radiation wavelength $\lambda = 1064$ nm.

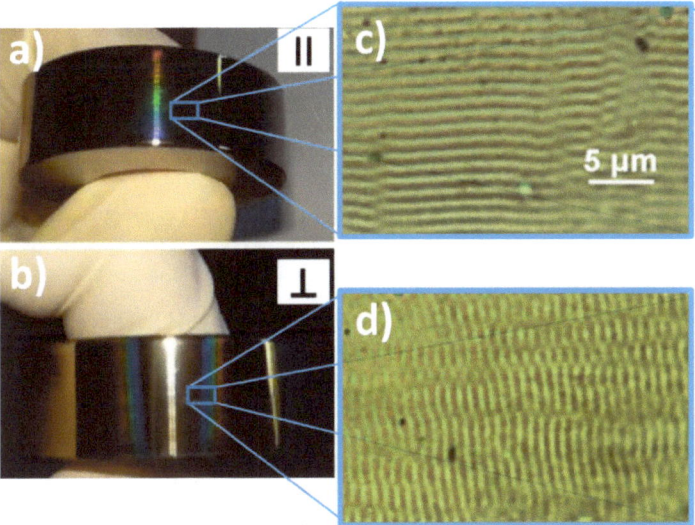

Figure 3. The images of the cylindrical shape steel samples irradiated with a Nd:YAG laser with laser-induced surface structures parallel (∥) and perpendicular (⊥) to the direction of rotation (**a,b**), optical microscope images of the laser-induced periodic surface structures (LIPSS) (**c,d**). The scale bar of 5 μm applies to both optical images in (**c,d**).

The AFM study of the samples allowed us to determine more accurately the shape and surface roughness of the LIPSS prior to studying their tribological properties. The AFM scans are presented in Figure 4 for non-irradiated (a) and laser-irradiated samples with LIPSS (b,c). It can be noted that, when irradiating the sample with the Nd:YAG laser, LIPSS consisting of periodically arranged hills and valleys with a period Λ of around 1 μm were formed on the surface. As can be seen from Figure 4d line profile, the average value of the peak-to-valley-height difference was ~ 20 nm for the LIPSS structures. A slight surface polishing had occurred, which was evidenced by reduced roughness root mean square values Ra from 29 nm to 13 nm for the non-irradiated and irradiated sample, respectively.

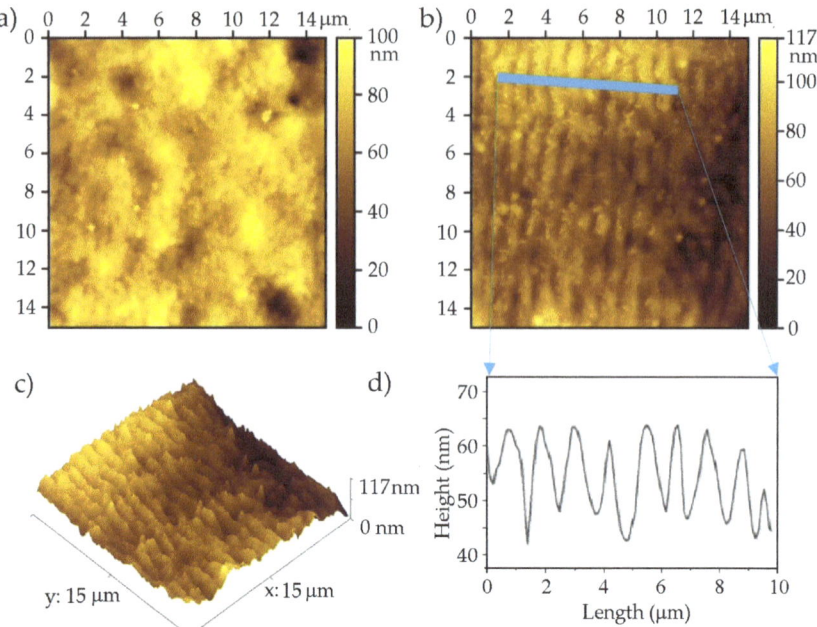

Figure 4. Atomic force microscopy (AFM) topography imaging of the cylindrical shape C45-grade steel surface: (**a**) non-irradiated, (**b**) 2D, (**c**) a 3D representation, and (**d**) line profile of laser-induced periodic surface structures (LIPSS) obtained through irradiation with a Nd:YAG laser.

The FE-SEM image of the irradiated C45-grade steel roller sample is shown in Figure 5. A comparison of the AFM (Figure 4b) and FE-SEM (Figure 5a) images showed that the structure was typical for such experiments. However, energy dispersive X-ray (EDX) analysis mapping (Figure 5b) revealed that oxides were formed on the hills of the LIPSS rather than in the valleys. Gachot et al. [37] conducted a study on oxide formation, morphology, and nano-hardness of steel that was patterned using the DLIP (direct laser interference patterning) method. This technique involves the interference of two laser beams to create a periodic structure in the micrometer range, with a periodicity of 18 μm and a depth of 8 μm. It was found that the laser-irradiated part of the steel had an oxide layer with thickness up to 15.4 nm. Results showed a significant increase in nano-hardness, from 2.2 GPa to 3.8 GPa, for laser-irradiated samples. It is well-known that metal oxides generally exhibit higher hardness compared to their corresponding pure metal forms [38]. It was concluded that both the oxide layer and the substrate structure played a significant role in this increased hardness. Figure 5c shows an EDS map of iron atoms which were evenly distributed over the sample surface. The SEM-EDS mapping revealed that the surface with LIPSS contains significant oxygen contents of up to 24 at.%. The oxygen atoms were distributed periodically on the hills of the LIPSS with a period of ~1 μm, which corresponded to that of the laser wavelength.

The room temperature Raman spectrum of the C45-grade steel sample with LIPSS is shown in Figure 6a. The spectrum consisted of a series of broad and weak peaks at 200 cm^{-1}, 320 cm^{-1}, 450 cm^{-1}, and 540 cm^{-1} and one strong peak at 668 cm^{-1}. Magnetite (Fe_3O_4) or hematite (Fe_2O_3) were formed as a result of the interaction between the Nd:YAG laser radiation and the C45-grade steel sample in air, as suggested by the possible peaks overlap [39,40]. According to [39], the modes at 193 and 538 cm^{-1} have a T_{2g} symmetry, while modes at 306 and 668 cm^{-1} are associated with E_g and A_{1g} ones. In comparison, the T_{2g} mode between 450 and 500 cm^{-1} was almost absent. Detailed descriptions of modes in the Raman spectrum of magnetite can be found in [39,41]. The Raman intensity maps

(Figure 6b) showed that the structure consisted of periodically arranged and oxidized hills and valleys, which were almost not oxidized. These findings correlate well with the SEM-EDS measurements. The Raman signal for C45-grade steel without LIPSS was weak and did not reveal any significant bands (not presented here).

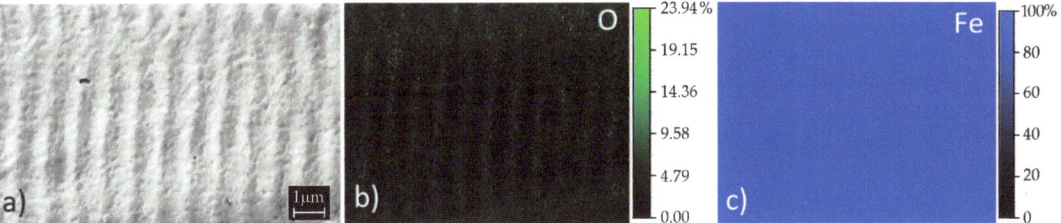

Figure 5. Field emission scanning electron microscopy (FE-SEM) image of the C45-grade steel sample with LIPSS formed by Nd:YAG laser radiation (**a**) and elemental energy dispersive X-ray (EDX) analysis maps of oxygen (**b**) and iron (**c**) concentrations.

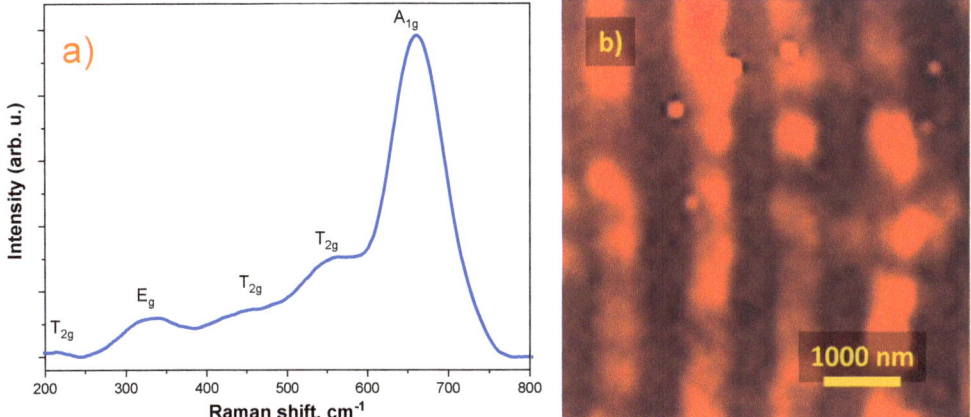

Figure 6. Typical Raman spectrum of a Nd:YAG laser-irradiated C45-grade steel sample with LIPSS (**a**) and Raman mapping of periodically arranged and oxidized hills (**b**).

In addition to the Raman analysis, XPS measurements were carried out to identify the composition of the iron oxide phase. The XPS Fe 2p spectrum of the irradiated C45-grade steel roller sample is shown in Figure 7. The presence of both iron oxidation states could be observed. Iron peaks Fe 2p_3/2 at 710.2 eV and Fe 2p_1/2 at 723.2 eV with a peak splitting of 13 eV could be attributed to the Fe^{2+} oxidation state. The peaks at Fe 2p_3/2 at 711.7 eV and Fe 2p_1/2 at 72.3 eV with a peak splitting of 13.6 eV corresponded to Fe^{3+} oxidation states. Multiple Fe satellite signals could also be observed. The concentrations of the Fe^{2+} and Fe^{3+} oxidation states were 38.6% and 61.4%, respectively. The XPS O 1s spectra can be observed in the Figure 7b. Maximum at 530.3 eV can be assigned to the lattice oxygen (Fe_3O_4) [42,43]. A significant amount of surface hydroxide groups can be observed at 532.1 eV. Hydroxyl groups form the first few monolayers of surface contamination [44] and can be attributed to ambient conditions. A comparison of the Raman (Figure 6a) and XPS (Figure 7) spectra showed that a magnetite phase and traces of a hematite phase were both present on the surface of the samples. According to these analyses, it was concluded that magnetite was the dominant oxide present on the surface of the samples.

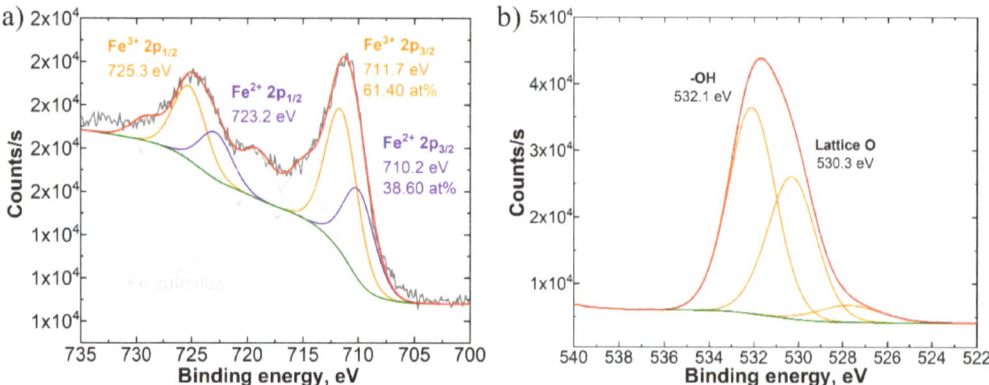

Figure 7. X-ray photoelectron spectroscopy (XPS) Fe 2p (**a**) and O 1s (**b**) spectra for a Nd:YAG laser-irradiated C45-grade steel sample with laser-induced periodic surface structures (LIPSS).

3.2. Evaluation of Tribological Properties

The schedule of the tribological testing is presented in Figure 2. The results of the friction torque variation during the first test cycle (20.000 m) at steady test modes (speed n = 400, 600, 800 rpm; load F = 300, 450, 600 N) are provided in Figure 8.

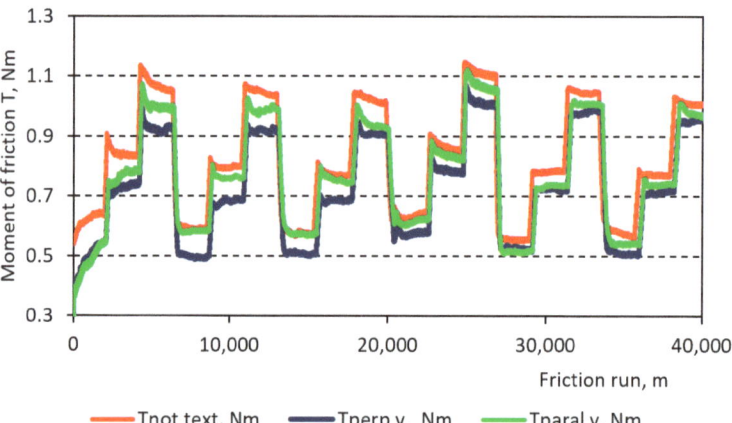

Figure 8. Variation of the friction torque T during the tests: T_{not}—surface without texture, T_{perp}—perpendicularly textured surface, and T_{paral}—parallel textured surface.

Our results revealed that the largest change in the friction torque occurred during the first stage of the test cycle (n = 400 rpm). The change in the friction torque when using the other modes was smaller during this test cycle, and in most cases, it was almost constant. However, this was exclusively due to the adaptation of the friction surfaces to each other during the initial stage of the friction pair testing. We see that in all cases, the friction pair with the non-textured surface had the highest friction torque. Table 1 shows the values of the average friction torque and energy consumed per 1 km of friction path for the different test variants. The table shows that in most cases, as the speed increased, the friction torque decreased, since an oil wedge was more readily formed when the speed increased. Conversely, as the load increased, the lubrication conditions deteriorated, and the friction torque increased.

Table 1. Mean values of friction torque T and energy consumption E per 1 km of friction path.

Test Parameters			Mean Values of Friction Torque T, Nm			Energy Consumption E, kJ per 1 km of Friction Path		
Test Cycle	Speed, rpm	Load, N	T_{not}	T_{perp}	T_{paral}	E_{not}	E_{perp}	E_{paral}
First	400	300	0.62	0.49	0.48	35.37	28.12	27.26
		450	0.84	0.73	0.77	48.17	41.55	43.77
		600	1.07	0.93	1.00	61.32	53.03	57.38
	600	300	0.59	0.50	0.58	33.94	28.51	33.26
		450	0.80	0.68	0.76	45.60	38.86	43.60
		600	1.05	0.92	0.99	59.83	52.52	56.57
	800	300	0.58	0.51	0.58	33.02	29.08	32.91
		450	0.77	0.68	0.76	44.22	39.08	43.30
		600	1.02	0.91	0.95	58.50	52.04	54.10
Second	400	300	0.63	0.58	0.62	36.23	32.92	35.15
		450	0.87	0.79	0.84	49.49	45.09	48.00
		600	1.11	1.02	1.07	63.60	58.00	61.26
	600	300	0.56	0.53	0.52	31.77	30.06	29.43
		450	0.78	0.72	0.73	44.74	41.32	41.89
		600	1.05	0.98	1.01	59.83	56.00	57.60
	800	300	0.58	0.51	0.54	33.02	29.14	30.96
		450	0.77	0.71	0.74	44.10	40.62	42.28
		600	1.01	0.95	0.98	57.64	54.39	55.93

The patterns of energy consumption (Table 1) showed that the reduction in the comparative energy consumption during the 1 km friction run in the first test cycle due to texturing was much higher than in the second one.

Two main factors could be distinguished when evaluating the influence of LIPSS on the energy consumption of the friction process and wear resistance of the surface: (1) an oxidized layer was formed on the surface of LIPSS hills; (2) LIPSS formed the periodic comb-striated relief on the surface of the metallic specimens. The influence of the relief on the friction and wear parameters was especially clear when the LIPSS were formed perpendicular to the direction of motion. In this case, texturing could reduce the energy consumption by up to 20% per km, which had a positive effect on the wear resistance on the whole friction pair. The wear resistance could potentially increase by more than twice compared to the friction pair with LIPSS formed parallel (||) to the direction of motion and the untextured roller (Figure 9).

As shown through the testing of laser-textured surfaces, the increase in wear resistance may have been due to the oil reservoir effect and hydrodynamic lifting effect provided by the LIPSS grooves/valleys (Figure 4d) between them which were regularly distributed in the contact zone and could act as integrated pressure pockets. These pockets could also act as traps for abrasion products, leaving a free medium between the surface of the roller and the coating [45,46]. A more continuous and durable boundary lubrication layer between the friction surfaces was provided as a result of the greater wettability of the surface and the formed grooves.

However, in the second test cycle, the texturing effect on the wear resistance was lower and reached up to 10%. This was explained by the fact that under high load conditions (450 and 600 N), the textured surface was more vulnerable to wear and therefore, the texturing effect became lower due to the wear of the texturizing structures compared to the effect obtained during the first test cycle.

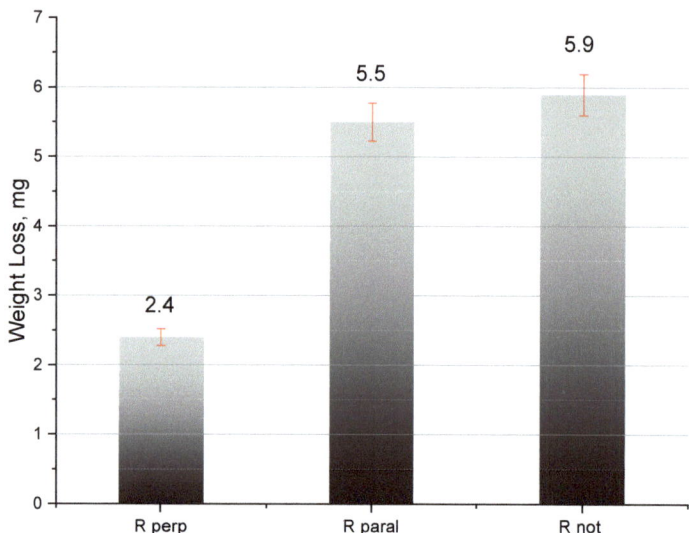

Figure 9. Total weight loss of friction pair bodies: R_{perp}–roller textured perpendicular to the direction of motion, R_{paral}–textured parallel to the direction of motion, and R_{not}–roller without textures.

Gachot et al. [33] presented a review paper stating that texturized surfaces could reduce the friction coefficient under boundary lubrication conditions only in the case of the texture being able to induce a hydrostatic force when the dimple was passed over and the normal reaction force counterbalanced the load and formed a continuous lubrication film. However, under mixed lubrication conditions, the shallow dimples and groove lines of texturized surfaces in most cases could induce hydrodynamic pressure and reduce the friction. The results of our tests showed that the surface texturized perpendicular to the sliding direction could create the conditions required for the formation of a lubrication film. The linear hills formed by LIPSS, when they were perpendicular to the sliding direction, created oil reservoirs in the valleys formed between the linear hills (also perpendicular to the sliding direction). This influenced the formation of the oil film and at the same time the lubrication mode in the friction pair. Our friction measurement results (Figure 8) showed that, when the LIPSS were perpendicular to the direction of movement, the coefficient of friction was around $\mu = 0.08 \ldots 0.09$, while for a non-textured surface $\mu = 0.1 \ldots 0.12$. The tops of LIPSS-formed hills were also strongly oxidized, as attested by the formation of a Fe_3O_4 layer. This surface property strengthened the boundary layer of the oil, making it more resistant to breach. Both features (formation of oil reservoirs and strengthening of the boundary oil layer due to better surface lubricity) contributed to the reduction of wear (Figure 9). Calculations of friction energy losses showed that up to 20 percent of the energy can be saved when using contact surfaces with LIPSS structures formed perpendicularly to the sliding direction. Such structures formed by precise LIPSS surface treatments could be used to produce low energy-consuming sub-micrometric textures of friction surfaces for precise components of implants, robots, and other precise elements. Such a LIPSS technology could also be used for haptic kinaesthetic communication technologies, sensory systems, etc. [23].

The Figure 10 shows FESEM images with different magnification of the non-irradiated (a,b) and laser-irradiated samples with laser-induced periodic surface structures (LIPSS) parallel (c,d) and perpendicular (e,f) to the direction of rotation after wear tests.

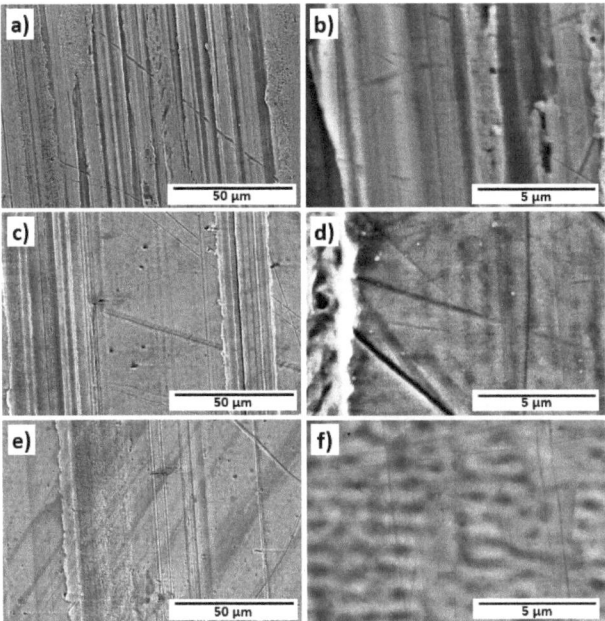

Figure 10. FESEM images of the non-irradiated samples (**a,b**) and the laser-irradiated samples with laser-induced periodic surface structures (LIPSS) parallel (**c,d**) and perpendicular (**e,f**) to the direction of rotation after wear tests.

The surface that was most affected was untextured (a,b). The surface textured parallel (c,d) and perpendicular (e,f) to the direction of movement was also damaged during the tests, but the damage is smaller compared to the non-textured surface (a,b). In all cases abrasion wear with comparably deep scratches is the prevailing wear mechanism in the case of our tests. It is also seen that the perpendicular LIPSS processing of the surface causes lower wear scratches. That could be related to the superior lubrication of the surfaces with perpendicular scars which preserves the lubricant from extrusion out of the contact zone, keeps the lubrication film stable, and strengthens the hydrodynamic effect. Textured friction surfaces are characterized by the presence of areas with intact or slightly damaged surfaces. These intact zones resulted in lower friction losses during testing.

4. Conclusions

In this study, a pulsed Nd:YAG laser with an intensity of $I = 870$ MW/cm^2 was employed to irradiate the lateral surface of a roller fabricated from C45 grade steel. The objective was to induce the formation of Laser-Induced Periodic Surface Structures (LIPSS). Two distinct sets of samples were created, each exhibiting LIPSS oriented perpendicular or parallel to the rotational motion direction of the roller. SEM-EDS mapping of the samples revealed that the surfaces with LIPSS contained significant oxygen concentrations of up to 24 at.%. The oxygen atoms were periodically distributed on the hills of the LIPSS with a period of ~1 µm, which corresponded to that of the laser wavelength used. Micro-Raman and XPS analyses revealed that a magnetite phase and traces of a hematite phase were both present on the surface of the samples. Raman mapping revealed that oxides were formed and distributed periodically (~1 µm) on the surface of the C45-grade steel. At the same time, morphological changes which were induced on the steel roller friction surface by laser radiation by forming LIPSS perpendicular to the direction of motion, led to a comparative reduction in energy consumption of up to 20% and more than twofold increase in wear resistance. Thus, laser texturing forming LIPSS perpendicular to the direction of motion could be recommended for friction pairs operating under low load conditions.

Author Contributions: Conceptualization, P.O., L.G., J.P., Y.H. and M.S.; methodology, P.O., L.G., J.P., R.D., D.W., M.I., J.L., J.K., R.R., R.K. and M.S.; validation, P.O., L.G., J.P. and M.S.; formal analysis, P.O., L.G., J.P., M.R., R.D., D.W., M.I., J.L. and M.S.; investigation, P.O., L.G., J.P., M.R., R.D., D.W., M.I., J.L., J.K., R.K., Y.H. and M.S.; resources, M.S.; data curation, P.O., L.G., J.P., R.D., D.W., M.I., J.L. and M.S.; writing—original draft preparation, P.O., L.G., J.P., M.R., R.D., M.I., J.L., J.K., R.R., Y.H. and M.S.; writing—review and editing, P.O., L.G., J.P., J.K., R.R. and M.S.; visualization, P.O., L.G., J.P., M.R., R.D., D.W., M.I., J.L., J.K., R.K. and M.S.; supervision, P.O., L.G., J.P. and M.S.; project administration P.O., L.G., J.P., Y.H. and M.S.; funding acquisition, P.O., L.G., J.P., Y.H. and M.S. All authors have read and agreed to the published version of the manuscript.

Funding: Liga Grase was supported by the European Regional Development Fund within the Activity 1.1.1.2 "Post-doctoral Research Aid" of the Specific Aid Objective 1.1.1 "To increase the research and innovative capacity of scientific institutions of Latvia and the ability to attract external financing, investing in human resources and infrastructure" of the Operational Programme "Growth and Employment" (No. 1.1.1.2/VIAA/4/20/638). This research was partially supported by the Bal-tic-German University Liaison Office, the German Academic Exchange Service (DAAD), with funds from the Foreign Office of the Federal Republic of Germany.

Institutional Review Board Statement: Not applicable.

Informed Consent Statement: Not applicable.

Data Availability Statement: The data presented in this study are available on request from the corresponding author.

Conflicts of Interest: The authors declare no conflict of interest.

References

1. Lu, P.; Wood, R.J.K.; Gee, M.G.; Wang, L.; Pfleging, W. The Use of Anisotropic Texturing for Control of Directional Friction. *Tribol. Int.* **2017**, *113*, 169–181. [CrossRef]
2. Maggiore, E.; Mirza, I.; Dellasega, D.; Tommasini, M.; Ossi, P.M. Sliding on Snow of Aisi 301 Stainless Steel Surfaces Treated with Ultra-Short Laser Pulses. *Appl. Surf. Sci. Adv.* **2022**, *7*, 100194. [CrossRef]
3. Zhuo, Y.; Xiao, S.; Amirfazli, A.; He, J.; Zhang, Z. Polysiloxane as Icephobic Materials—The Past, Present and the Future. *Chem. Eng. J.* **2021**, *405*, 127088. [CrossRef]
4. Ripamonti, F.; Furlan, V.; Savio, A.; Demir, A.G.; Cheli, F.; Ossi, P.; Previtali, B. Dynamic Behaviour of Miniature Laser Textured Skis. *Surf. Eng.* **2020**, *36*, 1250–1260. [CrossRef]
5. Ling, E.J.Y.; Uong, V.; Renault-Crispo, J.-S.; Kietzig, A.-M.; Servio, P. Reducing Ice Adhesion on Nonsmooth Metallic Surfaces: Wettability and Topography Effects. *ACS Appl. Mater. Interfaces* **2016**, *8*, 8789–8800. [CrossRef]
6. Florian, C.; Kirner, S.V.; Krüger, J.; Bonse, J. Surface Functionalization by Laser-Induced Periodic Surface Structures. *J. Laser Appl.* **2020**, *32*, 022063. [CrossRef]
7. Kumar, V.; Verma, R.; Kango, S.; Sharma, V.S. Recent Progresses and Applications in Laser-Based Surface Texturing Systems. *Mater. Today Commun.* **2021**, *26*, 101736. [CrossRef]
8. Mao, B.; Siddaiah, A.; Liao, Y.; Menezes, P.L. Laser Surface Texturing and Related Techniques for Enhancing Tribological Performance of Engineering Materials: A Review. *J. Manuf. Process.* **2020**, *53*, 153–173. [CrossRef]
9. Birnbaum, M. Semiconductor Surface Damage Produced by Ruby Lasers. *J. Appl. Phys.* **1965**, *36*, 3688–3689. [CrossRef]
10. Bonse, J.; Gräf, S. Ten Open Questions about Laser-Induced Periodic Surface Structures. *Nanomaterials* **2021**, *11*, 3326. [CrossRef]
11. Vorobyev, A.Y.; Makin, V.S.; Guo, C. Periodic Ordering of Random Surface Nanostructures Induced by Femtosecond Laser Pulses on Metals. *J. Appl. Phys.* **2007**, *101*, 034903. [CrossRef]
12. Jia, T.Q.; Chen, H.X.; Huang, M.; Zhao, F.L.; Qiu, J.R.; Li, R.X.; Xu, Z.Z.; He, X.K.; Zhang, J.; Kuroda, H. Formation of Nanogratings on the Surface of a ZnSe Crystal Irradiated by Femtosecond Laser Pulses. *Phys. Rev. B* **2005**, *72*, 125429. [CrossRef]
13. San-Blas, A.; Martinez-Calderon, M.; Buencuerpo, J.; Sanchez-Brea, L.M.; del Hoyo, J.; Gómez-Aranzadi, M.; Rodríguez, A.; Olaizola, S.M. Femtosecond Laser Fabrication of LIPSS-Based Waveplates on Metallic Surfaces. *Appl. Surf. Sci.* **2020**, *520*, 146328. [CrossRef]
14. Gao, Y.-F.; Yu, C.-Y.; Han, B.; Ehrhardt, M.; Lorenz, P.; Xu, L.-F.; Zhu, R.-H. Picosecond Laser-Induced Periodic Surface Structures (LIPSS) on Crystalline Silicon. *Surf. Interfaces* **2020**, *19*, 100538. [CrossRef]
15. Cubero, A.; Martínez, E.; Angurel, L.A.; de la Fuente, G.F.; Navarro, R.; Legall, H.; Krüger, J.; Bonse, J. Effects of Laser-Induced Periodic Surface Structures on the Superconducting Properties of Niobium. *Appl. Surf. Sci.* **2020**, *508*, 145140. [CrossRef]
16. Heitz, J.; Reisinger, B.; Fahrner, M.; Romanin, C.; Siegel, J.; Svorcik, V. Laser-Induced Periodic Surface Structures (LIPSS) on Polymer Surfaces. In Proceedings of the 2012 14th International Conference on Transparent Optical Networks (ICTON), IEEE, Coventry, UK, 2–5 July 2012; pp. 1–4.

17. Rodríguez-Beltrán, R.I.; Paszkiewicz, S.; Szymczyk, A.; Rosłaniec, Z.; Nogales, A.; Ezquerra, T.A.; Castillejo, M.; Moreno, P.; Rebollar, E. Laser Induced Periodic Surface Structures on Polymer Nanocomposites with Carbon Nanoadditives. *Appl. Phys. A Mater. Sci. Process.* **2017**, *123*, 717. [CrossRef]
18. Bonse, J.; Krüger, J.; Höhm, S.; Rosenfeld, A. Femtosecond Laser-Induced Periodic Surface Structures. *J. Laser Appl.* **2012**, *24*, 042006. [CrossRef]
19. Kasischke, M.; Maragkaki, S.; Volz, S.; Ostendorf, A.; Gurevich, E.L. Simultaneous Nanopatterning and Reduction of Graphene Oxide by Femtosecond Laser Pulses. *Appl. Surf. Sci.* **2018**, *445*, 197–203. [CrossRef]
20. Bonse, J.; Gräf, S. Maxwell Meets Marangoni—A Review of Theories on Laser-Induced Periodic Surface Structures. *Laser Photonics Rev.* **2020**, *14*, 2000215. [CrossRef]
21. Bonse, J. Quo Vadis LIPSS?—Recent and Future Trends on Laser-Induced Periodic Surface Structures. *Nanomaterials* **2020**, *10*, 1950. [CrossRef]
22. Fuentes-Edfuf, Y.; Sánchez-Gil, J.A.; Florian, C.; Giannini, V.; Solis, J.; Siegel, J. Surface Plasmon Polaritons on Rough Metal Surfaces: Role in the Formation of Laser-Induced Periodic Surface Structures. *ACS Omega* **2019**, *4*, 6939–6946. [CrossRef]
23. Costa, H.L.; Schille, J.; Rosenkranz, A. Tailored Surface Textures to Increase Friction—A Review. *Friction* **2022**, *10*, 1285–1304. [CrossRef]
24. Müller, F.A.; Kunz, C.; Gräf, S. Bio-Inspired Functional Surfaces Based on Laser-Induced Periodic Surface Structures. *Materials* **2016**, *9*, 476. [CrossRef]
25. Žemaitis, A.; Mimidis, A.; Papadopoulos, A.; Gečys, P.; Račiukaitis, G.; Stratakis, E.; Gedvilas, M. Controlling the Wettability of Stainless Steel from Highly-Hydrophilic to Super-Hydrophobic by Femtosecond Laser-Induced Ripples and Nanospikes. *RSC Adv.* **2020**, *10*, 37956–37961. [CrossRef] [PubMed]
26. Florian, C.; Déziel, J.L.; Kirner, S.V.; Siegel, J.; Bonse, J. The Role of the Laser-Induced Oxide Layer in the Formation of Laser-Induced Periodic Surface Structures. *Nanomaterials* **2020**, *10*, 147. [CrossRef]
27. Öktem, B.; Pavlov, I.; Ilday, S.; Kalaycıoğlu, H.; Rybak, A.; Yavaş, S.; Erdoğan, M.; Ilday, F.Ö. Nonlinear Laser Lithography for Indefinitely Large-Area Nanostructuring with Femtosecond Pulses. *Nat. Photonics* **2013**, *7*, 897–901. [CrossRef]
28. Bonse, J.; Kirner, S.V.; Griepentrog, M.; Spaltmann, D.; Krüger, J. Femtosecond Laser Texturing of Surfaces for Tribological Applications. *Materials* **2018**, *11*, 801. [CrossRef]
29. Florian, C.; Wonneberger, R.; Undisz, A.; Kirner, S.V.; Wasmuth, K.; Spaltmann, D.; Krüger, J.; Bonse, J. Chemical Effects during the Formation of Various Types of Femtosecond Laser-Generated Surface Structures on Titanium Alloy. *Appl. Phys. A Mater. Sci. Process.* **2020**, *126*, 266. [CrossRef]
30. Kirner, S.V.; Slachciak, N.; Elert, A.M.; Griepentrog, M.; Fischer, D.; Hertwig, A.; Sahre, M.; Dörfel, I.; Sturm, H.; Pentzien, S.; et al. Tribological Performance of Titanium Samples Oxidized by Fs-Laser Radiation, Thermal Heating, or Electrochemical Anodization. *Appl. Phys. A* **2018**, *124*, 326. [CrossRef]
31. Gachot, C.; Rosenkranz, A.; Reinert, L.; Ramos-Moore, E.; Souza, N.; Müser, M.H.; Mücklich, F. Dry Friction Between Laser-Patterned Surfaces: Role of Alignment, Structural Wavelength and Surface Chemistry. *Tribol. Lett.* **2013**, *49*, 193–202. [CrossRef]
32. Prodanov, N.; Gachot, C.; Rosenkranz, A.; Mücklich, F.; Müser, M.H. Contact Mechanics of Laser-Textured Surfaces. *Tribol. Lett.* **2013**, *50*, 41–48. [CrossRef]
33. Gachot, C.; Rosenkranz, A.; Hsu, S.M.; Costa, H.L. A Critical Assessment of Surface Texturing for Friction and Wear Improvement. *Wear* **2017**, *372–373*, 21–41. [CrossRef]
34. Zhang, H.; Komvopoulos, K. Scale-Dependent Nanomechanical Behavior and Anisotropic Friction of Nanotextured Silicon Surfaces. *J. Mater. Res.* **2009**, *24*, 3038–3043. [CrossRef]
35. Le, H.; Penchev, P.; Henrottin, A.; Bruneel, D.; Nasrollahi, V.; Ramos-de-Campos, J.A.; Dimov, S. Effects of Top-Hat Laser Beam Processing and Scanning Strategies in Laser Micro-Structuring. *Micromachines* **2020**, *11*, 221. [CrossRef]
36. Maragkaki, S.; Skaradzinski, C.A.; Nett, R.; Gurevich, E.L. Influence of Defects on Structural Colours Generated by Laser-Induced Ripples. *Sci. Rep.* **2020**, *10*, 53. [CrossRef]
37. Rosenkranz, A.; Reinert, L.; Gachot, C.; Aboufadl, H.; Grandthyll, S.; Jacobs, K.; Müller, F.; Mücklich, F. Oxide Formation, Morphology, and Nanohardness of Laser-Patterned Steel Surfaces. *Adv. Eng. Mater.* **2015**, *17*, 1234–1242. [CrossRef]
38. Takeda, M.; Onishi, T.; Nakakubo, S.; Fujimoto, S. Physical Properties of Iron-Oxide Scales on Si-Containing Steels at High Temperature. *Mater. Trans.* **2009**, *50*, 2242–2246. [CrossRef]
39. Shebanova, O.N.; Lazor, P. Raman Spectroscopic Study of Magnetite ($FeFe_2O_4$): A New Assignment for the Vibrational Spectrum. *J. Solid State Chem.* **2003**, *174*, 424–430. [CrossRef]
40. Dar, M.I.; Shivashankar, S.A. Single Crystalline Magnetite, Maghemite, and Hematite Nanoparticles with Rich Coercivity. *RSC Adv.* **2014**, *4*, 4105–4113. [CrossRef]
41. Verble, J.L. Temperature-Dependent Light-Scattering Studies of the Verwey Transition and Electronic Disorder in Magnetite. *Phys. Rev. B* **1974**, *9*, 5236–5248. [CrossRef]
42. Hao, C.; Gao, T.; Yuan, A.; Xu, J. Synthesis of Iron Oxide Cubes/Reduced Graphene Oxide Composite and Its Enhanced Lithium Storage Performance. *Chin. Chem. Lett.* **2021**, *32*, 113–118. [CrossRef]
43. Rajan, A.; Sharma, M.; Sahu, N.K. Assessing Magnetic and Inductive Thermal Properties of Various Surfactants Functionalised Fe_3O_4 Nanoparticles for Hyperthermia. *Sci. Rep.* **2020**, *10*, 15045. [CrossRef] [PubMed]

44. Tanuma, S.; Powell, C.J.; Penn, D.R. Calculations of Electron Inelastic Mean Free Paths. V. Data for 14 Organic Compounds over the 50-2000 EV Range. *Surf. Interface Anal.* **1994**, *21*, 165–176. [CrossRef]
45. Borghi, A.; Gualtieri, E.; Marchetto, D.; Moretti, L.; Valeri, S. Tribological Effects of Surface Texturing on Nitriding Steel for High-Performance Engine Applications. *Wear* **2008**, *265*, 1046–1051. [CrossRef]
46. Gualtieri, E.; Borghi, A.; Calabri, L.; Pugno, N.; Valeri, S. Increasing Nanohardness and Reducing Friction of Nitride Steel by Laser Surface Texturing. *Tribol. Int.* **2009**, *42*, 699–705. [CrossRef]

Disclaimer/Publisher's Note: The statements, opinions and data contained in all publications are solely those of the individual author(s) and contributor(s) and not of MDPI and/or the editor(s). MDPI and/or the editor(s) disclaim responsibility for any injury to people or property resulting from any ideas, methods, instructions or products referred to in the content.

Article

Enhancing Photovoltaic Performance and Stability of Perovskite Solar Cells through Single-Source Evaporation and CsPbBr₃ Quantum Dots Incorporation

Yuanzhe Kou *, Jianxiao Bian *, Xiaonan Pan and Jinchang Guo

School of Intelligent Manufacturing, Longdong University, Qingyang 745000, China; panxiaonan007@163.com (X.P.); guojinchang@lut.edu.cn (J.G.)
* Correspondence: kyzabc_1@tom.com (Y.K.); jxbian@ldxy.edu.cn (J.B.); Tel.: +86-152-9343-4444 (J.B.)

Abstract: This study investigates the potential of inorganic perovskite $CsPbBr_3$ as a photovoltaic material, highlighting its superior stability compared to that of organic–inorganic hybrid perovskite materials. Conventional methods for preparing $CsPbBr_3$ perovskite films, such as the two-step method and the dual-source thermal evaporation method, face challenges in obtaining high-purity films due to the decomposition of precursor films and the formation of multiple heterogeneous phases. To address this issue, we synthesized $CsPbBr_3$ powder material using thermal evaporation deposition, which effectively suppressed decomposition and the formation of heterogeneous phases. Consequently, we achieved uniform and dense $CsPbBr_3$ perovskite films. By incorporating energy-band engineering modification with $CsPbBr_3$ quantum dots (QDs), the all-inorganic perovskite solar cells (PSCs) attained a power conversion efficiency (PCE) of 7.01% under standard solar illumination conditions. The device PCE remained at 93% of its initial efficiency under 30% relative humidity conditions for over 100 days, showcasing its durability. The developed method produced an average grain size of 800 nm, resulting in a smooth and uniform film surface, thereby demonstrating the method's high repeatability. Additionally, the optimized PSCs exhibited a high open-circuit voltage (V_{OC}) with the champion device reaching a V_{OC} of 1.38 V and a PCE of 7.01%. This research presents a robust, efficient, and cost-effective approach for fabricating high-quality all-inorganic PSCs.

Keywords: perovskite solar cells; thermal evaporation deposition; quantum dots; all-inorganic

1. Introduction

Organic–inorganic hybrid perovskite solar cells (PSCs) have experienced remarkable advancements from 2009 to 2023, demonstrating immense potential for commercial and military applications [1–6]. The champion power conversion efficiency (PCE) in PSCs has reached as high as 25.7% by optimizing perovskite film properties and charge transfer capacity [7–13]. However, hybrid organic–inorganic PSCs still suffer from limited stability due to the inherent instability of hybrid perovskite materials in air conditions (oxygen-rich, moisture, heat, and light) [14,15], hindering further development. Therefore, exploring high-stability photovoltaic materials and designing novel device structures are crucial for promoting future applications and commercialization. To circumvent the instability of organic–inorganic hybrid perovskite materials, researchers have focused on stable all-inorganic halide perovskites of the type $CsPbX_3$ (X = I, Br, Cl, or mixed halides) for use in PSCs to enhance their stability. All-inorganic materials, including $CsPbI_3$ [16], $CsPbI_2Br$ [17], $CsPbIBr_2$ [18], and $CsPbBr_3$ [19], have been extensively studied due to their outstanding optical performance and enhanced stability [20–22]. Among these inorganic materials, $CsPbBr_3$ exhibits the best stability, providing protection against heat and moisture [23]. Kumar and colleagues demonstrated that $CsPbBr_3$ quantum dots are especially promising for improving the performance of photovoltaic devices due to their size-dependent energy gaps, high photoluminescence quantum yields, and advantageous band alignments [24].

Developing an efficient method for preparing inorganic CsPbBr$_3$ perovskite films is essential for achieving high film quality and device performance. Historically, solution processes have been employed to fabricate most CsPbBr$_3$ PSCs [25–27]. Nonetheless, CsPbBr$_3$ films prepared via solution engineering still exhibit minor degradation under UV irradiation. The solution process likely contributes to the imperfect photostability of CsPbBr$_3$ and results in high concentrations and low resistance to ultraviolet rays. Although solution-based processes do not require expensive vacuum equipment, they struggle to produce consistent, large-area, high-quality CsPbBr$_3$ films. Vacuum thermal evaporation (VTE), a mature technique commonly used in the coating industry, offers the ability to deposit multiple thin films on large areas, yielding excellent uniformity and flatness. However, VTE has been relatively underexplored for fabricating PSCs, especially compared to solution processes. Ma et al. utilized dual-source vacuum thermal evaporation to fabricate high-efficiency PSCs [28], demonstrating that vapor-deposited films can achieve uniformity. More recently, Chen et al. employed two-source VTE to fabricate high-performance CsPbBr$_3$-based organic solar cells with efficiencies of 14.03% [29]. Dual-source VTE has also been applied to fabricate other PSCs [30–35]. Bolink et al. successfully used MAI, CsBr, FAI, and PbI$_2$ as evaporation sources to prepare triple-cation Cs$_{0.5}$FA$_{0.4}$MA$_{0.1}$Pb(I$_{0.83}$Br$_{0.17}$)$_3$ perovskite films and manufacture solar cells with an efficiency of 16% [36].

In multi-source thermal evaporation, the raw material evaporation rate ratio significantly impacts the deposited film stoichiometry and PSC efficiency. Factors such as vacuum chamber pressure, crucible heating power, and the amount and distribution of evaporated material in the crucible all influence the evaporation rate ratio. Maintaining the correct evaporation rate of raw material throughout the process is challenging due to constantly changing experimental conditions, and adjusting and controlling the ratio of raw material evaporation rate is both difficult and time-consuming.

In this study, we employed a single crucible, single-source VTE method to deposit high-quality CsPbBr$_3$ thin films, which proved to be a simple and effective approach. We synthesized high-purity CsPbBr$_3$ material powder, compressed it into tablets, and placed it in a quartz crucible. The synthesis of CsPbBr$_3$ powder in our study was adapted from the method reported by Zhang et al. [37], with some modifications to optimize the process for our experimental setup. Zhang and colleagues have demonstrated a facile and efficient approach for the synthesis of highly luminescent CsPbBr$_3$ perovskite powder. We investigated the effects of evaporation rate and CsPbBr$_3$ film thickness on film quality and solar cell performance. Furthermore, we spin-coated a layer of quantum dots (QDs) on the TiO$_2$ electron transport layer to optimize energy level matching, thereby enhancing electron extraction capability and PCE [38]. Through continuous optimization, we successfully synthesized CsPbBr$_3$ thin films and fabricated a stable CsPbBr$_3$ solar cell with an efficiency of 7.01%. We achieved a PCE of 7.01% using the single-source thermal evaporation method. It is worth noting that some previous studies have reported higher PCE values for CsPbBr$_3$ perovskite solar cells. For example, Zhang et al. reported a PCE of 10.97%. Despite these higher PCE values, our method offers several advantages in terms of simplicity, reproducibility, and uniformity of the CsPbBr$_3$ films [39].

2. Experiment Section

2.1. Materials Synthesis

Lead bromide (PbBr$_2$), cesium bromide (CsBr), and tetrabutyl titanate were procured from Macklin Company. Spiro-OMeTAD was obtained from Xi'an Polymer Light Technology Corp. All other chemicals were purchased from Sigma-Aldrich. In this study, chemicals were used as received without further purification.

CsPbBr$_3$ powder was synthesized by dissolving 5 mmol of CsBr in 10 mL of an aqueous solution and 5 mmol of PbBr$_2$ in 20 mL of hydrogen bromide (48 wt% in an aqueous solution). The solutions were mixed in a 100 mL brown round-bottom flask and stirred vigorously at 0 °C for 12 h. Subsequently, the mixture was transferred to a 100 mL beaker, and 50 mL of absolute ethanol was added to obtain a yellow precipitate. The yellow precipitate was then subjected to rotary evaporation at a constant temperature of 60 °C for 4 h. The powder was dissolved in an aqueous solution and recrystallized in absolute ethanol three times. The yellow powder was dried at 60 °C in a vacuum drying oven for 12 h.

In accordance with previous reports, the preparation of CsPbBr$_3$ quantum dots (QDs) was facilitated by washing the precipitates with toluene through centrifugation at 9000 rpm for 20 min [40]. This washing process was repeated three times to eliminate as many organic ligands on the surface as possible. Finally, the precipitates were redispersed in ethyl acetate, yielding a stable colloidal solution.

2.2. Device Fabrication

A fluorine-doped tin oxide (FTO) glass substrate was etched using zinc powder and 35% HCl, followed by ultrasonic cleaning in acetone, isopropanol, deionized water, and ethanol for 20 min each. The cleaned substrate was then air-dried. To remove any remaining organic residues on the surface, the FTO substrate was treated with oxygen plasma for 10 min. A dense TiO$_2$ layer was spin-coated using a 0.15 M solution of diisopropoxy titanium bis (acetylpyruvate) (75 wt.% in isopropanol) in 1-butanol at 2000 rpm for 30 s. The coated substrate was then sintered at 500 °C for 30 min.

A 1 mL solution of CsPbBr$_3$ quantum dots (QDs) was spin-coated onto the TiO$_2$ layer at 2000 rpm for 30 s. The CsPbBr$_3$ light absorption layer was deposited on the FTO/TiO$_2$/CsPbBr$_3$ QDs layer by thermally evaporating CsPbBr$_3$ material in a vacuum (Figure 1a). During the thermal deposition process, the deposition chamber was first evacuated to approximately 2×10^{-5} Pa. The distance between the CsPbBr$_3$ target and the FTO substrate was set at 30 cm. The substrate was then annealed in a nitrogen-protected glove box for 5 min at a temperature of 100 °C.

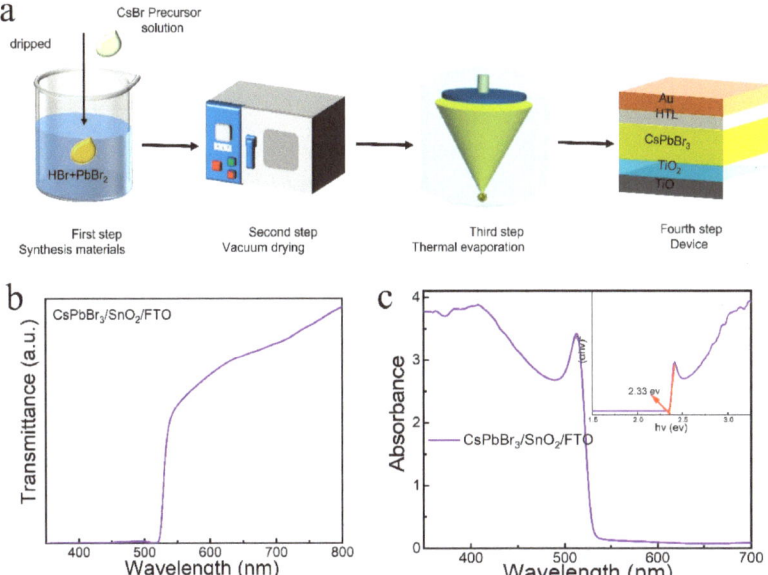

Figure 1. Schematic diagrams of (**a**) the synthesis steps of CsPbBr$_3$ powder and the single-source vacuum thermal evaporation deposition setup for CsPbBr$_3$ powder. (**b**) Transmission spectrum and (**c**) UV−Vis absorption spectra of CsPbBr$_3$ film on FTO/TiO$_2$/CsPbBr$_3$.

Subsequently, a Spiro-OMeTAD solution (50 mg of Spiro-MeOTAD, 22.5 µL of 4-tertbutylpyridine, and 22.5 µL of acetonitrile solution containing 170 mg·mL^{-1} of lithium bis-(trifluoromethylsulfonyl)imide in 1 mL of chlorobenzene) was deposited onto the surface of the CsPbBr$_3$ perovskite layer by spin-coating at 3000 rpm for 35 s in a nitrogen-protected glove box. Finally, gold electrodes with a thickness of 100 nm were deposited by thermal evaporation in a vacuum environment (5×10^{-4} Pa).

2.3. Device Characterization

Scanning electron microscopy (SEM) was employed to observe the surface morphology of the film and the cross-sectional view of the device. Energy dispersive spectroscopy (EDS) spectra were acquired using a Nova_NanoSEM430 instrument (FEI, Hillsboro, OR, USA). The X-ray diffraction (XRD) pattern of the CsPbBr$_3$ film was obtained using a Rigaku D/max 2550 X-ray diffractometer (Rigaku, Tokyo, Japan) with a monochromatic Cu target radiation source at a scan rate of 4°/min. Atomic force microscopy (AFM, 5500, Agilent, Santa Clara, CA, USA) was utilized to characterize the roughness of CsPbBr$_3$ films. Electrochemical impedance spectroscopy (EIS) measurements were performed using a CHI630E electrochemical analyzer (ChenHua, Shanghai, China). Absorption spectra were recorded using a UV-1800 spectrometer (Shimadzu, Tokyo, Japan). The optical properties of the CsPbBr$_3$ quantum dots and perovskite films were investigated using UV–Vis absorption spectroscopy and photoluminescence (PL) measurements. The absorption spectra were obtained using a UV–Vis spectrophotometer, while the PL spectra were recorded using a spectrofluorometer, following the procedures reported by Kumar et al. [41]. These techniques allowed us to determine the energy gap and emission properties of the CsPbBr$_3$ quantum dots, as well as to evaluate the charge carrier dynamics and recombination processes in the perovskite films. Current–voltage (I–V) measurements under 1-sun conditions (100 mW·cm^{-2} and AM 1.5 G radiation) were performed using an ABET Sun 2000 solar simulator system, with a Keithley 2400 digital source meter. The light intensity was calibrated using a reference silicon cell (RERA Solutions RR-1002, Shenzhen, China). The incident photocurrent conversion efficiency (IPCE) spectrum was recorded from 400 to 850 nm using a SolarCellScan100 system (ZOLIX, Beijing, China).

3. Results and Discussion

3.1. Structure and Morphology

As illustrated in the preparation Schematic diagrams in Figure 1a, the synthesis process of CsPbBr$_3$ powder involves mixing CsBr in an aqueous solution and PbBr$_2$ in a hydrogen bromide solution at a molar ratio of 1:1, resulting in a yellow-phase CsPbBr$_3$ powder. The yellow powder is then vacuum-dried. The obtained CsPbBr$_3$ powder is deposited onto the TiO$_2$/FTO substrate using vacuum thermal evaporation deposition techniques. The deposited films are yellow in color and exhibit a wide range of uniformity. The CsPbBr$_3$ thin film demonstrates relatively transparent properties with a high average visible light transmittance of 77% in the wavelength range from 550 to 800 nm (Figure 1b) [42]. Additionally, the deposited CsPbBr$_3$ films display strong absorption spectra from 300 to 550 nm in their UV–Vis absorbance spectra, indicating that single-source thermal deposition technology is an effective method for fabricating CsPbBr$_3$ films. The inset in Figure 1c reveals that the optical bandgap of the CsPbBr$_3$ film is 2.33 eV. In addition to the previously reported substrates, we have also prepared CsPbBr$_3$ perovskite films on glass and FTO (fluorine-doped tin oxide) substrates. The preparation procedure was the same as described earlier for the other substrates. The use of glass and FTO substrates allows us to investigate the effect of different substrate materials on the optical and morphological properties of the perovskite films. We have performed UV–Vis absorption spectroscopy and photoluminescence (PL) measurements on the CsPbBr$_3$ perovskite films deposited on both glass and FTO substrates. The absorption and PL spectra of the films on these substrates are presented in Figure 2. The absorption edge and PL peak positions for the CsPbBr$_3$ films on glass and FTO substrates are similar to those observed for the other substrates, indicating that the substrate material has minimal impact on the optical properties of the perovskite films.

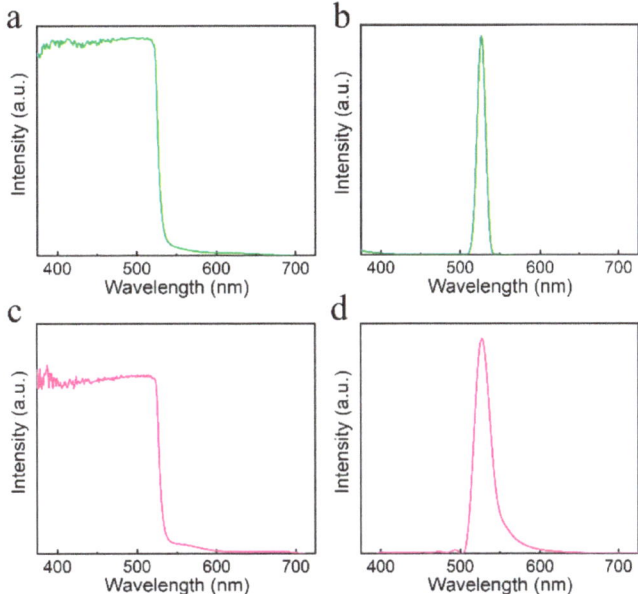

Figure 2. (**a,c**) UV−Vis absorption and (**b,d**) PL spectra of CsPbBr$_3$ film on FTO or glass substrate.

As depicted in Figure 3a, the evaporation rate does not alter the cubic phase of CsPbBr$_3$; however, it significantly influences the crystallinity and crystal orientation. Lower evaporation rates result in higher crystallinity and crystal orientation. The morphology of the CsPbBr$_3$ perovskite films plays a crucial role in determining the performance of photovoltaic devices. As observed in Figure 3b, films prepared at lower evaporation rates (0.5 nm/s) exhibit dense and large perovskite grains, which are desirable for high-performance devices. These characteristics reduce the number of grain boundaries where defects are typically located, leading to improved charge transport and minimized charge recombination. On the other hand, films prepared at higher evaporation rates (1.5 or 2 nm/s) display a reduced grain size, which may negatively affect the device performance due to increased grain boundaries and defect sites, The reason behind these observations is that at low evaporation rates, CsPbBr$_3$ molecules have more time to attach to the substrate electrode surface, resulting in a uniform film with large grain size and high orientation. Conversely, high evaporation rates make it difficult for CsPbBr$_3$ molecules to attach to the substrate due to increased kinetic energy, causing coalescence and leading to the formation of cracks, defects, small grains, and low crystallinity in the CsPbBr$_3$ film. This is advantageous for high-performance devices, as dense and large-grained perovskite crystals reduce the boundaries where most defects are located within the grain area [43]. Conversely, when the film is prepared by evaporation at higher rates (1.5 or 2 nm/s), the grain size significantly decreases. CsPbBr$_3$ molecules can more easily attach to the substrate electrode surface at low evaporation rates, ultimately yielding a uniform CsPbBr$_3$ film with large grain size and high orientation. In contrast, a high evaporation rate makes it difficult for CsPbBr$_3$ molecules to attach to the substrate; the increased kinetic energy causes subsequent coalescence, which inevitably leads to cracks and defects in the CsPbBr$_3$ film, accompanied by small grains and low crystallinity [44]. Although slower deposition rates may seem beneficial for improving film quality, there is a limit to the benefits that can be obtained from this approach. Excessively slow deposition rates can lead to increased substrate contamination, formation of unwanted intermediate phases, and an inefficient fabrication process. In our study, we found that a deposition rate of 0.5 nm/s provided an optimal balance between film quality and fabrication efficiency, resulting in the best device performance. The EDS mapping of $CsPbBr_3$ films in Figure 3c confirms the presence and uniform distribution of Cs, Pb, and Br elements, indicating the successful formation of the $CsPbBr_3$ perovskite structure. This uniform elemental distribution is essential for achieving consistent optoelectronic properties across the film and ensuring high performance in photovoltaic devices. The EDS mapping also verifies the stoichiometric composition of the $CsPbBr_3$ perovskite films, which is crucial for maintaining the desired crystal structure and phase purity. Additionally, the SEM images of $CsPbBr_3$ films with different evaporation rates (0.5, 1.5, and 2 nm/s) are shown in Figure $3d_{1-3}$. The SEM image and AFM of the $CsPbBr_3$ film evaporated at the lowest rate of 0.5 nm/s exhibit the largest grain size (850 nm) and the smallest roughness (root mean square = 3.17 nm). As the evaporation rate increases, the grain size reduces to 450 nm, further validating the XRD analysis results. In addition, we investigated the effect of different deposition rates on the performance of $CsPbBr_3$ perovskite solar cells. We found that the slow deposition rate yielded the best performance in terms of PCE. The J–V curves for solar cells fabricated at different deposition rates are provided in Table 1, clearly showing the impact of the deposition rate on the device performance.

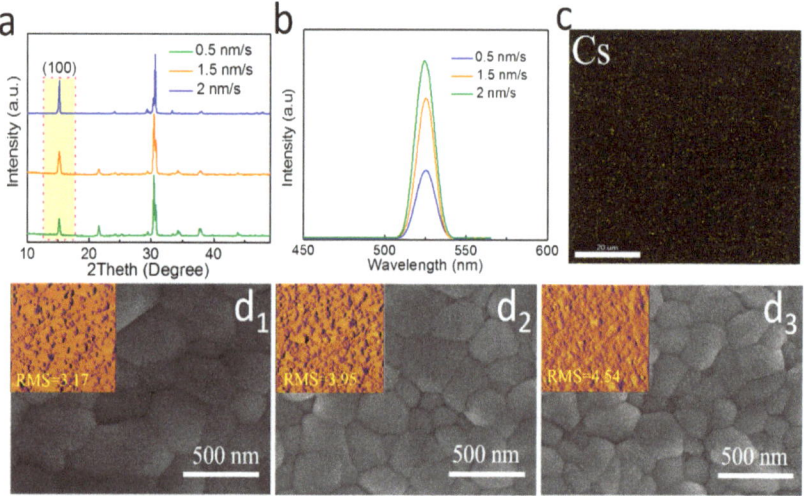

Figure 3. (**a**) XRD patterns of the deposited $CsPbBr_3$ films with different evaporation rates. (**b**) PL spectra and (**c**) EDS mapping, and (**d**$_{1-3}$) Surface SEM images and inserted picture is AFM height sensor image of the deposited $CsPbBr_3$ films with different evaporation rates: (**d**$_1$) 0.5 nm/s (**d**$_2$) 1.5 nm/s and (**d**$_3$) 2 nm/s.

Table 1. Performance Metrics of $CsPbBr_3$ Perovskite Solar Cell at different deposition rates.

Speed (nm/s)	J_{SC} (mA·cm^{-2})	V_{OC} (V)	FF (%)	PCE (%)
0.2	5.65	1.24	74.24	5.22
0.5	5.87	1.29	70.91	5.37
1	5.01	1.17	74.72	4.38
2	4.89	1.11	69.35	3.76

3.2. Device Performance

Figure 4a displays the device structure of CsPbBr$_3$ PSC, which consists of FTO/TiO$_2$/CsPbBr$_3$ QDs/CsPbBr$_3$/Spiro-OMeTAD/Au layers. In this structure, the CsPbBr$_3$ film serves as the full-inorganic light-absorbing layer, CsPbBr$_3$ QDs are employed to passivate the perovskite and TiO$_2$ contact surfaces, Spiro-OMeTAD is used as the hole transport layer (HTL), and Au electrodes are coated as the anode. To prepare the sample for cross-sectional SEM analysis, we first mechanically cleaved the CsPbBr$_3$ PSC device using a sharp scalpel blade. The device was carefully fractured along the edge, creating a cross-sectional surface that exposed the internal structure of the various layers. The cleaved sample was then mounted on an SEM sample holder using conductive carbon tape, ensuring that the cross-sectional surface was facing upwards. The sample was subsequently coated with a thin layer of gold using a sputter coater to enhance the conductivity and prevent charging effects during SEM imaging. Finally, the cross-sectional SEM image of the CsPbBr$_3$ PSC device was obtained (Figure 4b), illustrating the uniform deposition of each layer. The thicknesses of the CsPbBr$_3$ QDs, TiO$_2$, CsPbBr$_3$, Spiro-OMeTAD, and Au layers are 50, 600, 100, and 100 nm, respectively. Figure 4c presents the energy levels of each functional layer in CsPbBr$_3$-based PSCs, with the energy levels of TiO$_2$, CsPbBr$_3$ QDs, CsPbBr$_3$, Spiro-OMeTAD, and Au obtained from the literature [45]. As illustrated in Figure 4c, the valence and conduction bands of CsPbBr$_3$ QDs align well with those of the hole transport layer (HTL) and the electron transport layer (ETL), respectively. This band alignment enables efficient charge separation and transport across the interfaces: When photons are absorbed by the CsPbBr$_3$ QDs, electron–hole pairs (excitons) are generated. Due to the favorable band alignment, the photogenerated holes can easily transfer to the HTL, while the electrons can efficiently migrate to the ETL. The efficient charge transport minimizes the accumulation of charge carriers at the interfaces, which reduces the likelihood of non-radiative recombination losses. Additionally, the energy levels of the CsPbBr$_3$ QDs help to ensure that the built-in electric field within the device is strong enough to drive charge carriers toward their respective transport layers, further improving charge extraction and reducing recombination losses. The J–V characteristics and relevant parameters, including PCE, are shown in Figure 4d–f, and Table 2 (device parameters for solar cells with various thicknesses) summarizes and compares short-circuit current (J_{SC}), open-circuit voltage (V_{OC}), and fill factor (FF).

The PSCs with a 600 nm film thickness exhibit superior photovoltaic performance compared to other film thicknesses. Notably, the PSC with a 600 nm film thickness demonstrates the highest PCE of 6.52%, with a J_{SC} of 6.15 mA·cm^{-2}, a V_{OC} of 1.35 V, and an FF of 78.53%. This PCE value is 21%, 8%, and 3% higher than the devices with film thicknesses of 400 nm, 500 nm, and 700 nm, respectively. As the CsPbBr$_3$ film thickness increases, PCE, J_{SC}, FF, and V_{OC} show a slight increase, reaching a maximum at a film thickness of 600 nm, followed by a decrease upon further thickness increment. Therefore, the optimal film thickness for PSCs is determined to be 600 nm, at which the PCE improves from 5.37% to 6.52% (Figure 4d). These results suggest that the appropriate thickness can provide both effective light absorption and reduced charge recombination capacities. Although our PCE value might not be the highest among the reported studies, our single-source thermal evaporation method provides a solid foundation for future research to further enhance device performance. The simplicity, reproducibility, and uniformity of our method offer a promising route for synthesizing high-quality CsPbBr$_3$ perovskite films. Future studies could focus on optimizing the perovskite film morphology, thickness, and crystallinity, as well as improving the interfaces between the various layers of the device. Additionally, incorporating passivation techniques to reduce defects and non-radiative recombination could further enhance the PCE of our CsPbBr$_3$ perovskite solar cells.

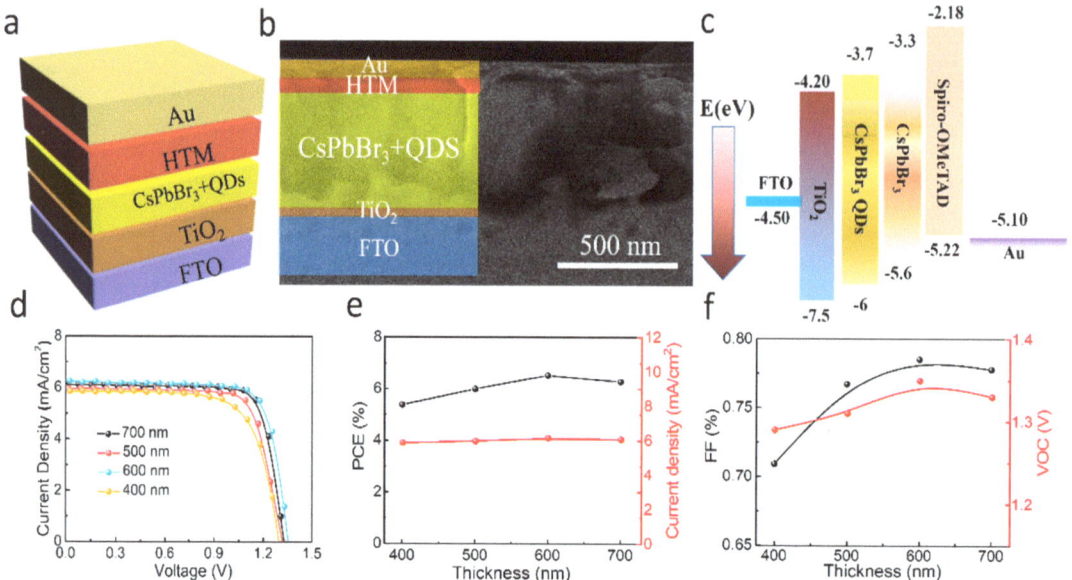

Figure 4. (**a**) Schematic device structure; (**b**) The cross-sectional SEM view; (**c**) Energy levels of each functional layer of CsPbBr$_3$ PSC; (**d**) J−V curves of PSC with different film thickness; (**e**) PCE, J$_{SC}$; (**f**) FF and V$_{OC}$ parameter distributions of PSCs.

Table 2. Device parameters for solar cells of various thicknesses and with QDs.

Thickness	J$_{SC}$ (mA·cm^{-2})	V$_{OC}$ (V)	PCE (%)	FF (%)
400	5.87	1.29	70.91	5.37
500	5.96	1.31	76.72	5.99
600 Without QDs	6.15	1.35	78.53	6.52
700	6.06	1.33	77.79	6.27
700 With QDs	6.42	1.38	79.12	7.01

Based on the aforementioned results, the PSCs were fabricated using a 600 nm thick film. Figure 5a presents the absorption, photoluminescence (PL)/excitation spectra, and SEM images of CsPbBr$_3$ quantum dots (QDs). An excitonic absorption peak at 510 nm and a green emission peak centered at 525 nm are observed for the CsPbBr$_3$ QDs under 365 nm excitation. We can observe that the CsPbBr$_3$ quantum dots exhibit an absorption band edge at approximately 2.3 eV. This value corresponds to the energy gap of the quantum dots, which is a crucial parameter affecting their optical and electronic properties. The energy gap of 2.3 eV indicates that the CsPbBr$_3$ quantum dots absorb light in the visible region of the electromagnetic spectrum, which is consistent with their green emission color. The observed energy gap also plays a significant role in the photovoltaic performance of the perovskite solar cell, as it determines the range of photon energies that can be effectively absorbed and converted into electrical energy by the device. The incorporation of CsPbBr$_3$ quantum dots in our perovskite solar cell contributes to enhanced performance in the UV region. This is due to their size-dependent properties, such as high photoluminescence quantum yields and strong absorption in the UV region, which enable efficient photon harvesting and charge carrier generation. As a result, the presence of CsPbBr$_3$ quantum dots in our device improves the overall efficiency by increasing the photocurrent and power conversion efficiency in the UV region. The J−V curves and PCEs of the CsPbBr$_3$ PSCs with and without QDs are displayed in Figure 5b and Table 2, respectively. The CsPbBr$_3$ PSC with QDs exhibits a higher PCE than the device without QDs, increasing from

6.52% to 7.01%. These results indicate that the QDs enhance the device's carrier extraction capability. Figure 5c depicts the incident photocurrent conversion efficiency (IPCE) spectra of the CsPbBr$_3$ PSCs over the wavelength range of 350 to 600 nm. The device with QDs demonstrates a higher IPCE than the device without QDs within the 350 to 510 nm range, thereby converting ultraviolet rays into usable green light and effectively improving the utilization rate of ultraviolet rays. Figure 5d reveals that under AM 1.5 sunlight irradiation, the device exhibits good stability. After 100 days of irradiation, the PCEs of PSCs with QDs and without QDs can maintain their initial values, accounting for more than 93% and 88%, respectively. Owing to the strong UV transfer ability of the CsPbBr$_3$ QDs, the long-term stability of the proposed device is significantly improved. Table 3 presents the performance metrics of multiple CsPbBr$_3$ perovskite solar cell devices fabricated in this study. The results demonstrate the reproducibility and consistency of our device fabrication process. The average PCE across all devices is 7%, with a standard deviation of 0.05%, showcasing the reliability and potential of our CsPbBr$_3$ perovskite solar cells for practical applications. In summary, the results and discussion demonstrate that the CsPbBr$_3$ PSCs with QDs exhibit enhanced photovoltaic performance, improved carrier extraction capability, and increased utilization of ultraviolet rays. Additionally, the device displays excellent long-term stability, making it a promising candidate for solar energy conversion applications. In our current study, we achieved a PCE of 7.01% using the single-source thermal evaporation method. Although some previous studies have reported PCE values over 8%, our method offers several advantages in terms of simplicity, reproducibility, and uniformity of the CsPbBr$_3$ films. Nevertheless, we acknowledge that there is room for improvement in the device's performance. Future studies could focus on optimizing the perovskite film morphology, thickness, and crystallinity, as well as improving the interfaces between the various layers of the device. Additionally, incorporating passivation techniques to reduce defects and non-radiative recombination could further enhance the PCE of our CsPbBr$_3$ perovskite solar cells.

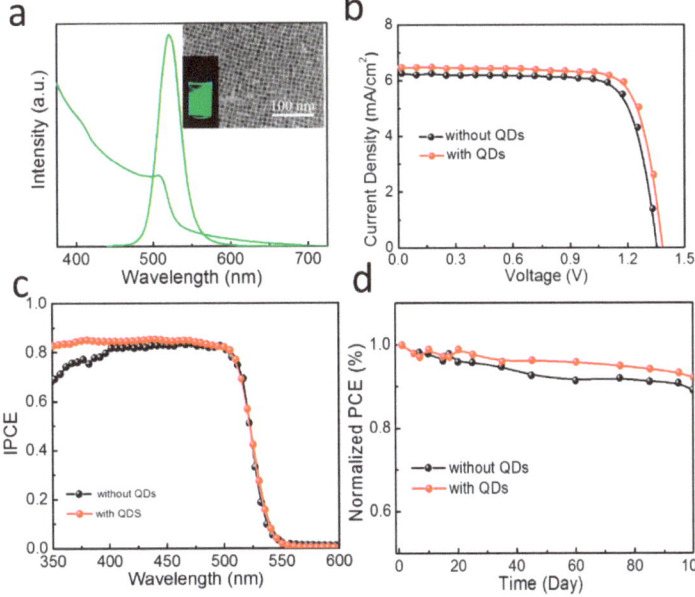

Figure 5. (**a**) Absorption and emission spectra of CsPbBr$_3$ QDs; (**b**) J−V curve; (**c**) IPCE spectra; (**d**) the long-term stability for PSCs with QDs or without QDs.

Table 3. Performance Metrics of CsPbBr3 Perovskite Solar Cell Devices.

No.	J_{SC} (mA·cm^{-2})	V_{OC} (V)	FF (%)	PCE (%)
1	6.42	1.38	79.12	7.01
2	6.38	1.35	81.04	6.98
3	6.45	1.37	79.21	7.00
4	6.32	1.38	79.68	6.95

4. Conclusions

In this work, we have successfully demonstrated a single-source thermal evaporation method for synthesizing CsPbBr$_3$ perovskite solar cells. We investigated the effects of evaporation rate and film thickness on the properties of the CsPbBr$_3$ films, and found that the optimal conditions led to uniform and reproducible films with desirable optoelectronic properties. Our method offers several advantages, including simplicity, reproducibility, and uniformity of the CsPbBr$_3$ films. The resulting CsPbBr$_3$ perovskite solar cells exhibited a power conversion efficiency (PCE) of 7.01% and good photostability. Although the stability improvement may not be significant compared to previous reports, our study provides a solid foundation for further exploration and optimization of CsPbBr$_3$ perovskite solar cells using our single-source thermal evaporation method.

Author Contributions: Formal analysis: Y.K.; investigation: Y.K.; project administration: J.B.; writing—original draft preparation: Y.K., J.B., X.P. and J.G.; writing—review and editing: Y.K., J.B., X.P. and J.G. All authors have read and agreed to the published version of the manuscript.

Funding: This study was financially supported by the Youth Science and Technology Fund of Gansu Province (Grant No. 21JR1RM340). College Teacher Innovation Fund of Gansu Provincial Department of Education (Grant No. 2023B-206).

Institutional Review Board Statement: Not applicable.

Informed Consent Statement: Not applicable.

Data Availability Statement: Data is contained within the article.

Conflicts of Interest: The authors declare no conflict of interest.

References

1. Kojima, A.; Teshima, K.; Shirai, Y.; Miyasaka, T. Organometal Halide Perovskites as Visible-Light Sensitizers for Photovoltaic Cells. *J. Am. Chem. Soc.* **2009**, *131*, 6050–6051. [CrossRef]
2. Im, J.-H.; Lee, C.R.; Lee, J.-W.; Park, S.W.; Park, N.G. 6.5% efficient perovskite quantum-dot-sensitized solar cell. *Nanoscale* **2011**, *3*, 4088–4093. [CrossRef] [PubMed]
3. Ummadisingu, A.; Steier, L.; Seo, J.Y.; Matsui, T.; Abate, A.; Tress, W.; Gratzel, M. The effect of illumination on the formation of metal halide perovskite films. *Nature* **2017**, *545*, 208–212. [CrossRef]
4. Yang, W.S.; Noh, J.H.; Jeon, N.J.; Kim, Y.C.; Ryu, S.; Seo, J.; Seok, S. High-performance Photovoltaic Perovskite Layers Fabricated through Intramolecular Exchange. *Science* **2015**, *348*, 1234–1237. [CrossRef] [PubMed]
5. Zhao, Y.; Heumueller, T.; Zhang, J.; Luo, J.; Kasian, O.; Langner, S.; Kupfer, C.; Liu, B.; Zhong, Y.; Elia, J.; et al. A bilayer conducting polymer structure for planar perovskite solar cells with over 1400 hours operational stability at elevated temperatures. *Nat. Energy* **2021**, *7*, 144–152. [CrossRef]
6. Zhao, Y.; Yavuz, I.; Wang, M.; Weber, M.H.; Xu, M.; Lee, J.H.; Tan, S.; Huang, T.; Meng, D.; Wang, R.; et al. Suppressing ion migration in metal halide perovskite via interstitial doping with a trace amount of multivalent cations. *Nat. Mater.* **2022**, *21*, 1396–1402. [CrossRef]
7. Im, J.-H.; Jang, I.-H.; Pellet, N.; Grätzel, M.; Park, N.-G. Growth of CH$_3$NH$_3$PbI$_3$ cuboids with controlled size for high-efficiency perovskite solar cells. *Nat. Nanotechnol.* **2014**, *9*, 927–932. [CrossRef] [PubMed]
8. Hao, F.; Stoumpos, C.C.; Cao, D.H.; Chang, R.P.H.; Kanatzidis, M.G. Lead-free solid-state organic–inorganic halide perovskite solar cells. *Nat. Photonics* **2014**, *8*, 489–494. [CrossRef]
9. Burschka, J.; Pellet, N.; Moon, S.-J.; Humphry-Baker, R.; Gao, P.; Nazeeruddin, M.K.; Grätzel, M. Sequential deposition as a route to high-performance perovskite-sensitized solar cells. *Nature* **2013**, *499*, 316–319. [CrossRef] [PubMed]
10. Jeon, N.J.; Noh, J.H.; Yang, W.S.; Kim, Y.C.; Ryu, S.; Seo, J.; Seok, S.I. Compositional engineering of perovskite materials for high-performance solar cells. *Nature* **2015**, *517*, 476–480. [CrossRef]

11. Bella, F.; Griffini, G.; Correa-Baena, J.-P.; Saracco, G.; Grätzel, M.; Hagfeldt, A.; Turri, S.; Gerbaldi, C. Improving efficiency and stability of perovskite solar cells with photocurable fluoropolymers. *Science* **2016**, *354*, 203–206. [CrossRef] [PubMed]
12. Cho, H.; Jeong, S.-H.; Park, M.-H.; Kim, Y.-H.; Wolf, C.; Lee, C.-L.; Heo, J.H.; Sadhanala, A.; Myoung, N.; Yoo, S.; et al. Overcoming the electroluminescence efficiency limitations of perovskite light-emitting diodes. *Science* **2015**, *350*, 1222–1225. [CrossRef] [PubMed]
13. Xing, G.; Mathews, N.; Sun, S.; Lim, S.S.; Lam, Y.M.; Grätzel, M.; Mhaisalkar, S.; Sum, T.C. Long-Range Balanced Electron and Hole-Transport Lengths in Organic-Inorganic CH3NH3PbI3. *Science* **2013**, *342*, 344–347. [CrossRef] [PubMed]
14. Mei, A.; Li, X.; Liu, L.; Ku, Z.; Liu, T.; Rong, Y.; Xu, M.; Hu, M.; Chen, J.; Yang, Y.; et al. A hole-conductor—free, fully printable mesoscopic perovskite solar cell with high stability. *Science* **2014**, *345*, 295–298. [CrossRef]
15. Jørgensen, M.; Norrman, K.; Krebs, F.C. Stability/degradation of polymer solar cells. *Sol. Energy Mater. Sol. Cells* **2008**, *92*, 686–714. [CrossRef]
16. Hu, Y.; Bai, F.; Liu, X.; Ji, Q.; Miao, X.; Qiu, T.; Zhang, S. Bismuth Incorporation Stabilized α-CsPbI3 for Fully Inorganic Perovskite Solar Cells. *ACS Energy Lett.* **2017**, *2*, 2219–2227. [CrossRef]
17. Lau, C.F.J.; Zhang, M.; Deng, X.; Zheng, J.; Bing, J.; Ma, Q.; Kim, J.; Hu, L.; Green, M.A.; Huang, S.; et al. Strontium-Doped Low-Temperature-Processed CsPbI$_2$Br Perovskite Solar Cells. *ACS Energy Lett.* **2017**, *2*, 2319–2325. [CrossRef]
18. Liang, J.; Zhao, P.; Wang, C.; Wang, Y.; Hu, Y.; Zhu, G.; Ma, L.; Liu, J.; Jin, Z. CsPb$_{0.9}$Sn$_{0.1}$IBr$_2$ Based All-Inorganic Perovskite Solar Cells with Exceptional Efficiency and Stability. *J. Am. Chem. Soc.* **2017**, *139*, 14009–14012. [CrossRef] [PubMed]
19. Begum, R.; Parida, M.R.; Abdelhady, A.L.; Murali, B.; Alyami, N.M.; Ahmed, G.H.; Hedhili, M.N.; Bakr, O.M.; Mohammed, O.F. Engineering Interfacial Charge Transfer in CsPbBr$_3$ Perovskite Nanocrystals by Heterovalent Doping. *J. Am. Chem. Soc.* **2017**, *139*, 731–737. [CrossRef]
20. Protesescu, L.; Yakunin, S.; Bodnarchuk, M.I.; Krieg, F.; Caputo, R.; Hendon, C.H.; Yang, R.X.; Walsh, A.; Kovalenko, M.V. Nanocrystals of Cesium Lead Halide Perovskites (CsPbX$_3$, X = Cl, Br, and I): Novel Optoelectronic Materials Showing Bright Emission with Wide Color Gamut. *Nano Lett.* **2015**, *15*, 3692–3696. [CrossRef]
21. Swarnkar, A.; Marshall, A.R.; Sanehira, E.M.; Chernomordik, B.D.; Moore, D.T.; Christians, J.A.; Chakrabarti, T.; Luther, J.M. Quantum dot–induced phase stabilization of α-CsPbI$_3$ perovskite for high-efficiency photovoltaics. *Science* **2016**, *354*, 92–95. [CrossRef]
22. Sanehira, E.M.; Marshall, A.R.; Christians, J.A.; Harvey, S.P.; Ciesielski, P.N.; Wheeler, L.M.; Schulz, P.; Lin, L.Y.; Beard, M.C.; Luther, J.M. Enhanced mobility CsPbI$_3$ quantum dot arrays for record-efficiency, high-voltage photovoltaic cells. *Sci. Adv.* **2017**, *3*, 4204. [CrossRef] [PubMed]
23. Kulbak, M.; Gupta, S.; Kedem, N.; Levine, I.; Bendikov, T.; Hodes, G.; Cahen, D. Cesium Enhances Long-Term Stability of Lead Bromide Perovskite-Based Solar Cells. *J. Phys. Chem. Lett.* **2016**, *7*, 167–172. [CrossRef]
24. Kumar, A.; Swami, S.K.; Sharma, R.; Yadav, S.; Singh, V.N.; Schneider, J.J.; Sinha, O.P.; Srivastava, R. A study on structural, optical, and electrical characteristics of perovskite CsPbBr$_3$ QD/2D-TiSe$_2$ nanosheet based nanocomposites for optoelectronic applications. *Dalton Trans.* **2022**, *51*, 4104–4112. [CrossRef] [PubMed]
25. Panigrahi, S.; Jana, S.; Calmeiro, T.; Nunes, D.; Martins, R.; Fortunato, E. Imaging the Anomalous Charge Distribution Inside CsPbBr$_3$ Perovskite Quantum Dots Sensitized Solar Cells. *ACS Nano* **2017**, *11*, 10214–10221. [CrossRef]
26. Li, Y.; Duan, J.; Zhao, Y.; Tang, Q. All-inorganic bifacial CsPbBr$_3$ perovskite solar cells with a 98.5%-bifacial factor. *Chem Commun.* **2018**, *54*, 8237–8240. [CrossRef] [PubMed]
27. Hoffman, J.B.; Zaiats, G.; Wappes, I.; Kamat, P.V. CsPbBr$_3$ Solar Cells: Controlled Film Growth through Layer-by-Layer Quantum Dot Deposition. *Chem. Mater.* **2017**, *29*, 9767–9774. [CrossRef]
28. Ma, Q.; Huang, S.; Wen, X.; Green, M.A.; Ho-Baillie, A.W.Y. Hole Transport Layer Free Inorganic CsPbIBr$_2$ Perovskite Solar Cell by Dual Source Thermal Evaporation. *Adv. Energy Mater.* **2016**, *6*, 1502202. [CrossRef]
29. Chen, W.; Zhang, J.; Xu, G.; Xue, R.; Li, Y.; Zhou, Y.; Hou, J.; Li, Y. A Semitransparent Inorganic Perovskite Film for Overcoming Ultraviolet Light Instability of Organic Solar Cells and Achieving 14.03% Efficiency. *Adv. Mater.* **2018**, *30*, e1800855. [CrossRef] [PubMed]
30. Frolova, L.A.; Anokhin, D.V.; Piryazev, A.A.; Luchkin, S.Y.; Dremova, N.N.; Stevenson, K.J.; Troshin, P.A. Highly Efficient All-Inorganic Planar Heterojunction Perovskite Solar Cells Produced by Thermal Coevaporation of CsI and PbI$_2$. *J. Phys. Chem. Lett.* **2017**, *8*, 67–72. [CrossRef]
31. Zhang, X.; Jin, Z.; Zhang, J.; Bai, D.; Bian, H.; Wang, K.; Sun, J.; Wang, Q.; Liu, S.F. All-Ambient Processed Binary CsPbBr$_3$-CsPb$_2$Br$_5$ Perovskites with Synergistic Enhancement for High-Efficiency Cs-Pb-Br-Based Solar Cells. *ACS Appl. Mater. Interfaces* **2018**, *10*, 7145–7154. [CrossRef]
32. Wei, Z.; Perumal, A.; Su, R.; Sushant, S.; Xing, J.; Zhang, Q.; Tan, S.T.; Demir, H.V.; Xiong, Q. Solution-processed highly bright and durable cesium lead halide perovskite light-emitting diodes. *Nanoscale* **2016**, *8*, 18021–18026. [CrossRef]
33. Wang, Q.; Zhang, X.; Jin, Z.; Zhang, J.; Gao, Z.; Li, Y.; Liu, S.F. Energy-Down-Shift CsPbCl$_3$:Mn Quantum Dots for Boosting the Efficiency and Stability of Perovskite Solar Cells. *ACS Energy Lett.* **2017**, *2*, 1479–1486. [CrossRef]
34. Wang, P.; Zhang, X.; Zhou, Y.; Jiang, Q.; Ye, Q.; Chu, Z.; Li, X.; Yang, X.; Yin, Z.; You, J. Solvent-controlled growth of inorganic perovskite films in dry environment for efficient and stable solar cells. *Nat. Commun.* **2018**, *9*, 2225. [CrossRef]

35. Teng, P.; Han, X.; Li, J.; Xu, Y.; Kang, L.; Wang, Y.; Yang, Y.; Yu, T. Elegant Face-Down Liquid-Space-Restricted Deposition of CsPbBr$_3$ Films for Efficient Carbon-Based All-Inorganic Planar Perovskite Solar Cells. *ACS Appl. Mater. Interfaces* **2018**, *10*, 9541–9546. [CrossRef]
36. Gil-Escrig, L.; Momblona, C.; La-Placa, M.G.; Boix, P.P.; Sessolo, M.; Bolink, H.J. Vacuum Deposited Triple-Cation Mixed-Halide Perovskite Solar Cells. *Adv. Energy Mater.* **2018**, *8*, 1703506. [CrossRef]
37. Zhang, X.; Wang, W.; Xu, B.; Liu, S.; Dai, H.; Bian, D.; Chen, S.; Wang, K.; Sun, X.W. Thin film perovskite light-emitting diode based on CsPbBr 3 powders and interfacial engineering. *Nano Energy* **2017**, *37*, 40–45. [CrossRef]
38. Liu, C.; Hu, M.; Zhou, X.; Wu, J.; Zhang, L.; Kong, W.; Li, X.; Zhao, X.; Dai, S.; Xu, B.; et al. Efficiency and stability enhancement of perovskite solar cells by introducing CsPbI$_3$ quantum dots as an interface engineering layer. *NPG Asia Mater.* **2016**, *26*, 2686–2694. [CrossRef]
39. Tong, G.; Chen, T.; Li, H.; Qiu, L.; Liu, Z.; Dang, Y.; Song, W.; Ono, L.K.; Jiang, Y.; Qi, Y. Phase transition induced recrystallization and low surface potential barrier leading to 10.91%-efficient CsPbBr$_3$ perovskite solar cells. *Nano Energy* **2019**, *65*, 104015. [CrossRef]
40. Moyen, E.; Kanwat, A.; Cho, S.; Jun, H.; Aad, R.; Jang, J. Ligand removal and photo-activation of CsPbBr$_3$ quantum dots for enhanced optoelectronic devices. *Nanoscale* **2018**, *10*, 8591–8599. [CrossRef]
41. Kumar, A.; Swami, S.K.; Singh, V.N.; Gupta, B.K.; Sinha, O.P.; Srivastava, R. Ethylcellulose-Encapsulated Inorganic Lead Halide Perovskite Nanoparticles for Printing and Optoelectronic Applications. *Part. Part. Syst. Charact.* **2022**, *39*, 2100250. [CrossRef]
42. Chang, X.; Li, W.; Zhu, L.; Liu, H.; Geng, H.; Xiang, S.; Liu, J.; Chen, H. Carbon-Based CsPbBr$_3$ Perovskite Solar Cells: All-Ambient Processes and High Thermal Stability. *ACS Appl. Mater. Interfaces* **2016**, *8*, 33649–33655. [CrossRef] [PubMed]
43. Bai, D.; Zhang, J.; Jin, Z.; Bian, H.; Wang, K.; Wang, H.; Liang, L.; Wang, Q.; Liu, S.F. Interstitial Mn^{2+}-Driven High-Aspect-Ratio Grain Growth for Low-Trap-Density Microcrystalline Films for Record Efficiency CsPbI$_2$Br Solar Cells. *ACS Energy Lett.* **2018**, *3*, 970–978. [CrossRef]
44. Li, J.; Gao, R.; Gao, F.; Lei, J.; Wang, H.; Wu, X.; Li, J.; Liu, H.; Hua, X.; Liu, S. Fabrication of efficient CsPbBr$_3$ perovskite solar cells by single-source thermal evaporation. *J. Alloys Compd.* **2019**, *818*, 152903. [CrossRef]
45. Chen, C.; Li, H.; Jin, J.; Cheng, Y.; Liu, D.; Song, H.; Dai, Q. Highly enhanced long time stability of perovskite solar cells by involving a hydrophobic hole modification layer. *Nano Energy* **2017**, *32*, 165–173. [CrossRef]

Disclaimer/Publisher's Note: The statements, opinions and data contained in all publications are solely those of the individual author(s) and contributor(s) and not of MDPI and/or the editor(s). MDPI and/or the editor(s) disclaim responsibility for any injury to people or property resulting from any ideas, methods, instructions or products referred to in the content.

Article

Investigating the Effects of Geometrical Parameters of Re-Entrant Cells of Aluminum 7075-T651 Auxetic Structures on Fatigue Life

Amir Ghiasvand [1], Alireza Fayazi Khanigi [2], John William Grimaldo Guerrero [3], Hamed Aghajani Derazkola [4,*], Jacek Tomków [5,*], Anna Janeczek [5] and Adrian Wolski [5]

[1] Department of Mechanical Engineering, University of Tabriz, Tabriz 5166616471, Iran
[2] Department of Materials Science and Engineering, Tarbiat Modares University, Tehran 1435685553, Iran
[3] Departamento de Energía, Univesidad de la Costa, Barranquilla 080001, Colombia
[4] Department of Mechanics, Design and Industrial Management, University of Deusto, Avda Universidades 24, 48007 Bilbao, Spain
[5] Faculty of Mechanical Engineering and Ship Technology, Gdansk University of Technology, 11/12 Gabriela Narutowicza Str., 80-229 Gdansk, Poland
* Correspondence: h.aghajani@deusto.es (H.A.D.); jacek.tomkow@pg.edu.pl (J.T.)

Abstract: In this study, the effects of two geometrical parameters of the re-entrant auxetic cells, namely, internal cell angle (θ) and H/L ratio in which H is the cell height, and L is the cell length, have been studied on the variations of Poisson's ratio and fatigue life of Aluminum 7075-T6 auxetic structures. Five different values of both the H/L ratio and angle θ were selected. Numerical simulations and fatigue life predictions have been conducted through the use of ABAQUS (version 2022) and MSC Fatigue (version 11.0) software. Results revealed that increases in both the H/L ratio and angle θ improved the average value of Poisson's ratio. Increasing the H/L ratio from 1 to 1.4 and θ from 50° to 70° increased the values of Poisson's ratio, respectively, 7.7% and 80%. In all angles, increasing the H/L values decreased the fatigue life of the structures significantly. Furthermore, in all H/L values, an increment in θ caused a reduction in fatigue life. The effects of H/L and θ parameters on fatigue life were dominant in the low cycle fatigue regime. Results also showed that the H/L ratio parameter had greater influence as compared to the θ angle, and the structures with higher auxeticity experienced higher fatigue resistance. It was found that the auxetic property of the structure has a direct relationship with the fatigue resistance of the structure. In all samples, structures with greater auxetic property had higher fatigue resistance.

Keywords: auxetic structures; re-entrant cell; Poisson's ratio; fatigue life; aluminum 7075-T6

1. Introduction

Poisson's ratio is one the most important parameters for describing the mechanical behavior of materials undergoing various loadings. It defines the pattern of deformation and stress state within the material perpendicular to the loading direction. The value of Poisson's ratio for a wide range of materials is positive; however, based on some thermodynamics considerations and strain energies in theory of elasticity, this quantity for some homogeneous isotropic solid materials can vary between 0.5 and −1. Theoretically, there are materials with negative Poisson ratios [1–3].

In recent decades, structures with negative Poisson ratios have been introduced named auxetic structures. According to Figure 1, these materials show different behaviors under various kinds of loadings compared to other materials which have positive Poisson ratios, so that when the structure undergoes a tension in longitudinal direction, it experiences a positive strain in lateral direction as well. The Poisson ratio of the structures can change with any variation in part geometry, properties, stiffness, and matrix characterization. The

auxetic phenomenon is independent from the size of the component [4–6]. These characteristics are associated with the internal geometry of the part and with the deformation that occurs while undergoing an off-axis tension. The auxetic phenomenon has been seen in nano scale in materials such as silicates, in microscopic scale such as foams and polymers, and in macro scale such as re-entrant, chiral, and ant-chiral systems. Some auxetic materials can often be found in nature including cubic elemental metals, cristobalites, biological tissues, and bones [7].

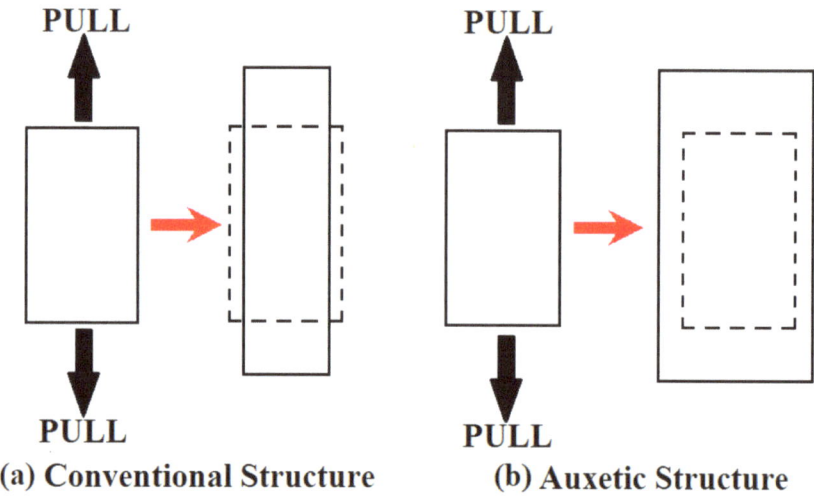

Figure 1. The behavior of conventional and auxetic structure subjected to a tension load.

A negative Poisson ratio is an abnormal property that can have significant effects on the mechanical properties of the structures such as corrosion resistance or roughness, compression, shear strength, and other aspects of the dynamic performance of the structure [8–10]. The auxetic structures show better practical performances as compared with those traditional structures which are widely used in different sectors such as automotive industries. A unique characteristic of an auxetic structure is its ability of shrinkage when undergoing the compression loads, enabling improvements in shear resistance, energy absorption, fracture toughness, and fatigue strength of the auxetic structure [11].

Although there has been an extensive number of studies dealing with the auxetic structures so far, most of the investigations have been concentrated on the behavior or performance of the structures subjected to the quasi-static or impact loads and their ability for energy absorption. Investigations regarding the fatigue strength of the auxetic structures can rarely be found in the literature. Choi et al. [12] studied the fracture toughness of two different copper foams with both positive and negative Poisson's ratios. They evidenced that the fracture toughness of the auxetic structure was higher than that with positive Poisson's ratio. Necember et al. [13] studied the effects of geometrical parameters of chiral auxetic cell on the mechanical properties and deformation patterns of the structure undergoing multiaxial loading conditions. They concluded that the geometrical parameters affected noticeably the mechanical response of the auxetic structure. Kramberger et al. [14] studied the impact of auxetic cell geometrical factors on the fracture behavior of the auxetic structures fabricated from Al 7075-T6 alloy. They showed that the cell geometry was the main factor in determining the crack propagation trajectory of the auxetic structure. Necember et al. [15] conducted numerical analysis and experimental tests to study the influence of re-entrant cell orientation on the fatigue crack initiation and crack growth in auxetic structures. They showed that the crack initiation occurred in the cell edges with maximum stress magnitudes. Bezazi et al. [16] studied the behavior of two different polyurethane foam structures with positive and negative Poisson ratios undergoing cyclic loads. They reported that the auxetic foam possessed higher fatigue strength compared to that with a positive Poisson ratio. Francesconi et al. [17] studied fatigue strength and crack propagation in two aluminum structures with both positive and negative Poisson ratios, numerically and experimentally. Their test results showed that the aluminum auxetic structure had greater fatigue life compared to the normal aluminum structure at similar load levels. Necember et al. [18] predicted low cycle fatigue lives in two different auxetic structures with re-entrant and chiral cells through conducting numerical analysis. They evidenced that the auxetic structures with chiral cells possessed higher fatigue life compared to those with re-entrant cells. Necember et al. [19] investigated fatigue behavior of the re-entrant and rotated re-entrant auxetic specimens made of Al-alloy 7075-T651. Two geometric layouts of the base unit cell were analyzed (re-entrant and rotated re-entrant structure). For the fatigue life calculation, the strain life approach was used, based on the Coffin–Manson model with a Morrow mean stress correction. The comparison between computational and experimental results regarding fatigue-life curves and observed fatigue failure path showed a reasonable agreement. Necember et al. [20] presented the experimental and computational analysis for determining the fatigue life of the auxetic cellular structures made of aluminum alloys. The experimental and computational results showed that the chiral and re-entrant auxetic structures have significantly different rigidity. Compared to the chiral auxetic specimen, the re-entrant auxetic specimen demonstrated approximately ten times higher rigidity at almost the same relative porosity (chiral structure 67%, reentrant structure 70%). Lvov et al. [21] proposed a three-dimensional auxetic structure and analyzed the mechanical characteristics using static and low-cycle compression tests. According to the tests results, auxetic structures are able to withstand cyclic loads longer than reverse non-auxetic cellular structures. Buckling at sufficiently low strain causes stresses along the rods that form an auxetic structure. Necembe et al. [22] investigated the fatigue behavior of the re-entrant auxetic structures made of the aluminum alloy AA 7075-T651. The influence of the unit cell orientation on the crack path and fatigue life was studied using experimental and computational approaches. The experimental and computational results showed that the unit cell's orientation has a minor influence on the fatigue life of both analyzed auxetic structures, but impact on the direction of the fatigue failure path significantly. Ulbin et al. [23] studied the effects of fillet radius of the auxetic cell on the fatigue life of the auxetic structures subjected to the cyclic loads. Based on the variations in the fillet radius, five different samples with negative Poisson ratios were examined. Results revealed that in decreasing the Poisson ratio, the fatigue life increased. Michalski and Strek [24] studied the high-cycle fatigue phenomenon in re-entrant auxetic and hexagonal honeycomb cells. They reported that re-entrant auxetic cell experienced higher fatigue lives as compared to hexagonal honeycomb cell structure.

Since the number of works regarding fatigue life of auxetic structures are very limited, further investigations on fatigue strength of these structures are necessary. More specifically, there have been no comprehensive investigations in the literature dealing with the effects of geometrical features of the re-entrant auxetic cells on fatigue life of the structures. It has been mentioned in the research literature that resistance to fatigue loading of auxetic structures has a direct relationship with the auxeticity property of the structure, but the parametric investigation in this field has not been conducted so far. It is considered that in the available literature, no research has discussed the effect of geometrical parameters of the re-entrant cell on the fatigue life of the structure. Therefore, in the current research, these two effective parameters (H/L ratio and θ) are used, and the effect of these parameters on the auxeticity property and fatigue life is investigated. This study intends to examine the effects of geometrical parameters of re-entrant auxetic cells on both Poisson's ratio and fatigue life of the Al 7075-T6 auxetic structures. To do so, numerical simulations and fatigue life predictions were performed through the use of ABAQUS and MSC Fatigue FE software. The implemented method was validated through appropriate case studies and the experimental test data available in the literature.

2. Numerical Simulations

2.1. Subsection

The stress analysis of the auxetic structures was performed through ABAQUS FEA package. To this end, first, the re-entrant auxetic structures were designed as shown in Figure 2.

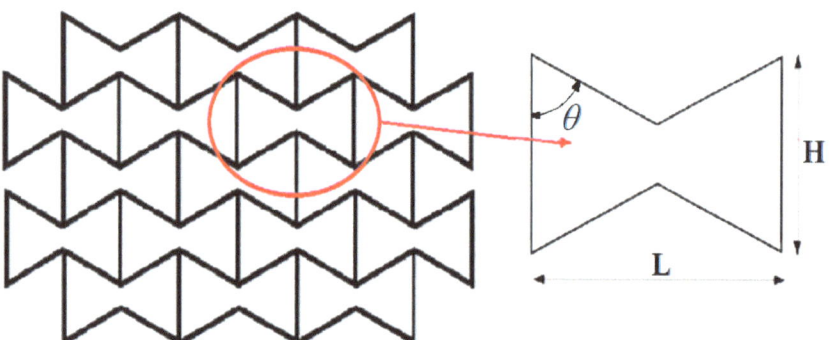

Figure 2. Schematic view of a re-entrant auxetic cell.

In a re-entrant auxetic cell, L is the length, H is the height, and θ is the internal angle of the cell. In this study, the effects of two geometrical parameters, namely, H/L and θ, were examined on the Poisson ratio and fatigue life of auxetic structures. Each mentioned factor was studied in five different levels. Table 1 shows all studied factors and their levels. In order to investigate comprehensive interactions between the two mentioned factors, a full factorial design was used in the numerical simulations. In total, 50 auxetic samples were designed and modelled. Among the models, 25 samples were utilized for examining the variations in Poisson ratios and the rest were used for fatigue life predictions. For all samples, the cell length L was set to be 10 mm, and the thickness of the structure's depth was considered to be 2 mm. It is worth mentioning that the strut thickness parameter was designed in such a way that the structure's mass for all samples was the same, enabling to have a better comparison between the results.

Table 1. Studied factors and levels.

Level	θ (Degrees)	H/L
1	50	1
2	55	1.1
3	60	1.2
4	65	1.3
5	70	1.4

Samples with different dimensions were utilized to assess the Poisson ratios and fatigue lives. To obtain Poisson ratios, samples with the dimensions of 60 × 60 mm were modelled, while to predict fatigue lives, the auxetic structures which have been introduced in Ref. [25] were employed. In order to apply a uniform load to the whole auxetic structures, thin films were placed at both ends of the structures. This enables the load to be distributed in the central part of the structures. The width of these thin films was constant and considered to be 3 mm for all models. The schematic view of the samples' geometry used for assessing Poisson's ratio and fatigue lives are illustrated, respectively, in Figure 3a,b.

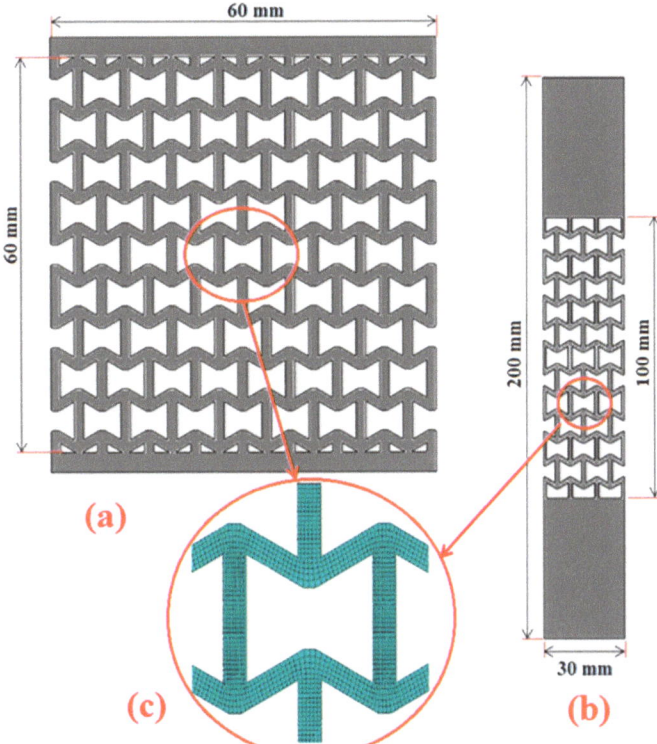

Figure 3. The schematic view of the auxetic samples for (**a**) static test modelling, (**b**) fatigue test modelling, and (**c**) mesh pattern of a typical cell.

The two-dimensional plain stress CPS4R Lagrangian elements were used in numerical modelling and simulations. A mesh-sensitivity analysis was performed to obtain an optimum mesh size. Based on this analysis, the ratio of mesh size to cell thickness was selected to be 0.2. Figure 3c illustrates the meshed model and mesh pattern in a re-entrant auxetic cell.

2.2. Material Properties, Loading, and Boundary Conditions

The aluminum 7075-T6 alloy was used for the auxetic structures undergoing static and cyclic loadings. The mechanical properties of the aluminum 7075-T6 alloy which has been extracted from the literature and assigned to the ABAQUS and MSC Fatigue software are listed in Table 2. In the simulations for performing the static and fatigue analyses, two different steps were carried out. In the first step, ABAQUS software was employed to assess the Poisson ratios of the auxetic structures subjected to static loads. To do so, the geometrical conditions, material properties, and loading type were assigned to the software. Then, all 25 auxetic structures were subjected to a displacement-controlled loading until the tensile strain reached to 0.1. After that, each sample was numerically analyzed, and the initial Poisson ratios, as well as the variations in the average magnitudes of the Poisson ratios with the strain values, were calculated. In the second step, the remaining 25 fatigue sample models were subjected to cyclic loads with the load ratio of 0.1. Fatigue life predictions for each auxetic structure were conducted at five different strain levels. Stress analysis was carried out in ABAQUS software, and the results of the stress states were imported to the MSC Fatigue software to predict fatigue lives. The authors of [25] chose the same material and experimental fatigue tests. It should be noted that the aforementioned reference was used only for the verification stage. Considering that in auxetic structures, the main factor affecting the auxetic properties of the structure is the geometrical parameters of the cells that make up the structure. In re-entrant cells, the geometrical parameters of the H/L ratio, the thickness of the cell arm, and θ are three geometrical parameters, and by changing each of these parameters, the structure's response against loading changes drastically.

Table 2. Mechanical properties of Al7075-T651 used in static and fatigue analysis [25].

Properties	Static	Fatigue
Young Modulus (GPa)	68.9	-
Poisson Ratio	0.33	-
Yield Stress (MPa)	539	-
Ultimate Tensile Stress (MPa)	596	-
Elongation at The Break (%)	0.12	-
Fatigue Strength Coefficient (MPa)	-	1145
Fatigue Ductility Coefficient	-	0.0686
Fatigue Strength Exponent	-	−0.0048
Fatigue Ductility Exponent	-	−0.3605

According to the amount of loading of the samples and the presence of plastic behavior in the auxetic structures under the load investigated in this research, the bilinear elastic–plastic physical material model has been used to define the behavior of the material. The parameters required to develop this model are given in Table 2. It should be noted that the hardening is also considered as Isotropic. Each of the fatigue analyzes carried out in the present research is performed as follows:

1. In the first stage, stress analysis has been performed in Abaqus software. In the stress analysis stage, considering that the loading has exceeded the elastic limit, it is necessary to repeat the loading cycles until the shape of the stress cycles becomes stable. Therefore, to stabilize each of the models, 50 cycles have been simulated so that the stress becomes stable and can be used in the next step;
2. In the second stage, the results of the stabilized stress analysis have been imported into the fatigue life estimation software, and the fatigue life estimation for each model has been performed by the fatigue model mentioned in the text of the article and the parameters presented in Table 2.

2.3. Fatigue Life Prediction

Fatigue fracture can take place at any number of cyclic reversals depending on the load levels and other test conditions. High cycle fatigue (HCF) and low cycle fatigue (LCF) regimes are usually defined based on a transition fatigue life (N_t). Normally, reversals lower than 10^4 are classified as LCF and greater than 10^4 cycles are categorized as HCF [26]. In HCF, elastic strains are dominant since the plastic strain magnitudes are very small and can be negligible as opposed to LCF conditions. The strain-based approaches are usually utilized to predict fatigue life of structures based on the strain components and their ranges. This approach is suitable to predict the fatigue life of metallic parts especially in low cycle fatigue regime. An appropriate strategy to perform the fatigue analysis is to consider both elastic and plastic strain components and develop the method based on their magnitudes. The Manson–Coffin criterion has been developed based on the above-mentioned concept [27].

$$\frac{\Delta \varepsilon}{2} = \frac{\sigma'_f}{E}(2N_f)^b + \varepsilon'_f(2N_f)^c \qquad (1)$$

In this equation, $\Delta \varepsilon$ is the strain range, E is the elastic modulus, and N_f is the number of cycles to failure. Furthermore, σ'_f, ε'_f, b, and c are, respectively, the fatigue strength coefficient, fatigue ductility coefficient, fatigue strength exponent, and fatigue ductility exponent. It is worth mentioning that, in the Manson–Coffin criterion, the effects of mean stress (σ_m) has not been included. A number of criteria have been proposed to consider the effects of mean stress [28]. In a criterion which has been proposed by Smith–Watson–Topper (SWT) for each mean stress condition, the product of maximum stress (σ_{max}) and strain amplitude (ε_a) has been considered to be equal to that of completely reversed condition at a same number of cycles to failure. The SWT criterion can be expressed as follows [29]:

$$\sigma_{max} \frac{\Delta \varepsilon}{2} = \frac{(\sigma'_f)^2}{E}(2N_f)^{2b} + \sigma'_f \varepsilon'_f (2N_f)^{b+c} \quad (2)$$

It has been proven that the SWT parameter led to appropriate results for a wide range of aluminum alloys [30]. However, this criterion can lead to erroneous results for loading conditions with negative mean stresses. In this study, the selected material is an aluminum alloy [31], and mean stresses are non-zero positive values; therefore, it seems that the SWT criterion is a suitable method for fatigue life predictions [32]. The fatigue parameters of Al 7075-T6 alloy required for the SWT criterion are summarized in Table 2.

3. Verification Procedure

One of the necessities and prerequisites of any numerical models such as finite element methods is implementing and conducting the validation process.

To ensure the accuracy and correctness of the numerical models, the validation results need to be reported prior to present other results of the simulations. In this study, to validate the results of simulations, a comparison made between the experimental data available in the literature and the numerical results of two different specimens undergoing both static and cyclic loads. To this end, some numerical models of the standard fatigue specimens, as well as the auxetic fatigue samples similar to those used in Ref. [25], were created and the results were compared with associated fatigue test data. The dimensions of the standard fatigue sample were in accordance with the ASTM-E606 standard. The geometry of the studied fatigue samples for validation process are shown in Figure 4.

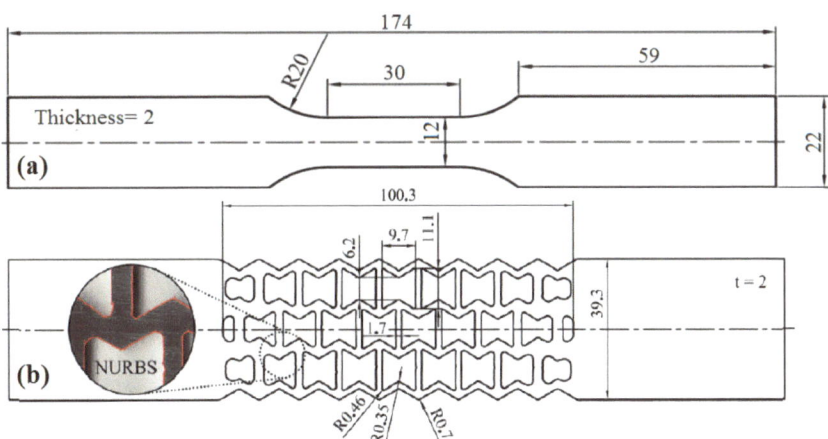

Figure 4. Dimensions of fatigue specimens: (**a**) standard fatigue sample and (**b**) auxetic fatigue sample (All dimensions are in mm) [25].

4. Results and Discussions

4.1. Results of Case Studies for Validation

Figure 5a shows the fatigue life and strain contours of the standard fatigue sample at 0.01 strain level. Figure 5b depicts the numerically obtained strain versus fatigue life curve of the standard fatigue sample, and compares the numerical results with experimental test data. The results show that the maximum strain component (E11) was in the sample's center, near 0.01. Similarly, to maximum strain, the maximum fatigue life predicted in the center of the sample was 234 repeats. Figure 5c presents the simulation results of strain distribution and fatigue life of auxetic sample. Figure 5d compares the numerically obtained fatigue lives of the auxetic fatigue samples with experimental fatigue test data. As seen in Figure 5b,d, the numerically obtained fatigue lives are in good agreement with the experimental fatigue test data. This accuracy between the results proves the reliability of the procedure used for other case studies.

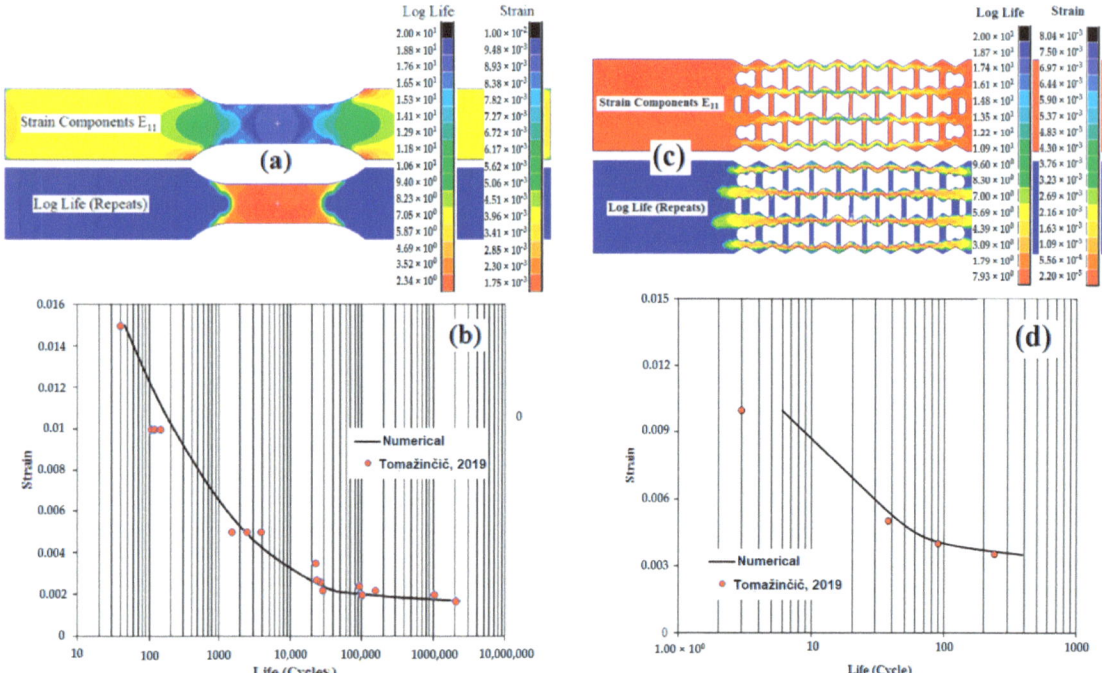

Figure 5. Case study of the standard fatigue sample: (**a**) the contour of strain and fatigue lives in logarithmic scale and (**b**) the strain versus fatigue life curve. (**c**) The contour of strain and fatigue life in logarithmic scale and (**d**) the strain versus fatigue life curve of auxetic sample [25].

4.2. Effects of H/L Parameter on Poisson's Ratio

The influence of variation of height to the length of the re-entrant cell has been studied on Poisson ratios of different auxetic structures. Figure 6 shows the variations of the average of the Poisson ratios of the auxetic structures versus H/L ratios. As seen, any geometrical changes in the re-entrant cell led to a relatively slight variation in Poisson ratios. In the case of constant cell's length, increasing in cell's height resulted in an improvement in the Poisson ratio of the structure. As an average, an increment in H/L ratio from 1 to 1.4 led to 7.7% increase in the average value of the Poisson ratio.

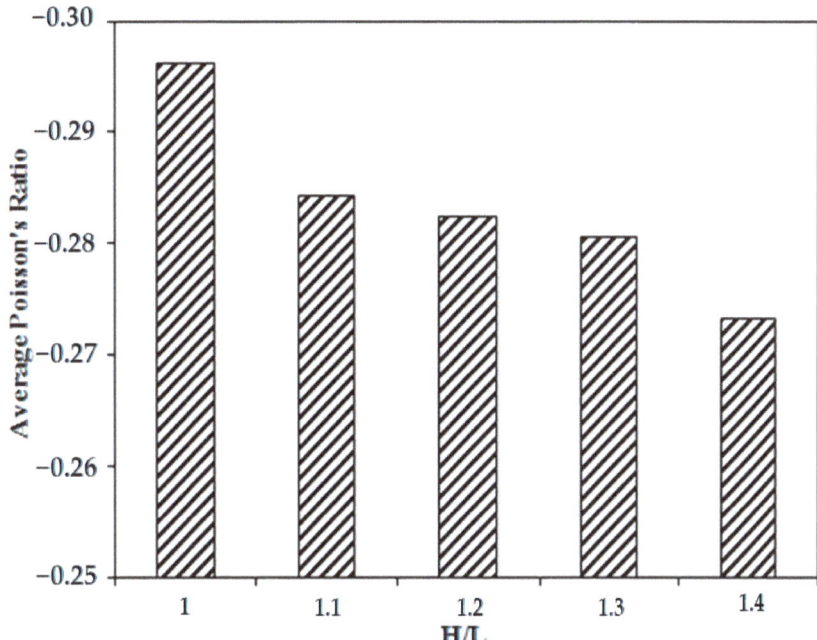

Figure 6. The variations of the average of the Poisson ratios of the auxetic structures versus H/L ratios.

The behavior of auxetic cells subjected to different loadings and deformation patterns vary with the cell's dimensional ratios [30,31]. In general, with increasing the cell's height, the number of cells along the loading direction decrease, leading to an increase in the length of longitudinal arms of the cells and a reduction in transverse displacements. As an example, a deformation pattern in an auxetic structure in $\theta = 50°$ and various H/L values is shown in Figure 7. As can be seen in Figure 7, in structures with a larger H/L ratio, the amount of displacement in the direction perpendicular to the loading has decreased. By reducing the transverse displacement of the sample under the same axial loading, the Poisson ratio of the structure decreases. The changes in Poisson ratio in three H/L ratios of 1.1, 1.2, and 1.3 are small, but when this ratio reaches 1.4, the increase in Poisson ratio occurs with a larger slope.

4.3. Effects of Cell's Angle (θ) on Poisson's Ratio

Figure 8 illustrates the average value of Poisson's ratio of the re-entrant auxetic structures versus the cell's angle.

A variation in cell's angle results in a significant change in the structure's Poisson ratio. An increase in θ leads to an elevation in the average value of Poisson's ratio and a noticeable reduction in the structure's auxeticity. The highest and lowest values of the average of Poisson's ratios occurred, respectively, at $\theta = 70°$ and $\theta = 50°$. Increasing θ from 50° to 70° resulted in 80% improvement in Poisson's ratio. Moreover, increasing θ would lead to change the cell's shape from a re-entrant cell to a square honeycomb shape, resulting in a reduction in the structure's auxeticity. Figure 9 shows the variations in deformation patterns of the auxetic structures in different cell angles.

Figure 7. Deformation pattern in an auxetic structure in $\theta = 50°$ and various H/L values.

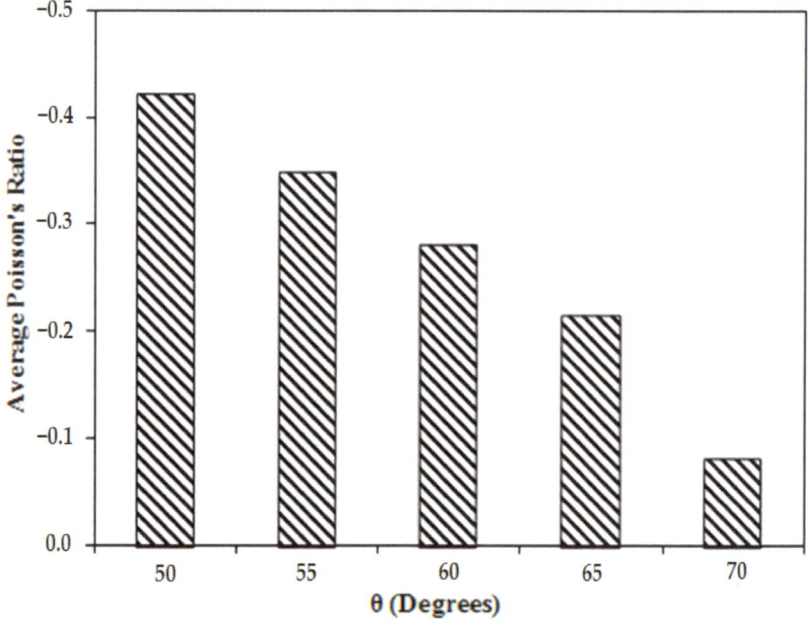

Figure 8. The average value of Poisson's ratio of the re-entrant auxetic structures versus the cell's angle.

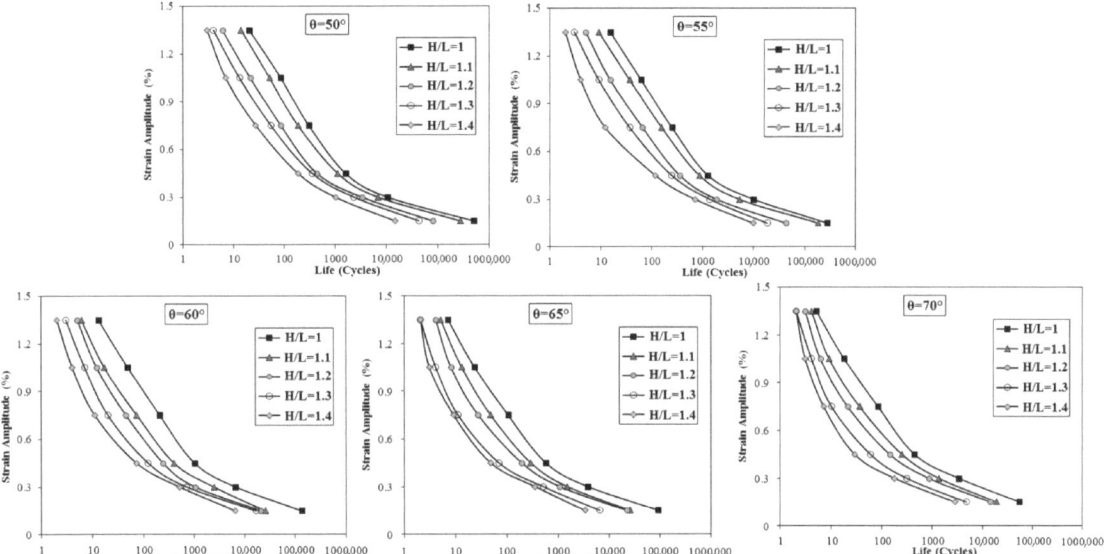

Figure 9. Effects of H/L ratio on the fatigue resistance of the auxetic structures in different cell's angle θ.

According to Figure 10, with the increase in θ of the auxetic cell, the overall shape of the cell undergoes geometrical changes, and the auxetic structure under axial loading shows behavior similar to the typical honeycomb structures. As the angle θ increases, the load shifts from the transverse arms to the longitudinal arms of the cell, and this causes local buckling in the longitudinal arms. With local buckling of these longitudinal arms, the amount of transverse positive displacement of the structure under tensile load is reduced. As can be seen in Figure 9, the auxetic structure with angle θ equal to 70, which was the largest angle examined, showed almost the same deformation pattern as the square honeycomb structure.

4.4. Effects of H/L Ratio on Fatigue Life

For the studied models in this investigation, the fatigue strength of the structure was studied in five different stress levels. Figure 11 depicts the effects of H/L ratio on the fatigue resistance of the auxetic structures in different cell's angle (θ).

In all angles, the fatigue strength decreased significantly with increasing the H/L ratio. The variation in H/L value in LCF regime had a greater influence on fatigue strength compared to that of HCF regime, causing a larger reduction in fatigue life. Results revealed that the structure's auxeticity had a direct relationship with fatigue strength. The structures with higher auxeticity values possessed greater fatigue strength. Figure 11 shows stress contours and fatigue life contours in logarithmic scale in one cell at different H/L values. As evidenced in Figure 11, regardless of the cell's H/L value, the maximum stress occurred in the conjunction area of the internal cell's arm. It is obvious in Figure 11 that the stress value has a direct relationship with cell's H/L value. With increasing the H/L value, the maximum stress values and the stress concentration areas have been increased. As evidenced in fatigue life contour, fatigue crack initiation areas are exactly matched with stress concentration regions. The joints between the nodes of the re-entrant cell's arms are the most susceptible points for fatigue cracking. In the nodes connecting the transverse and longitudinal arms of the cell, load transfer occurs and this load transfer leads to more stress concentration in these areas. With the increase in stress concentration in these areas, the probability of fatigue crack formation under fatigue loading increases, which is also evident in the contours of the log of life shown in Figure 11. After the mentioned areas, the longitudinal arms of the cells are prone to fatigue cracks. In the longitudinal arms of the cell, the internal parts of the longitudinal spans have a shorter fatigue life than the peripheral areas, and cracks generally grow along the thickness from the inside to the outside of the cell wall.

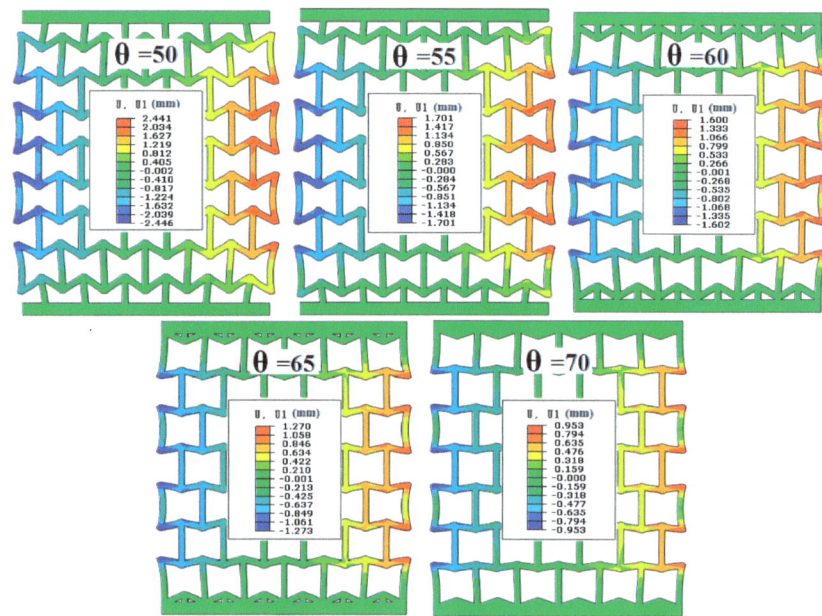

Figure 10. The variations in deformation patterns of the auxetic structures in different cell angles.

Figure 11. Stress contours and fatigue life contours in logarithmic scale in one cell at different H/L values.

4.5. Effects of Cell's Angle (θ) on Poisson's Ratio

In Figure 12, the effects of θ on the fatigue life of auxetic structures at different θ angles are shown.

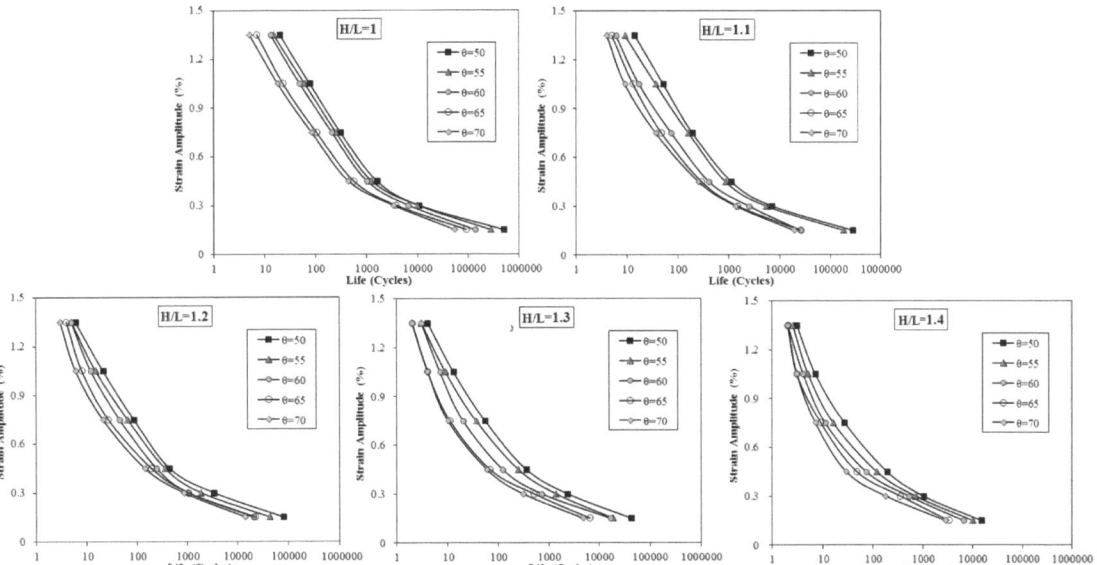

Figure 12. Stress contours and fatigue life contours in logarithmic scale in one cell at different H/L values.

Furthermore, in Figure 13, the stress contours and the log of life contours of the life of a cell representing the entire structure with different θ angles are displayed. According to Figure 12, in all different H/L ratios, the fatigue life of the structure decreases with the increase in cell angle θ. By comparing Figures 10 and 12, it can be seen that the changes in the H/L ratio have a stronger effect on the fatigue life than the changes in the θ angle. For the θ parameter, it is also observed that the changes in this parameter in the low cycle region have a stronger effect on the fatigue life of the structure and its increase in the larger strain ranges has caused a sharper decrease in the fatigue life.

According to Figure 13, regardless of θ of the cell, the highest stress occurred in the joint area of the middle sides of re-entrant auxetic cells. According to the contours of the log of life, the upper and lower nodes are the most prone areas to initiate fatigue damage. According to the stress contours, a strong stress concentration has occurred in the nodes connecting cells to adjacent cells due to the presence of shear stress. In all samples, in the connecting areas of the middle arms, the angle and orientation of the maximum stress areas have a phase difference of 45 degrees with the direction of loading.

By studying the effect of θ parameter on the fatigue life and Poisson's ratio of the structure, it was found that the auxetic property of the structure has a direct relationship with the fatigue resistance of the structure. In all samples, it was observed that structures with greater auxetic property have higher fatigue resistance.

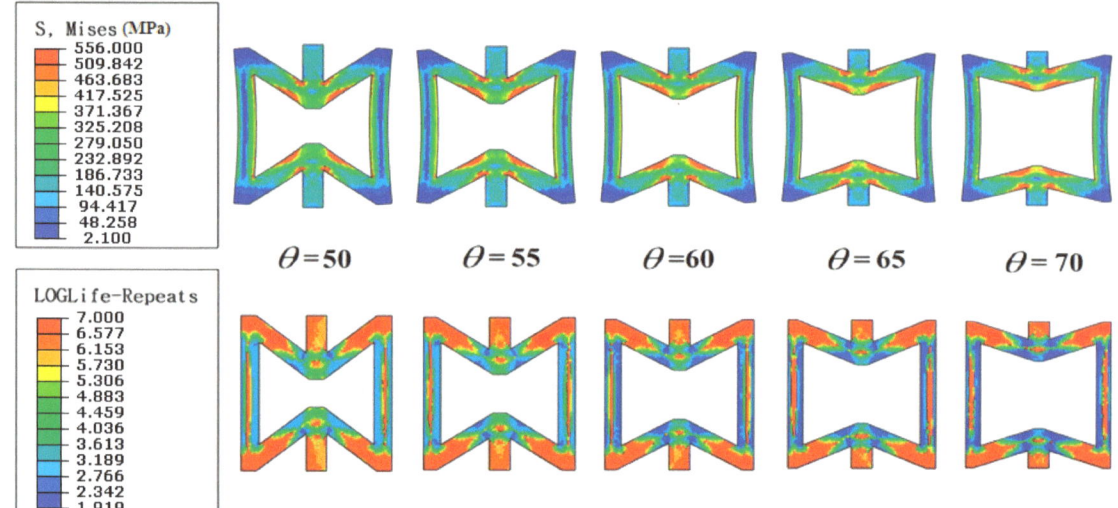

Figure 13. Stress contours and fatigue life contours in logarithmic scale in one cell at different cell's angle θ.

5. Conclusions

In this study, the influence of the θ angle and H/L ratio of the re-entrant auxetic cells on the variations in Poisson's ratio and fatigue life of the auxetic structures were numerically examined. Five different θ angles and five various H/L ratios were studied. The following conclusions were drawn out from the results:

1. Increasing the H/L ratios of the re-entrant cell relatively increased the average value of the Poisson ratio of the structure. As an average, the increasing H/L ratio from 1 to 1.4 elevated the Poisson ratio by 7.7%;
2. It was found that the parameter θ angle had a direct relationship with the Poisson ratio of the structure. Increasing θ from 50° to 70° elevated the value of the Poisson ratio by about 80%;
3. In general, with increasing the H/L ratio, the number of cells along the loading direction decreased, resulting in an enlargement of the longitudinal arms of the cells along the loading direction and a reduction in the ability of the lateral displacement of the structure. Moreover, with increasing θ angle, the re-entrant cell's shape changed to square honeycomb cell, leading to a drop in the structure's auxeticity;
4. In all θ angles, with increasing H/L value, the fatigue life of the structure decreased significantly. The effect of the variation in H/L value in the LCF regime was greater than that in the HCF regime, resulting in greater reduction in fatigue life;
5. In all H/L values, it was evidenced that increasing θ angle of the re-entrant cell would decrease the fatigue strength of the auxetic structure. The impact of the variation in θ angle in the LCF regime was more influential than that in the HCF regime.
6. Overall, the effects of H/L parameter on fatigue life were greater than that of θ angle. Regardless of the values of H/L ratio or θ angle, the maximum stresses occurred in the conjunction of the cell's arms. According to the fatigue analysis, the crack initiation areas were the stress concentration regions of the internal cells of the auxetic structure.

Author Contributions: Conceptualization, A.G., A.F.K., J.W.G.G., H.A.D. and J.T.; methodology, A.G., A.F.K. and J.W.G.G.; software, A.G. and A.F.K.; validation, A.G. and A.F.K.; formal analysis, A.G., A.F.K., J.W.G.G., H.A.D., J.T., A.J. and A.W.; investigation, A.G., A.F.K., J.W.G.G. and H.A.D.; resources, A.G., A.F.K., J.W.G.G., H.A.D. and J.T.; data curation, A.G., A.F.K., J.W.G.G. and H.A.D.; writing—original draft preparation, A.G. and A.F.K.; writing—review and editing, A.G., A.F.K., J.W.G.G., H.A.D., J.T., A.W. and A.J. All authors have read and agreed to the published version of the manuscript.

Funding: This research received no external funding.

Institutional Review Board Statement: Not applicable.

Informed Consent Statement: Not applicable.

Data Availability Statement: Not applicable.

Conflicts of Interest: The authors declare no conflict of interest.

References

1. Gao, Q.; Ge, C.; Zhuang, W.; Wang, L.; Ma, Z. Crashworthiness analysis of double-arrowed auxetic structure under axial impact loading. *Mater. Des.* **2019**, *161*, 22–34. [CrossRef]
2. Imbalzano, G.; Linforth, S.; Ngo, T.; Lee, P.; Tran, P. Blast resistance of auxetic and honeycomb sandwich panels: Comparisons and parametric designs. *Compos. Struct.* **2018**, *183*, 242–261. [CrossRef]
3. Ren, X.; Shen, J.; Tran, P.; Ngo, T.; Xie, Y. Design and characterisation of a tuneable 3D buckling-induced auxetic metamaterial. *Mater. Des.* **2018**, *139*, 336–342.
4. Lim, T.-C. *Auxetic Materials and Structures*; Springer: Berlin/Heidelberg, Germany, 2015.
5. Wang, Z.; Zulifqar, A.; Hu, H. *Auxetic Composites in Aerospace Engineering, Advanced Composite Materials for Aerospace Engineering*; Elsevier: Amsterdam, The Netherlands, 2016; pp. 213–240.
6. Munteanu, L.; Dumitriu, D.; Donescu, Ș.; Chiroiu, V. On the complexity of the auxetic systems. In *Proceedings of the European Computing Conference*; Springer: Berlin/Heidelberg, Germany, 2009; pp. 631–636.
7. Lakes, R. Negative-Poisson's-ratio materials: Auxetic solids. *Annu. Rev. Mater. Res.* **2017**, *47*, 63–81. [CrossRef]
8. Albag, O. *Auxetic Materials, Material Balance*; Springer: Berlin/Heidelberg, Germany, 2021; pp. 65–74.
9. Khare, E.; Temple, S.; Tomov, I.; Zhang, F.; Smoukov, S. Low fatigue dynamic auxetic lattices with 3D printable, multistable, and tuneable unit cells. *Front. Mater.* **2018**, *5*, 45. [CrossRef]
10. Nasim, M.S.; Etemadi, E. Three dimensional modeling of warp and woof periodic auxetic cellular structure. *Int. J. Mech. Sci.* **2018**, *136*, 475–481. [CrossRef]
11. Essassi, K.; Rebiere, J.-L.; El Mahi, A.; Ben Souf, M.A.; Bouguecha, A.; Haddar, M. Experimental and analytical investigation of the bending behaviour of 3D-printed bio-based sandwich structures composites with auxetic core under cyclic fatigue tests. *Compos. Part A Appl. Sci. Manuf.* **2020**, *131*, 105775. [CrossRef]
12. Choi, J.; Lakes, R. Non-linear properties of metallic cellular materials with a negative Poisson's ratio. *J. Mater. Sci.* **1992**, *27*, 5375–5381. [CrossRef]
13. Nečemer, B.; Glodež, S.; Novak, N.; Kramberger, J. Numerical modelling of a chiral auxetic cellular structure under multiaxial loading conditions. *Theor. Appl. Fract. Mech.* **2020**, *107*, 102514. [CrossRef]
14. Kramberger, J.; Nečemer, B.; Glodež, S. Assessing the cracking behavior of auxetic cellular structures by using both a numerical and an experimental approach. *Theor. Appl. Fract. Mech.* **2019**, *101*, 17–24. [CrossRef]
15. Nečemer, B.; Kramberger, J.; Vuherer, T.; Glodež, S. Fatigue crack initiation and propagation in re-entrant auxetic cellular structures. *Int. J. Fatigue* **2019**, *126*, 241–247. [CrossRef]
16. Bezazi, A.; Scarpa, F. Mechanical behaviour of conventional and negative Poisson's ratio thermoplastic polyurethane foams under compressive cyclic loading. *Int. J. Fatigue* **2007**, *29*, 922–930. [CrossRef]
17. Francesconi, L.; Baldi, A.; Dominguez, G.; Taylor, M. An investigation of the enhanced fatigue performance of low-porosity auxetic metamaterials. *Exp. Mech.* **2020**, *60*, 93–107. [CrossRef]
18. Nečemer, B.; Klemenc, J.; Glodež, S. The computational LCF-analyses of chiral and Re-entrant auxetic structure using the direct cyclic algorithm. *Mater. Sci. Eng. A* **2020**, *789*, 139618. [CrossRef]
19. Nečemer, B.; Kramberger, J.; Glodež, S. Fatigue crack growth in the re-entrant auxetic structure. *Procedia Struct. Integr.* **2022**, *39*, 34–40. [CrossRef]
20. Nečemer, B.; Klemenc, J.; Zupanič, F.; Glodež, S. Modelling and predicting of the LCF-behaviour of aluminium auxetic structures. *Int. J. Fatigue* **2022**, *156*, 106673. [CrossRef]
21. Lvov, V.; Senatov, F.; Stepashkin, A.; Veveris, A.; Pavlov, M.; Komissarov, A. Low-cycle fatigue behavior of 3D-printed metallic auxetic structure. *Mater. Today Proc.* **2020**, *33*, 1979–1983. [CrossRef]
22. Nečemer, B.; Vuherer, T.; Glodež, S.; Kramberger, J. Fatigue behaviour of re-entrant auxetic structures made of the aluminium alloy AA7075-T651. *Thin-Walled Struct.* **2022**, *180*, 109917. [CrossRef]

23. Ulbin, M.; Borovinšek, M.; Vesenjak, M.; Glodež, S. Computational Fatigue Analysis of Auxetic Cellular Structures Made of SLM AlSi10Mg Alloy. *Metals* **2020**, *10*, 945. [CrossRef]
24. Michalski, J.; Strek, T. *Fatigue Life of Auxetic Re-entrant Honeycomb Structure, International Scientific-Technical Conference Manufacturing*; Springer: Berlin/Heidelberg, Germany, 2019; pp. 50–60.
25. Tomažinčič, D.; Nečemer, B.; Vesenjak, M.; Klemenc, J. Low-cycle fatigue life of thin-plate auxetic cellular structures made from aluminium alloy 7075-T651. *Fatigue Fract. Eng. Mater. Struct.* **2019**, *42*, 1022–1036. [CrossRef]
26. Schijve, J. *Fatigue of Structures and Materials*; Springer Science & Business Media: Berlin/Heidelberg, Germany, 2001.
27. Niesłony, A.; el Dsoki, C.; Kaufmann, H.; Krug, P. New method for evaluation of the Manson–Coffin–Basquin and Ramberg–Osgood equations with respect to compatibility. *Int. J. Fatigue* **2008**, *30*, 1967–1977. [CrossRef]
28. Suresh, S. *Fatigue of Materials*; Cambridge University Press: Cambridge, UK, 1998.
29. Li, J.; Sun, Q.; Zhang, Z.; Qiao, Y.-J.; Liu, J. A modification of Smith-Watson-Topper damage parameter for fatigue life prediction under non-proportional loading. *Fatigue Fract. Eng. Mater. Struct.* **2012**, *35*, 301–316. [CrossRef]
30. Tamadon, A.; Pons, D.J.; Sued, K.; Clucas, D. Internal Flow Behaviour and Microstructural Evolution of the Bobbin-FSW Welds: Thermomechanical Comparison between 1xxx and 3xxx Aluminium Grades. *Adv. Mater. Sci.* **2021**, *21*, 40–64. [CrossRef]
31. Mohan, D.G.; Tomków, J.; Gopi, S. Induction Assisted Hybrid Friction Stir Welding of Dissimilar Materials AA5052 Aluminium Alloy and X12Cr13 Stainless Steel. *Adv. Mater. Sci.* **2021**, *21*, 17–30. [CrossRef]
32. Tu, X.; Shahba, A.; Shen, J.; Ghosh, S. Microstructure and property based statistically equivalent RVEs for polycrystalline-polyphase aluminum alloys. *Int. J. Plast.* **2019**, *115*, 268–292. [CrossRef]

Disclaimer/Publisher's Note: The statements, opinions and data contained in all publications are solely those of the individual author(s) and contributor(s) and not of MDPI and/or the editor(s). MDPI and/or the editor(s) disclaim responsibility for any injury to people or property resulting from any ideas, methods, instructions or products referred to in the content.

Article

Effect of Al$_2$O$_3$ Content on High-Temperature Oxidation Resistance of Ti$_3$SiC$_2$/Al$_2$O$_3$

Yuhang Du [1], Qinggang Li [2,*], Sique Chen [2], Deli Ma [1], Baocai Pan [2], Zhenyu Zhang [1] and Jinkai Li [2]

[1] Shandong Provincial Key Laboratory of Preparation and Measurement of Building Materials, Jinan 250022, China
[2] School of Material Science and Engineering, University of Jinan, Jinan 250022, China
* Correspondence: mse_liqg@ujn.edu.cn

Abstract: Considering the lack of an effective anti-oxidation protective layer for the oxidation process of Ti$_3$SiC$_2$, an in situ synthesis of Ti$_3$SiC$_2$ and Al$_2$O$_3$ was designed. Thermally stable Al$_2$O$_3$ was used to improve the high-temperature oxidation resistance of Ti$_3$SiC$_2$. Samples without TiC were selected for the oxidation test, and the oxidation morphology and weight gain curves of the oxidized surface in air at 1400 °C are reported. The change in the oxidation behavior occurred 4 h after oxidation. The addition of Al$_2$O$_3$ changed the composition of the oxide layer and compensated for the lack of a dense protective layer during Ti$_3$SiC$_2$ oxidation. Moreover, after 4 h of oxidation, the newly generated Al$_2$TiO$_5$ and the composite layer formed by diffusion were the main reasons for the large difference in the final weight gain between the two sets of samples.

Keywords: Ti$_3$SiC$_2$; Al$_2$O$_3$; high temperature; oxidation resistance

1. Introduction

Ceramic is a type of material with unique characteristics, such as high-temperature oxidation resistance, high strength, and elastic stiffness, but it has inherent brittleness and low machinability [1–4]. A special group of materials in the ceramic family are the MAX-phase ceramics, which have a hexagonal structure and combined metal-ceramic properties. MAX-phase materials have broad application prospects owing to their excellent properties. However, compared with conventional ceramics, the hardness and high-temperature oxidation resistance of MAX-phase materials are lower, which significantly limits their application in the engineering field. Therefore, it is necessary to improve their mechanical properties and high-temperature stability. Ti$_3$SiC$_2$ is a MAX-phase compound with a layered structure that is a promising candidate for high-temperature applications [5,6]. In addition to its simple machinability, this material has excellent properties, such as electrical conductivity, thermal conductivity, and thermal shock resistance [7,8]. As a typical MAX-phase material, Ti$_3$SiC$_2$ is a promising structural ceramic for high-temperature applications such as heating elements in high-temperature furnaces and fuel-combustion components in automobiles and aircraft engines [9,10].

Notably, the oxidation resistance of Ti$_3$SiC$_2$ is crucial and has been investigated extensively, whether in the application of high-temperature structural ceramics or connection materials for solid oxide fuel cells. The preferential oxidation behavior of Ti$_3$AlC$_2$ is different from that of Ti$_3$SiC$_2$, which has a continuous Al$_2$O$_3$ layer [11–13]. The antioxidation capacities of Ti$_3$SiC$_2$ require further improvement for its effective application.

Reinforcement phases, including TiC, SiC, c-BN, TiB$_2$, and ZrO$_2$, have been used to improve the mechanical properties and oxidation resistance of Ti$_3$SiC$_2$ [14–17]. Li et al. [18] prepared dense SiC/Ti$_3$Si(Al)C$_2$ composites using an in situ hot-pressing sintering method and reported that the oxide layers formed at 1200 and 1300 °C were divided into outer, middle and inner layers. To obtain high-purity Ti$_3$SiC$_2$, Xu et al. [19] demonstrated that the incorporation of a small amount of Al was beneficial for improving the purity of Ti$_3$SiC$_2$.

Moreover, the addition of Al was advantageous for improving the oxidation resistance of the composites [20,21]. Some researchers believe that the optical and electrical properties of alumina at high temperature have crucial application value and prospects for fusion technology [22–24].

Thus, the dense Al_2O_3 layer formed during the oxidation of Ti_3AlC_2 is the design inspiration for this study. Moreover, considering the lack of an effective anti-oxidation protective layer in the Ti_3SiC_2 oxidation process, thermally stable Al_2O_3 is selected as a reinforcement phase in this study to change the oxidation resistance of Ti_3SiC_2. A Ti_3SiC_2/Al_2O_3 composite is synthesized in situ using the hot-pressing sintering method, and the high-temperature oxidation resistance of the composite is reported. Therefore, this study aims to present a detailed investigation of the high-temperature oxidation resistance of Ti_3SiC_2/Al_2O_3.

2. Experimental Procedure

The volume capacity of 30%, 40% and 50% Al_2O_3 were added and the powders of Ti:Si:TiC:Al in the molar ratio of 1:1.2:2:0.3 were used to synthesize Ti_3SiC_2/Al_2O_3 composites. In situ synthesis of Ti_3SiC_2 and Al_2O_3 was designed, namely TSC70 (Ti_3SiC_2/30 vol.% Al_2O_3), TSC60 (Ti_3SiC_2/40 vol.% Al_2O_3), TSC50(Ti_3SiC_2/50 vol.% Al_2O_3). TiC (99.9% purity, average particle size 1 μm, Shanghai ST-Nano Technology Co., Ltd., Shanghai, China), Ti (99.9% purity, average particle size 1–3 μm, Shanghai ST-Nano Technology Co., Ltd., Shanghai, China), Al (99.9% purity, average particle size 50 nm, Shanghai ST-Nano Technology Co., Ltd., Shanghai, China), Si (99.9% purity, average particle size 1 μm, Shanghai ST-Nano Technology Co., Ltd., Shanghai, China) and Al_2O_3 (99.9% purity, average particle size 30 nm, Shanghai ST-Nano Technology Co., Ltd., Shanghai, China) were used as raw materials. The original powders were mixed into ethanol by ball-milling for 4 h. Then the slurry was dried in a drying oven at 40 °C for 6 h and then sieved under 100 mesh. The Ti_3SiC_2/Al_2O_3 composites were in situ fabricated by vacuum hot-press sintering (VVPgr-80-2200, Shanghai, China) at 1450 °C with an applied pressure of 30 MPa for 1.5 h (the vacuum degree was 6.71×10^{-3} MPa).

For the oxidation experiments, rectangular blocks of size 5 mm × 4 mm × 4 mm were cut using a cylindrical SiC blade. The surface was polished with SiC paper. Thereafter, the samples were ultrasonically cleaned with ethanol to remove surface impurities. Oxidation tests was carried out in an alumina tube furnace at 1400 °C and the samples were exposed for up to 20 h. After the alumina tube furnace was heated to the test temperature, the block Ti_3SiC_2/Al_2O_3 composites to be tested were placed in the furnace. Using an electronic balance of accuracy 1×10^{-7} kg, the difference in weight gain was calculated.

X-ray diffraction (XRD) (D8 ADVANCE, Bruker, Saarbrucken, Germany) was used to confirm the phases of the samples before and after oxidation. The microstructure of the oxidized samples was observed by scanning electron microscopy (SEM) (FEI QUANTA FEG 250, Hillsboro, OR, USA) with energy dispersive X-ray spectrum (EDS).

3. Results and Discussion

The phase components of the Ti_3SiC_2/Al_2O_3 composite were characterized by X-ray diffraction (XRD). Figure 1 shows the XRD patterns of TSC70, TSC60 and TSC50 before oxidation. As shown in Figure 1, Ti_3SiC_2 and Al_2O_3 did not generate Ti–Al compounds, indicating that the composite degree between Ti_3SiC_2 and Al_2O_3 was appropriate. This plays an important role in the subsequent investigation of the high-temperature oxidation resistance of the Ti_3SiC_2/Al_2O_3 composite. Notably, several TiC peaks were observed in TSC70, but none were detected in TSC50 and TSC60. Under the same sintering conditions, the peak intensities of Al_2O_3 increased with increasing Al_2O_3 content, corresponding with the change in the volumetric fraction added during synthesis. This phenomenon indicates that the selection of the raw material was successful. Sun et al. [25] reported that the oxidation rate of Ti_3SiC_2 is slower than that of TiC, and TiC is detrimental to the oxidation resistance of Ti_3SiC_2. Therefore, in the subsequent experiments, two sets of samples without TiC were selected for the oxidation test.

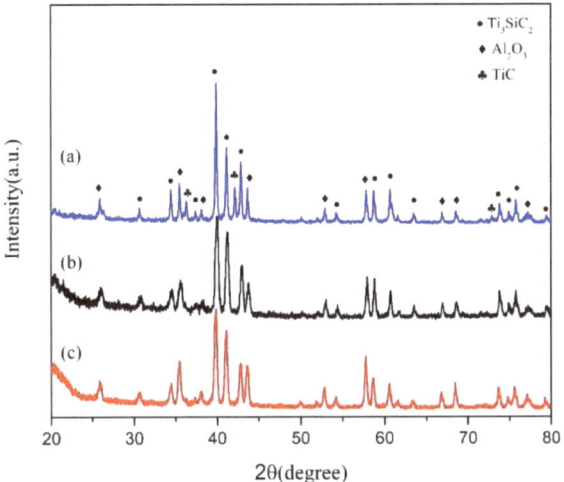

Figure 1. X-ray diffraction patterns of non-oxidizing materials: (a) TSC70, (b) TSC60 and (c) TSC50.

Figures 2 and 3 show the XRD patterns of the TSC50 and TSC60 samples, respectively, oxidized at 1400 °C. After the oxidation of TSC50 for up to 4 h, XRD showed that the TiO_2 content was relatively low. The presence of matrix Ti_3SiC_2 and Al_2O_3 was caused by the short oxidation time, and a dense and continuous oxide layer was not formed. Thus, the exposure of the matrix to the surface was accompanied by a small amount of TiO_2. In contrast to TSC50, the oxidation products of TSC60 with less Al_2O_3 were different after 4 h of oxidation. The intensity of the Ti_3SiC_2 peaks in TSC60 were significantly reduced, and the oxidized surface was mainly composed of TiO_2, Al_2O_3, and newly generated Al_2TiO_5, indicating that a continuous and thin oxide layer was formed on the sample surface.

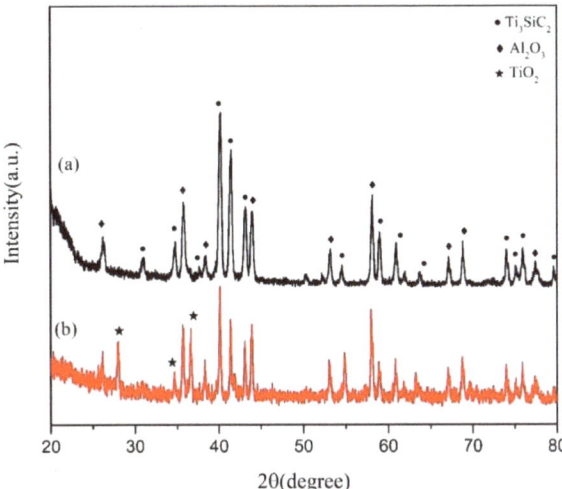

Figure 2. X-ray diffraction patterns of the TSC50 (a) before oxidation and (b) after oxidation at 1400 °C for 4 h.

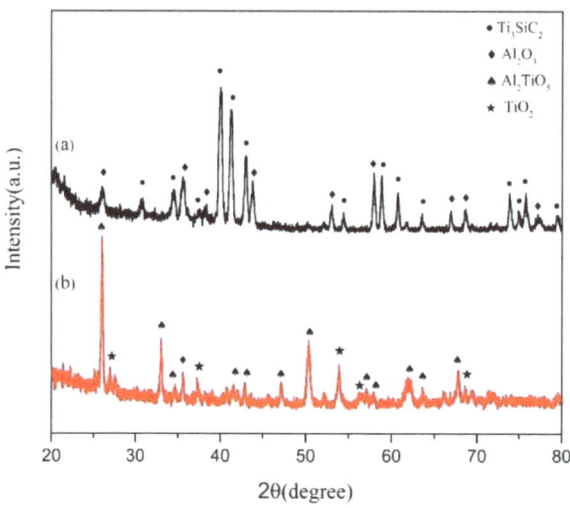

Figure 3. X-ray diffraction patterns of the TSC60 (a) before oxidation and (b) after oxidation at 1400 °C for 4 h.

As shown in Figure 4, the scanning electron microscopy (SEM) results show the surface morphology of the oxide layer after the oxidation of TSC50 and TSC60 for 4 h. Ti_3SiC_2 and Al_2O_3 remained on the surface of the TSC50 oxide layer. The morphology of the grains was massive and layered, and the pores were clearly observed. After oxidation, some grains were aggregated, which indicates oxide layer growth. The newly generated TiO_2 was connected to Ti_3SiC_2 with an evident layered structure, indicating its tendency to encapsulate Ti_3SiC_2. Conversely, after the oxidation of TSC60, the grain morphology on the oxide layer surface did not exhibit a lamellar structure. Although the grain size on the oxide layer surface of TSC60 was larger than that of TSC50, the grain morphology of TSC50 was more regular with a more distinct orientation. Stomata were observed on the surface of TSC50, and these pores may serve as channels for oxygen diffusion into the matrix. Compared with the sparsely oxidized surface of TSC50, TSC60 exhibited a state of mutual fusion and airtightness among the grains. Moreover, the fusion boundary of TSC60 could be clearly observed. Trace pores and cracks were observed on the oxidized surface of TSC60. The cause of the cracks may be the short oxidation time and the incomplete growth of the oxide layer. The difference in the oxidized surface morphology may be owing to the variation in the Al_2O_3 contents, which causes different degrees of oxidation between oxygen and Ti_3SiC_2 during the oxidation process.

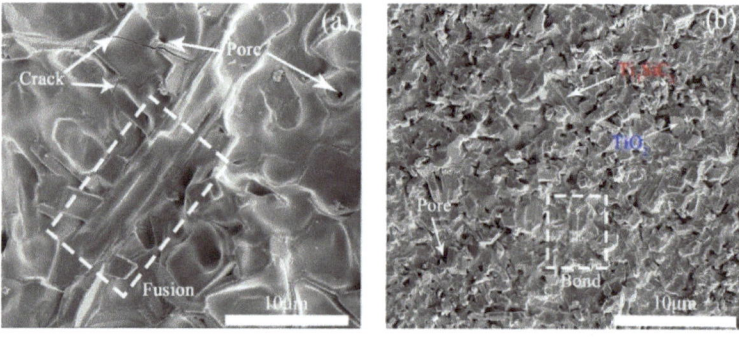

Figure 4. Surface morphology of the oxide layer after oxidation of (**a**) TSC60 and (**b**) TSC50 at 1400 °C for 4 h.

Figures 5 and 6 show the surface morphology and energy dispersive X-ray spectroscopy (EDS) results of TSC50 and TSC60 after 4h of oxidation. The oxidized surface formed a well-shaped crystal. EDS analysis indicated that the dense massive crystals on the oxidized surface of TSC50 (Figure 5) mainly contained Ti, Al, and O. Based on the types of elements and grain structures observed, the main components observed at points 1 and 2 in Figure 5 were identified as Al_2O_3 and TiO_2. Moreover, Si was not detected, indicating that SiO_2 was not present on the oxidized surface. Furthermore, a minor difference in the contents of Ti and Al was observed. Additionally, Figure 6 shows two types of crystal morphologies, in which the content of Al (the flat grain identified by point 1) was much higher than that of Ti. Based on the XRD results, the oxidized surface of TSC60 contained a small amount of Al_2TiO_5. Aluminium titanate has a plate-titanite-type crystal morphology with typical plate-like and blade-like crystals. Therefore, the grain indicated at point 1 in Figure 6 is proposed to be Al_2TiO_5. Simultaneously, the grain indicated at point 2 had a higher content of elemental Ti and appeared columnar, which is consistent with the crystal structure of rutile. Therefore, the EDS results were in good agreement with the XRD results.

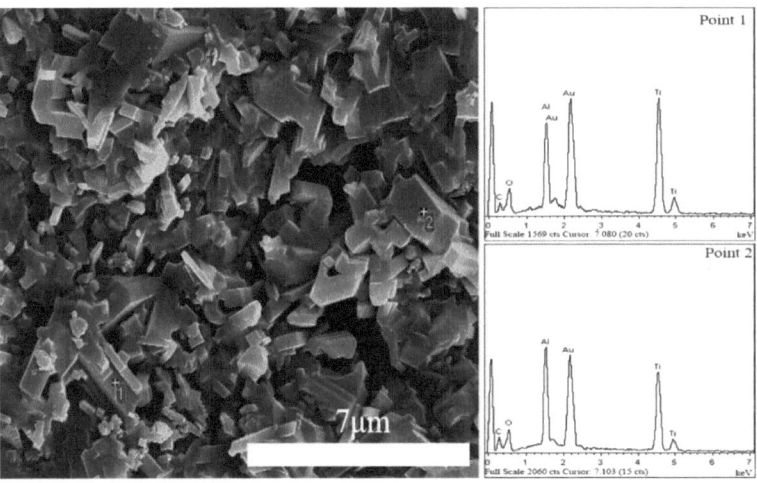

Figure 5. EDS analysis of TSC50 after oxidation at 1400 °C for 4 h.

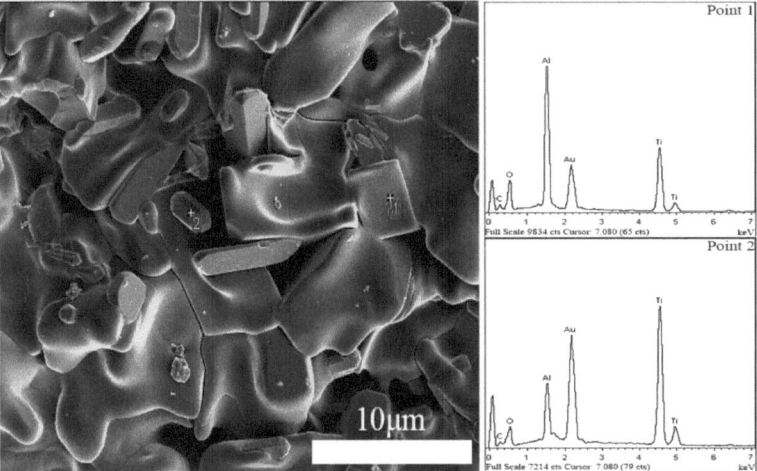

Figure 6. EDS analysis of TSC60 after oxidation at 1400 °C for 4 h.

Figure 7 shows the change in the weight gain per unit area of the two composites over time at a temperature of 1400 °C. During the oxidation period of 0–4 h, the weight gain per unit area decreased with the increasing Al_2O_3 content in the composites. However, when the oxidation time exceeded 4 h, the weight gain per unit area increased with the increasing Al_2O_3 content. With increasing time, the weight gain of TSC60 stabilized, indicating that TSC60 had transformed after 4 h of oxidation, thus reducing the degree of the subsequent oxidation processes. After 20 h of oxidation, the weight gain of TSC50 was 42.253×10^{-3} kg/m^2, whereas that of TSC60 was 29.411×10^{-3} kg/m^2, which is approximately 70% of the former. According to the XRD and EDS results, a mixed layer of Al_2TiO_5 and TiO_2 formed on the surface of TSC60 after 4 h of oxidation. Therefore, it is proposed that the presence of a mixed layer effectively reduces the rate of the subsequent oxidation of the composites. As a material with good thermal shock resistance and excellent high-temperature stability, Al_2TiO_5 played a significant role in reducing the oxidation rate of Ti_3SiC_2/Al_2O_3 in this study.

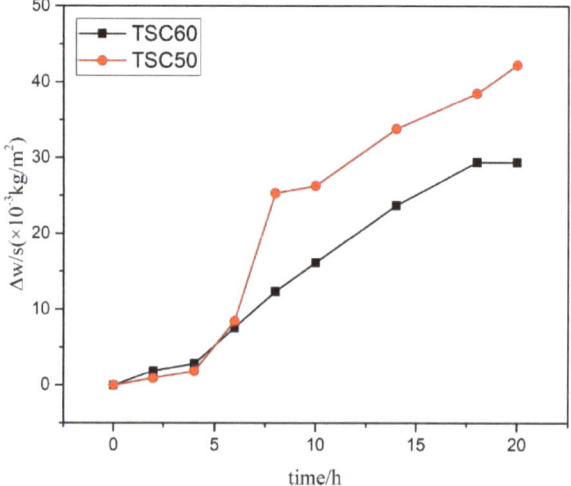

Figure 7. Weight gain per unit area of TSC50 and TSC60 at 1400 °C for 20 h.

As discussed, pores and cracks were present on the oxidized surface of Ti_3SiC_2/Al_2O_3, providing a diffusion channel for oxygen. Therefore, the oxidation of Ti_3SiC_2/Al_2O_3 was a diffusion-controlled process. Moreover, the oxidation of Ti_3SiC_2 was caused by the outward diffusion of Ti, Si and carbon, and the inward diffusion of oxygen. Although the oxidation of Ti_3SiC_2/Al_2O_3 was a diffusion-controlled process, the addition of Al_2O_3 changed the composition of the oxide layer. The cross-section of the oxide layer after 20 h of oxidation is shown in Figure 8. The thickness of the oxide layer of TSC60 was approximately 253 μm. Moreover, the oxide layer exhibited a silicone-free outer layer of approximately 40 μm. Based on the EDS results, the outer layers were clearly composed of Al_2TiO_5 and TiO_2. The presence of Si in the composite layer was detected, indicating that Si did not undergo external diffusion when it reached the composite layer, as shown in Figure 9. Thus, SiO_2 formed by silicon diffusion, and Al_2TiO_5 and Al_2O_3 compounded to form a glass phase, which increased the compactness of the composite layer and prevented the diffusion of Si and Ti [26]. As shown in Figure 9, a large amount of TiO_2 was present in the Al-deficient layer. Because the oxygen pressure in the outer layer was higher than that in the inner layer, Si gave priority to SiO gas generation. As the diffusion process progressed, SiO gas became a SiO_2 barrier that encapsulated TiO_2, resulting in a condition of Si enrichment; therefore, only a small amount of Al was detected [27]. In contrast, TSC50 did not have dense layers but it also had an Al-deficient layer of approximately 18 μm (Figure 10). At elevated temperatures and extended periods, such as 1400 °C and 20 h, the oxidation rate was high. Notably, this temperature is similar to the melting point of Si; thus, Si was more reactive. The Si content may be one of the reasons for the difference between the TSC60 and TSC50 oxide layers. Owing to its excellent high-temperature stability, Al_2O_3 did not decompose at this oxidation temperature; however, it reacted with other oxides to protect the matrix.

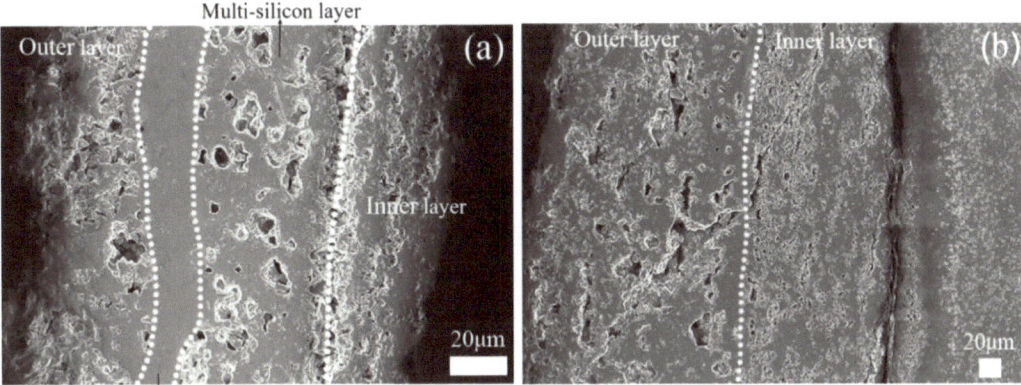

Figure 8. Cross-section of the oxide layer after oxidation of (**a**) TSC60 and (**b**) TSC50 at 1400 °C for 20 h.

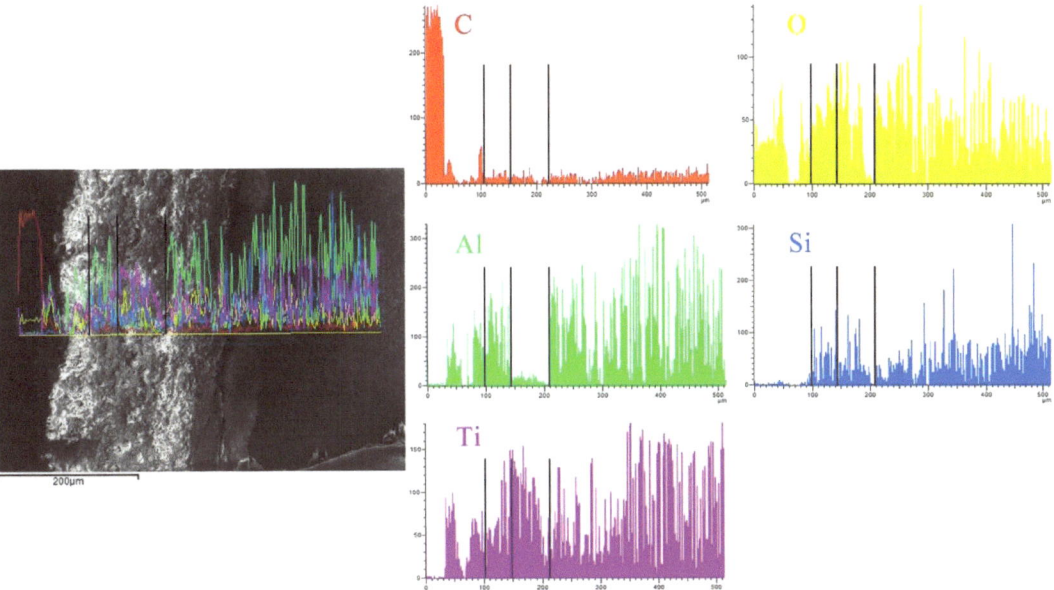

Figure 9. EDS line scanning results of TSC60 after oxidation at 1400 °C for 20 h.

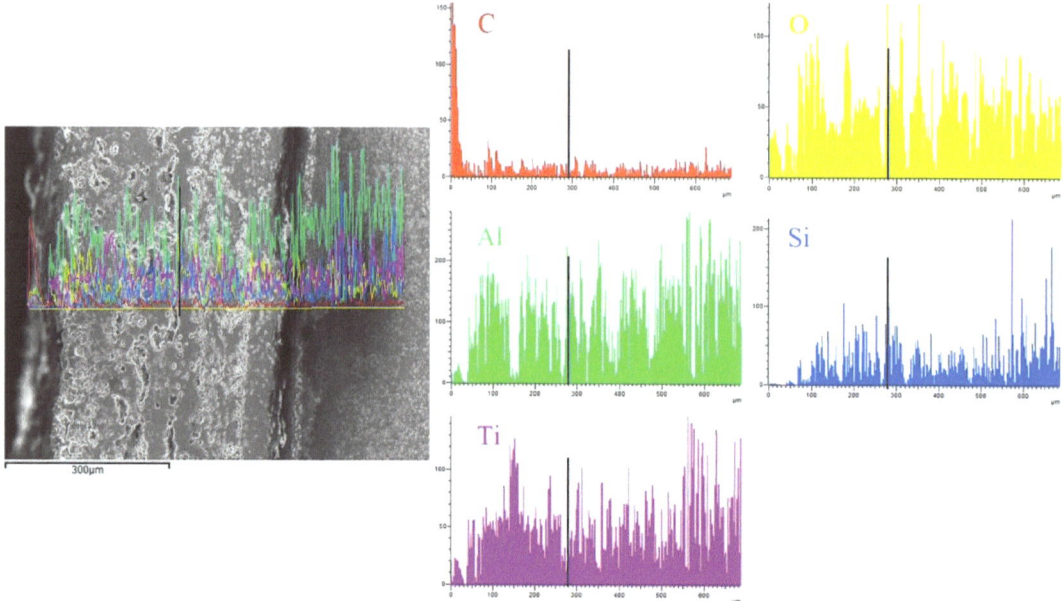

Figure 10. EDS line scanning results of TSC50 after oxidation at 1400 °C for 20 h.

Based on the above analysis, it is evident that Al_2O_3 improves the high-temperature oxidation resistance of Ti_3SiC_2. Compared with the studies reported by Zhang et al. [28] and Gao et al. [29], the oxide layer of Ti_3SiC_2/Al_2O_3 formed in this study was thinner. Moreover, the Al_2O_3 content was one of the factors influencing the oxidation resistance of the composite. Increasing the Al_2O_3 content did not always yield positive results. Thus, it is proposed that an optimal range of Al_2O_3 content exists, in which the high-temperature oxidation resistance of Ti_3SiC_2/Al_2O_3 is improved. This aspect will be investigated further in future research studies.

4. Conclusions

The compositional morphology and oxidation kinetics of Ti_3SiC_2/Al_2O_3 composites at 1400 °C were investigated in this study. The weight gain of the composite decreased with an increase in the volumetric content of Al_2O_3, indicating that Al_2O_3 addition delays the oxidation of the composite. However, when the oxidation time exceeded 4 h and continued until 20 h, the oxidation process accelerated, indicating that 4 h was the limit for the oxidation stability of the composite. After 4 h of oxidation, at 1400 °C, the surface of the composite with a high volume of Al_2O_3 exhibited more pores, facilitating the diffusion of oxygen into the matrix, which may have caused the acceleration of the oxidation process of the composites during extended periods of oxidation. The presence of Al_2TiO_5 on the oxidation surface of the composite with a low volume of Al_2O_3 may have reduced the rate of matrix oxidation and hindered the oxidation instability of the composites. Moreover, the composite oxide layer inhibited the diffusion-controlled process.

Author Contributions: Conceptualization, Q.L. and Y.D.; methodology, S.C.; software, D.M.; validation, Q.L., Y.D. and Z.Z.; formal analysis, Y.D.; investigation, B.P.; resources, Q.L.; data curation, J.L.; writing—original draft preparation, Q.L.; writing—review and editing, Y.D.; visualization, Y.D.; supervision, Q.L.; project administration, Q.L.; funding acquisition, Q.L. All authors have read and agreed to the published version of the manuscript.

Funding: The authors appreciate the financial support provided by the National Natural Science Foundation of China (Grant No. 51872118, 51701081), the Key Research and Development Program of Shandong Province (Grant No. 2019GGX104077, 2019RKB01018), the Shandong Provincial Natural Science Foundation, (Grant No. ZR2018PEM008, ZR2019MEM055). The project was supported by the State Key Laboratory of Advanced Technology for Materials Synthesis and Processing (Wuhan University of Technology). This work was financially supported by National Natural Science Foundation of China (51632003), the Taishan Scholars Program, and the Case-by-Case Project for Top Outstanding Talents of Jinan.

Institutional Review Board Statement: Not applicable.

Informed Consent Statement: Not applicable.

Data Availability Statement: Not applicable.

Conflicts of Interest: No conflict of interest exits in the submission of this manuscript, and the manuscript was approved by all authors for publication. I would like to declare on behalf of my co-authors that the work described was original research that has not been published previously, and is not under consideration for publication elsewhere, in whole or in part. All the authors listed have approved the manuscript that is enclosed.

References

1. Sun, Z.M. Progress in research and development on MAX phases: A family of layered ternary compounds. *Int. Mater. Rev.* **2011**, *56*, 143–166. [CrossRef]
2. Atazadeh, N.; Heydari, M.S.; Baharvandi, H.R.; Ehsani, N. Reviewing the effects of different additives on the synthesis of the Ti_3SiC_2 MAX phase by mechanical alloying technique. *Int. J. Refract. Met. Hard Mater.* **2016**, *61*, 67–78. [CrossRef]
3. Qin, J.; He, D. Phase stability of Ti_3SiC_2 at high pressure and high temperature. *Ceram. Int.* **2013**, *39*, 9361–9367. [CrossRef]
4. Dezellus, O.; Gardiola, B.; Andrieux, J.; Lay, S. Experimental evidence of copper insertion in a crystallographic structure of Ti_3SiC_2 MAX phase. *Scr. Mater.* **2015**, *104*, 17–20. [CrossRef]
5. Islak, B.Y.; Ayas, E. Evaluation of properties of spark plasma sintered Ti_3SiC_2 and Ti_3SiC_2/SiC composites. *Ceram. Int.* **2019**, *45*, 12297–12306. [CrossRef]
6. Liu, X.; Zhang, H.; Jiang, Y.; He, Y. Characterization and application of porous Ti_3SiC_2 ceramic prepared through reactive synthesis. *Mater. Des.* **2015**, *79*, 94–98. [CrossRef]
7. El Saeed, M.A.; Deorsola, F.A.; Rashad, R.M. Optimization of the Ti_3SiC_2 MAX phase synthesis. *Int. J. Refract. Met. Hard Mater.* **2012**, *35*, 127–131. [CrossRef]
8. Cai, Y.Z.; Cheng, L.F. Effect of positioning impregnation on the oxidation behaviour of Ti_3SiC_2/SiC functionally graded materials at 1400 °C. *J. Alloy. Compd.* **2018**, *742*, 180–190. [CrossRef]
9. Yang, J.S.; Zhang, X.Y. Fabrication of Ti_3SiC_2 powders using TiH_2 as the source of Ti. *Ceram. Int.* **2012**, *38*, 3509–3512. [CrossRef]
10. Zheng, L.-L.; Sun, L.-C.; Li, M.-S.; Zhou, Y.-C. Improving the high-temperature oxidation resistance of $Ti_3(SiAl)C_2$ by Nb-doping. *J. Am. Ceram. Soc.* **2011**, *94*, 3579–3586. [CrossRef]
11. Li, X.; Qian, Y.; Zheng, L.; Xu, J.; Li, M. Determination of the critical content of Al for selective oxidation of Ti_3AlC_2 at 1100 °C. *J. Eur. Ceram. Soc.* **2016**, *36*, 3311–3318. [CrossRef]
12. Li, X.; Zheng, L.; Qian, Y.; Xu, J.; Li, M. Breakaway oxidation of Ti_3AlC_2 during long-term exposure in air at 1100 °C. *Corros. Sci.* **2016**, *104*, 112–122. [CrossRef]
13. Gong, Y.; Tian, W.; Zhang, P.; Chen, J.; Zhang, Y.; Sun, Z. Slip casting and pressureless sintering of Ti_3AlC_2. *J. Adv. Ceram.* **2019**, *8*, 367–376. [CrossRef]
14. Qi, F.F.; Wang, Z. Improved mechanical properties of Al_2O_3 ceramic by in-suit generated Ti_3SiC_2 and TiC via hot pressing sintering. *Ceram. Int.* **2017**, *43*, 10691–10697. [CrossRef]
15. Shi, S.L.; Pan, W. Toughening of Ti_3SiC_2 with 3Y-TZP addition by spark plasma sintering. *Mater. Sci. Eng. A* **2007**, *447*, 303–306. [CrossRef]
16. Islak, B.Y.; Candar, D. Synthesis and properties of TiB_2/Ti_3SiC_2 composites. *Ceram. Int.* **2021**, *47*, 1439–1446. [CrossRef]
17. Zhang, J.; Wang, L.; Jiang, W.; Chen, L. High temperature oxidation behavior and mechanism of Ti_3SiC_2-SiC nanocomposites in air. *Compos. Sci. Technol.* **2008**, *68*, 1531–1538. [CrossRef]
18. Li, S.; Song, G.M.; Zhou, Y. A dense and fine-grained $SiC/Ti_3Si(Al)C_2$ composite and its high-temperature oxidation behavior. *J. Eur. Ceram. Soc.* **2012**, *32*, 3435–3444. [CrossRef]
19. Xu, X.; Ngai, T.L.; Li, Y. Synthesis and characterization of quarternary $Ti_3Si_{(1-x)}Al_xC_2$ MAX phase materials. *Ceram. Int.* **2015**, *41*, 7626–7631. [CrossRef]
20. Guedouar, B.; Hadji, Y. Oxidation behavior of Al-doped Ti_3SiC_2-20wt.%Ti_5Si_3 composite. *Ceram. Int.* **2021**, *47*, 33622–33631. [CrossRef]
21. Heider, B.; Scharifi, E.; Engler, T.; Oechsner, M.; Steinhoff, K. Influence of heated forming tools on corrosion behavior of high strength aluminum alloys. *Mater. Sci. Eng. Technol.* **2021**, *52*, 145–151. [CrossRef]

22. Popov, A.I.; Lushchik, A.; Shablonin, E.; Vasil'chenko, E.; Kotomin, E.A.; Moskina, A.M.; Kuzovkov, V.N. Comparison of the F-type center thermal annealing in heavy-ion and neutron irradiated Al_2O_3 single crystals. *Nucl. Instrum. Methods Phys. Res. Sect. B Beam Interact. Mater. At.* **2018**, *433*, 93–97. [CrossRef]
23. Averback, R.S.; Ehrhart, P.; Popov, A.I. Defects in ion implanted and electron irradiated Mgo and Al_2O_3. *Radiat. Eff. Defects Solids* **1995**, *136*, 169–173. [CrossRef]
24. Shablonin, E.; Popov, A.I.; Prieditis, G.; Vasil'chenko, E.; Lushchik, A. Thermal annealing and transformation of dimer F centers in neutron-irradiated Al_2O_3 single crystals. *J. Nucl. Mater.* **2021**, *543*, 152600. [CrossRef]
25. Sun, Z.; Zhou, Y.; Li, M. High temperature oxidation behavior of Ti_3SiC_2-based material in air. *Acta Mater.* **2001**, *49*, 4347–4353. [CrossRef]
26. Dong, X.; Wang, Y.; Wang, R.; Wang, X.; Li, Y. Study on Al_2TiO_5-SiO_2-Al_2O_3 composites. *Bull. Chin. Ceram. Soc.* **2008**, *27*, 649–653.
27. Zhang, H.B.; Shen, S.Y. Oxidation behavior of porous Ti_3SiC_2 prepared by reactive synthesis. *Trans. Nonferrous Met. Soc. China* **2018**, *28*, 1774–1783. [CrossRef]
28. Zhang, H.B.; Zhou, Y.C.; Bao, Y.W.; Li, M.S. Improving the oxidation resistance of Ti_3SiC_2 by forming a $Ti_3Si_{0.9}Al_{0.1}C_2$ solid solution. *Acta Mater.* **2004**, *52*, 3631–3637. [CrossRef]
29. Gao, H.; Benitez, R.; Son, W.; Arroyave, R.; Radovic, M. Structural, physical and mechanical properties of $Ti_3(Al_{1-x}Si_x)C_2$ solid solution with x = 0–1. *Mater. Sci. Eng. A* **2016**, *676*, 197–208. [CrossRef]

MDPI AG
Grosspeteranlage 5
4052 Basel
Switzerland
Tel.: +41 61 683 77 34

Coatings Editorial Office
E-mail: coatings@mdpi.com
www.mdpi.com/journal/coatings

Disclaimer/Publisher's Note: The statements, opinions and data contained in all publications are solely those of the individual author(s) and contributor(s) and not of MDPI and/or the editor(s). MDPI and/or the editor(s) disclaim responsibility for any injury to people or property resulting from any ideas, methods, instructions or products referred to in the content.

www.ingramcontent.com/pod-product-compliance
Lightning Source LLC
LaVergne TN
LVHW072349090526
838202LV00019B/2506